COOK MILK IN ANY FLAVOUR YOU LIKE.

OMMB

Tanager
Press

Published by ©The Ontario Milk Marketing Board, Mississauga, Ontario, Canada, L5N 2L8.

Includes index.

ISBN 1-895410-02-9

Designed by Durkin, Rodgers & Battaglia Inc.

Printed in Canada by Matthews, Ingham and Lake Inc.

The Ontario Milk Marketing Board gratefully acknowledges co-operation and assistance of the following:

Entrants in The Ontario Milk Marketing Board Recipe Contests; Dairy Bureau of Canada; United Dairy Industry Association; California Milk Advisory Board; Canadian Egg Marketing Agency; Ontario Turkey Producers Marketing Board; American Dairy Association and Foodland Ontario; Ms. B. Stern, Mrs. A. Gryfe, Mrs. K. Bush, Mrs. J. Rodmell, Sharon Dale and Joan Colnett.

Cook milk in any flavour you like/ Ontario
   Milk Marketing Board

    ISBN 1-895410-02-9

    1 . Cookery (Milk) . 2. Milk .
I . Ontario Milk Marketing Board .
TX759   641 . 6 ' 714 – dc20

This edition published by Tanager Press,
Mississauga, Ontario L5G 1S8

TABLE OF CONTENTS

INTRODUCTION   I

All About Milk	II-VI
Milk and Cooking	VII-IX
Metric Conversion Guides	X-XI
Cupboard and Food Storage Guides	XII-XIV
Glossary of Cooking Terms Used in this Cookbook	XV-XVII
Weight Volume Equivalents Guide	XVIII-XXI
Emergency Substitutions Guide	XXII-XXIII
Cookbook Re-order Form	XXV

RECIPES

Appetizers, Snacks & Beverages	1-43
Appetizers and Snacks	1-14
Beverages	15-43
Soups & Chowders	45-102
Main Meals	103-352
Beef & Veal	103-142
Cheese, Eggs & Pasta	143-212
Chicken	213-258
Fish & Shellfish	259-306
Lamb	307-312
Pork & Ham	313-340
Turkey	341-352
Vegetables, Side Dishes, Sauces & Dressings	353-408
Vegetables and Side Dishes	353-395
Sauces and Dressings	396-408
Baked Goods	409-484
Desserts	485-612

INDEX   613-637

INTRODUCTION

THE NEW EDITION

This is it. The definitive edition of a truly delicious Canadian bestseller. And the 600 tried and true recipes that make up the newest version of this essential cookbook have been reorganized so that whatever you need is right at your fingertips.

The Table of Contents is now organized into six familiar categories:

• APPETIZERS, SNACKS AND BEVERAGES

• SOUPS AND CHOWDERS

• MAIN MEALS

• VEGETABLES, SIDE DISHES, SAUCES AND DRESSING

• BAKED GOODS

• DESSERTS

The index couldn't be more complete. The recipes are listed alphabetically by name and cross-referenced by both main ingredients and category for your convenience.

We've also included valuable nutritional information on Milk, tips for cooking, measuring and storage, conversion and substitution charts, as well as a glossary of the cooking terms used in this cookbook.

A cookbook and reference guide that you can use day in, day out. That's why since its first edition *Cook Milk In Any Flavour You Like* has sold well over 150,000 copies.

All About Milk

MILK AND ITS NUTRIENTS

Throughout our lives, Milk in all its many varieties, provides us with important energy plus a total of 15 nutrients key to human health. The goodness in Milk contributes to both our present health and our future well-being.

What do all the nutrients in Milk actually do? That's an important question.

For the answer, see the chart below. You'll also discover exactly how much of the Recommended Daily Intake (RDI) of each nutrient Milk provides.

Important Nutrients	Definition	Percentage RDI for 250 mL* of 2% Milk
Vitamin A	Helps normal bone and tooth development. Promotes good night vision. Maintains the health of the skin and membranes.	11%
Vitamin D	Enhances calcium and phosphorus utilization in the formation and maintenance of healthy bones and teeth.	44%
Thiamin/ Vitamin B1	Releases energy from carbohydrate. Assists in normal growth and appetite.	8%
Riboflavin/ Vitamin B2	Maintains healthy skin and eyes. Maintains a normal nervous system. Releases energy to body cells during metabolism.	25%
Niacin	Helps normal growth and development. Maintains a normal nervous system and gastrointestinal tract.	10%
Vitamin B_6	Helps in many aspects of protein metabolism. Assists in the formation of red blood cells.	6%

ALL ABOUT MILK

Important Nutrients	Definition	Percentage RDI for 250 mL* of 2% Milk
Folacin	Contributes to red blood cell formation	6%
Vitamin B$_{12}$	Contributes to red blood cell formation. Helps maintain healthy nerve and gastrointestinal tissues.	45%
Pantothenate	Involved in the release of energy from carbohydrate as well as the breakdown and metabolism of fat.	11%
Calcium	Helps in formation and maintenance of strong bones and teeth. Promotes healthy nerve function and normal blood clotting.	29%
Phosphorus	Helps in formation and maintenance of strong bones and teeth.	22%
Magnesium	Assists in formation and maintenance of strong bones and teeth. Helps in energy metabolism and tissue formation.	14%
Zinc	Contributes to energy metabolism and tissue formation.	11%
Protein	Builds and repairs body tissues. Builds antibodies, the blood components which fight infection.	8.6g per serving
Carbohydrate	Supplies energy. Assists in the utilization of fats.	12g per serving

• 250mL is the metric equivalent of 1 cup/8 oz.

WHY IS MILK FORTIFIED WITH VITAMINS A AND D?

Our food, in general, does not contain much Vitamin D. In fact, those Canadians that don't consume Milk regularly are likely to have difficulty getting enough Vitamin D in their diet.

ALL ABOUT MILK

Milk is an ideal vehicle for Vitamin D because it contains the right combination of calcium and phosphorus, which together with the added Vitamin D, help assure healthy bones and teeth.

Whole Milk is also an important source of Vitamin A, a fat-soluble vitamin important for skin health. When Milk fat is reduced, as with skim Milk, the Vitamin A content is also lowered. Therefore, lowered fat Milks are fortified with Vitamin A to maintain the higher levels found in whole Milk.

MILK AND CANADA'S FOOD GUIDE

According to Canada's Food Guide, both adult men and women should consume *2 servings daily* from the Milk and Milk Products Food Group. Yet Canadian adults drink on average only $3/4$ of a cup (175 mL) of Milk per day.

One simple way to add more of the essential nutrients contributed to your diet by Milk and Milk Products is to cook with Milk whenever you can.

CANADA'S FOOD GUIDE

Variety
Choose different kinds of foods from each group in appropriate numbers of servings and portion sizes.

Energy Balance
Needs vary with age, sex and activity. Balance energy intake from foods with energy output from physical activity to control weight. Foods selected according to the Guide can supply 4000-6000 kJ (kilojoules) (1000-1400 kilocalories) per day. For additional energy, increase the number and size of servings from the various food groups and/or add other foods.

Moderation
Select and prepare food with limited amounts of fat, sugar and salt. If alcohol is consumed, use limited amounts.

ALL ABOUT MILK

Four Food Groups From Canada's Food Guide

Milk and Milk Products

Children up to 11 years   2-3 servings
Adolescents   3-4 servings
Pregnant and nursing   3-4 servings
   women
Adults   2 servings

Skim, 1%, 2%, whole, buttermilk, reconstituted dry or evaporated Milk may be used as a beverage or as the main ingredient in other foods. Cheese may also be chosen.

Some examples of one serving:
250 mL (1 cup) Milk
175 mL ($^3/_4$ cup) yogurt
45g (1 $^1/_2$ ounces) cheddar
   or process cheese

In addition, a supplement of Vitamin D is recommended when Milk is consumed which does not contain added Vitamin D.

Fruits and Vegetables

4-5 Servings
Include at least two vegetables.
Choose a variety of both vegetables and fruits - cooked, raw or their juices. Include yellow or green or green leafy vegetables.

Some examples of one serving:
125 mL ($^1/_2$ cup) vegetables or fruit -
   fresh, frozen or canned
1 medium-sized potato, carrot, tomato,
   peach, apple, orange or banana.
125 mL ($^1/_2$ cup) juice - fresh, frozen or
   canned

Breads and Cereals

3-5 Servings
Whole grain or enriched whole grain products are recommended.

Some examples of one serving:
1 slice bread
125 mL ($^1/_2$ cup) cooked cereal
175 mL ($^3/_4$ cup) ready-to-eat cereal
1 roll or muffin
125 to 175 mL ($^1/_2$ to $^3/_4$ cup) cooked
   rice, macaroni, spaghetti or noodles
$^1/_2$ hamburger or wiener bun

Meat, Fish, Poultry and Alternates

2 Servings

Some examples of one serving:
60 to 90 g (2-3 ounces) cooked lean
   meat, fish, poultry or liver
60 mL (4 tablespoons) peanut butter
250 mL (1 cup) cooked dried peas,
   beans or lentils
125 mL ($^1/_2$ cup) nuts or seeds
60 g (2 ounces) cheddar cheese
125 mL ($^1/_2$ cup) cottage cheese
2 eggs

ALL ABOUT MILK

THE MANY FACES OF MILK

Once consuming Milk, infants under 1 year should stick to whole Milk for growth and energy needs. After that, there's a wonderful assortment of Milk products to suit almost everyones' needs and tastes.

That's because all Milks have fundamentally the same nutrient content. All Milks contain the same 15 essential nutrients. The key variable is fat content.

Milk	Fat Content By Weight	Grams of Fat Per 250mL*	Total Calories per 250mL*
Whole Milk	3.25%	8.6	157
2% Milk	2%	5.0	129
2% Chocolate Milk	2%	5.3	189
1% Milk	1%	2.7	108
Buttermilk	0.1%-2%	2.0	105
Skim Milk	trace	0.5	91
Dry Instant Skim Milk	trace	0.5	91

* 250mL is the metric equivalent of 1 cup/8 oz.

In terms of specialized Milks, UHT (Ultra High Temperature) Milk is a whole or partly-skimmed Milk which has been heated to eliminate all organisms that might make it go off. UHT Milk can be stored at room temperature for up to 3 months without any appreciable loss of nutrients or taste. Once opened, use and store as you would any other fresh Milk.

LACTEEZE and LACTAID are lactose-reduced Milks suitable for people who are lactose-intolerant, allowing them to benefit from all the natural goodness that Milk contains.

Milk And Cooking

MILK MAKES GOOD FOOD BETTER

Nutrition, while significant, is not the only reason to cook with Milk. Milk's cooking advantages in terms of taste, texture, versatility, economy and speed are important too.

As you can see from the recipes in this cookbook – when it comes to preparing soups or sauces, dinners or desserts – Milk is a natural. Milk not only tenderizes a wide variety of meats, fish and vegetables, it helps produce moister cakes, creamier sauces and smoother soups.

Substituting Milk for water when preparing pre-packaged foods is another good strategy to ensure your family gets enough of Milk's goodness. Cake, muffin and quick bread mixes rise like a dream when you ignore package instructions and use Milk as the liquid additive. Or try using Milk instead of water to prepare hot chocolate mixes. One taste and you'll be hooked forever.

TIPS FOR COOKING WITH MILK

Baked Goods: Ingredients for baking need to be at or about room temperature to ensure even mixing and best results. Milk is no exception. Remove the chill from Milk by using your microwave or by letting the measured amount stand until it reaches room temperature. It takes only a short time to bring it to optimum baking temperature.

Custards and Puddings: When large amounts of Milk are used for baked desserts, such as custards and puddings, it is best to preheat or "scald" Milk before using to prevent scorching or protein formation on top.

Cream Cheese Mixtures: Mix Milk gradually into beaten cream cheese mixtures like cheesecakes or dips to prevent lumping.

Sauces: Stir Milk-based sauces constantly to ensure a smooth, velvety texture.

Mashed Potatoes: Warm the Milk you use for mashed potatoes so that they stay piping hot for serving.

Acidic Foods: If any tangy flavoured ingredients, such as tomato sauce or lemon juice, are to be added to a Milk-based sauce, be sure the sauce is thickened first with flour or corn starch to prevent curdling.

Milk And Cooking

MEASURING TIPS

To measure Milk for cooking or baking purposes be sure to use a clear glass measuring cup and take a reading at eye level. While it probably doesn't matter too much if your measurement is exact for a soup or chowder, accuracy will affect the outcome of more precise recipes, such as any baked goods.

Whether you use the Imperial (cups or spoons) or the Metric (millilitres) system, one rule is key. Stay with one system only, do not switch back and forth. Recipe conversion is not simple. Baked goods and sauces, in particular, are vulnerable to converting and rounding out amounts.

PRESERVING MILK'S QUALITY

Exceptional care is taken by dairy farmers and processors to ensure that the Milk you buy is of the highest quality. Storing it properly can preserve this quality and freshness right down to the last drop.

At store level:

• Always buy Milk well within the expiry date.

• Pick up your Milk near the end of your shopping, so that it remains cold for as long as possible.

• Purchase Milk from responsible grocers who rotate stock frequently and have low-light dairy cases. (Some nutrients may be reduced by excessive light exposure).

MILK AND COOKING

At home:

• Here the 3 C's apply - keep it Clean, Cold and Covered.

• Refrigerate as soon as possible. Milk stays fresher longer when it is kept cold. It is best stored at 40°F (4°C) inside the refrigerator rather than in the refrigerator door.

• Do not let Milk stand exposed to sunlight. This can destroy the B Vitamins and quickly spoil the flavour.

• To avoid contamination it's best not to transfer Milk to other containers. A freshly bought litre of Milk should not be mixed with one which has a different expiry date.

• Milk should be kept covered to protect it from dust, bacteria and food odours in the refrigerator. (An absorbed flavour alters the taste of Milk, but it is still safe to use.)

• Pasteurized Milk does not "sour" in the same way as raw, unpasteurized Milk. Do not use it once it has gone off. If a recipe calls for "sour" Milk it's better to start with fresh Milk and sour it yourself (see the Emergency Substitutions Guide on pages XXII-XXIII).

FREEZING MILK

Milk may be frozen for up to six weeks. Although freezing does not greatly affect the nutritive value, it may cause separation. That's why you should shake it gently before using.

It's also best to thaw Milk in the refrigerator. In a pinch, 1 litre of Milk may be thawed in a microwave at high power (100%) for 20 minutes, but this isn't recommended on a regular basis. As well, remember not to refreeze Milk after thawing.

Metric Conversion Guides

The following charts, tips and basic cooking information have been compiled to provide you with handy, simple reference guides. Ones which make your cooking as easy and uncomplicated as possible.

Oven Temperature Conversion Guide

Fahrenheit °F	Celcius°C
250°F	120°C
275°F	140°C
300°F	150°C
325°F	160°C
350°F	180°C
375°F	190°C
400°F	200°C
425°F	220°C
450°F	230°C
475°F	240°C
500°F	260°C

°C represents the metric equivalent of the corresponding °F

METRIC CONVERSION GUIDES

Cookware and Bakeware Equivalents Guide

Bakeware	Dimensions		Volume	
	Imperial	**Metric**	**Imperial**	**Metric**
Round layer cake pans	8x1½-inch 9x1⅜-inch	20x4 cm 22x3.5 cm	5 cups 6 cups	1.2L 1.5L
Square cake pans	8x8x2-inch 9x9x1¾ inch	20x20x5 cm 23x23x4.5 cm	8 cups 10 cups	2L 2.5L
Loaf pans	8½x4½ x 2½-inch 9x5x2½-inch	22x11x6 cm 23x13x6 cm	6 cups 8 cups	1.5L 2L
Tube pans	9x4-inch 10x4½-inch	23x10 cm 25x11 cm	12 cups 16 cups	3L 4L
Jelly roll pans	15½x10½ x ¾ inch 17½ x 11½ x ¾ inch	39x27x2 cm 45x29x2 cm	8 cups 12 cups	2L 3L
Baking dishes	11x7x1½-inch 12x8x1¾-inch 13x9x2-inch	28x18x4 cm 30x20x4.5 cm 33x23x5 cm	8 cups 12 cups 14 cups	2L 3L 3.5L
Pie plates	8x1½ inch 9x1½-inch 10x1¾-inch	20x4 cm 23x4 cm 25x4.5 cm	3 cups 4 cups 6 cups	750mL 1L 1.5L
Springform pans	8x2½-inch 9x2½-inch 10x2½-inch	20x6 cm 23x6 cm 25x6 cm	8 cups 10 cups 12 cups	2L 2.5L 3L

Reprinted with the permission of Canadian Living Magazine

CUPBOARD AND FOOD STORAGE GUIDES

Cupboard Storage Guide

• Store food in the coolest kitchen cabinets - not over range, refrigerator or exhaust fan.

• Keep foods in airtight containers.

• Choose fresh packages and avoid cans with swollen ends and dents.

• Date all products you are storing. With longer storage times than recommended here, flavours may deteriorate and nutrients are eventually lost.

Food	Time
Baking powder & Baking soda	18 months
B.B.Q, ketchup, chili sauces	1 month
Bouillon cubes	1 year
Canned Foods:	
Fish, fruit, vegetables, meat, poultry, soup	1 year
Cereals (ready-to-eat and cooked)	6 months
Chocolate squares	1 year
Coconut	1 year
Coffee: instant	6 months
vacuum packed	1 year
Flour (all types except whole wheat)	1 year

Food	Time
Fruit (dried)	6 months
Gelatin (flavoured and unflavoured)	18 months
Herbs & Spices	6 months
Honey	1 year
Jams & Jellies	1 year
Pasta	1 year
Rice: white	2 years
brown & wild	1 year
Sugar: granulated	2 years
brown	4 months
Tea: loose	6 months
bags	1 year

CUPBOARD AND FOOD STORAGE GUIDES

Refrigerator Storage Guide

- Keep refrigerator temperatures around 40°F (4°C).
- Wrap food well or place in air-tight containers to prevent drying out and transference of odours.
- Keep uncooked meat, fish and poultry in the coldest part of the refrigerator, fresh fruit and vegetables in crispers or loosely covered.
- All dairy products, leftovers and packaged foods should be tightly covered or wrapped in moisture-resistant wrap or foil.

Food	Time *Always check Best Before dates*
Butter	2 weeks
Buttermilk, sour cream, yogurt	2 weeks
Cheese:	
Cottage	5 days
Cream	2 weeks
Sliced	2 weeks
Whole piece	2 months
Cream	1 week
Eggs (in shell)	1 month
Milk	1 week
Fruit:	
Apples	1 month
Bananas, melons, peaches, pears	2 weeks
Cherries	3 days
Citrus	2 weeks
Vegetables:	
Asparagus	3 days

Food	Time *Always check Best Before dates*
Vegetables:	
Carrots, broccoli, brussel sprouts, green onions	5 days
Cauliflower, celery, green beans, peppers, tomatoes	1 week
Lettuce, spinach	5 days
Fresh Meat:	
Beef, lamb, pork, veal: chops/steaks	5 days
roasts	5 days
Ground Meat	2 days
Sausage & variety meats	2 days
Prepared Meats:	
Bacon, wieners	1 week
Ham: canned, unopened	6 months
sliced	3 days
whole	1 week
Luncheon meat	5 days
Fish & shellfish	1 day
Poultry	2 days

Cupboard And Food Storage Guides

Freezer Storage Guide

• Keep freezer at 0°F (-15°C).

• For longer than two weeks storage wrap food in moisture/vapour-proof materials or specially coated or laminated freezer paper. Properly wrapped and frozen foods will hold full flavour and nutrients for times listed below. After that, flavours may fade but food is still safe to eat.

• Never refreeze or eat any food with off odour or colour.

• Never refreeze Milk once it has thawed.

• Cheese which has been frozen is best used in recipes as it may crumble when thawed.

Food	Time
Milk	up to 6 weeks
Ice cream/sherbet	1 month
Butter	9 months
Bread/baked goods	3-6 months
Fish:	
Fatty fish	3 months
Lean fish	6 months
Flour, whole wheat	1 year
Fruit	1 year
Meat/Poultry:	
Beef: roasts & steaks	1 year
ground	4 months
Lamb/Veal:	
roasts & steaks	9 months

Food	Time
Meat/Poultry:	
Pork: roasts	8 months
chops	4 months
Chicken/Turkey:	
parts	6 months
whole	1 year
Duck, Goose & Turkey rolls	6 months
Shellfish:	
Breaded & cooked	3 months
Shrimp (raw)	1 year
Vegetables	8 months
Cheese:	
Dry curd, cottage, ricotta	2 weeks
Natural process	3 months
Cream (whipped)	1-2 months
Nuts	3 months

XV

GLOSSARY OF COOKING TERMS

For your convenience, here is a comprehensive list of the cooking terms used in this cookbook.

Bake To cook by free circulating hot, dry air. It is very important to preheat oven before using. For even cooking do not overcrowd oven.

Baste To moisten food while baking by pouring liquid or fat over it. A bulb baster is convenient for this purpose.

Batter A mixture of flour, liquid and other ingredients which can be beaten or stirred.

Beat To incorporate air into a mixture by over and over motion either by spoon, whisk or electric beater.

Blanch To immerse foods, briefly, in boiling water, followed by a quick cooling in cold water. Used to bring out the colour, loosen skins for peeling or to mellow flavours.

Blend To combine ingredients without excess mixing.

Boil To heat until bubbles constantly break on the surface.

Braise To simmer in a covered dish in a small amount of liquid. Ideal for tougher cuts of meat and firm fleshed fish.

Broil To cook with intense direct heat either on a grill or under a broiler. The high heat seals in the juices, browns the outside and keeps the food tender.

Brown To cook food quickly in a preheated oven broiler or hot frypan to "brown" the outside and seal in the juices.

Brush To coat lightly with liquid such as Milk, egg, etc. using a small brush.

Caramelize To heat dry sugar or foods containing sugar until light brown and a caramel colour.

Chill To place in a refrigerator until cold.

Chop To cut into pieces ranging from small (finely chopped) to large (coarsely chopped).

Combine To mix ingredients together.

Clarify To make butter clear by heating, separating and discarding solids.

Cool To allow mixture to come to room temperature.

Cream To mix until soft and fluffy. Usually applied to a mixture of fat and sugar.

Crush To press to extract juice.

Cube To cut food into cube-shaped pieces ranging in size from $1/4$ inch (6mm) to 1 inch (2.5cm).

Crimp To decorate the edge of a pie crust by pinching the dough together with the fingers.

GLOSSARY OF COOKING TERMS

Cut To combine butter or shortening with dry ingredients until mixture resembles coarse meal using 2 dinner knives or a pastry blender or the fingers.

Deglaze To add liquid to a pan in which food, meat or poultry has been cooked, stirring, scraping up and dissolving the browned bits in the bottom of the pan. The mixture is then heated to reduce the liquid to the desired consistency.

Dice To cut into small cubes ranging from $1/4$ inch (6mm) to $1/2$ inch (12mm).

Dough A mixture of liquid and flour stiff enough to be handled or kneaded.

Dredge To coat completely with flour or sugar.

Dust To sprinkle lightly with flour or sugar.

Eviscerate To remove internal organs from animals, fish or poultry.

Fillet To cut meat, chicken or fish from the bones.

Flake To break into small pieces, usually with a fork.

Fold To incorporate one ingredient with another without stirring or beating but instead by gently lifting from underneath with a rubber spatula.

Fry To cook food in hot fat in a frypan over direct heat until crisp and brown. Food is often dipped in flour or batter first.

Glaze To coat with syrup, thin frosting, jelly or jam.

Grate To change a solid food to fine particles by using a hand grater or food processor.

Grease To lightly coat a pan with some form of fat to prevent foods from sticking and to help browning.

Grind To use a mortar and pestle, food processor or meat grinder to transform a solid piece of food to fine pieces.

Julienne To cut fresh vegetables or other food into thin match-stick sized strips of, uniform length.

Knead To work dough with hands by folding it over on itself, pushing down and away with heels of the hands, turning down dough $1/4$ turn after each pushing and folding motion.

Line To cover the surface of a baking sheet or roasting pan with waxed or parchment paper or aluminium foil to prevent food from sticking.

Marinate To tenderize and flavour food by placing it in a seasoned liquid, usually a mixture of oil, lemon juice or vinegar and seasonings.

Melt To change solids to liquid by use of low heat. Commonly used in reference to butter and chocolate.

Mince To chop very finely.

Mix To combine ingredients by stirring.

Glossary Of Cooking Terms

Pare To use a thin knife to remove skin or rind from fruit or vegetables.

Peel To strip off outer coatings; e.g. oranges.

Pinch The amount of a dry ingredient you can hold between thumb and finger.

Poach To cook food gently in simmering liquid that does not boil.

Preheat To set oven or broiler to the desired temperature 5 to 10 minutes before use so that the desired temperature is reached before the food is put into the oven to cook.

Prick To pierce food or pastry with tines of a fork to prevent it from bursting or rising during baking.

Purée To mash solid food or pass it through a food mill or processor.

Reduce To thicken or concentrate a sauce by boiling down, which lessens the volume and intensifies the flavour.

Roast To cook by the free circulation of dry heat, often beginning with a very hot oven to seal in juices, then lowering heat to complete cooking.

Roll To spread pastry with a rolling pin to desired thickness.

Sauté To cook in a skillet in a small amount of fat. It is a quick process, so food should be thinly sliced and tender.

Scald To heat a liquid, usually Milk or cream, over low heat until just before it boils. Small bubbles will form around edge of pan.

Score To make very thin slashes to mark the surface of the food with lines.

Sear To brown the surface of food quickly with high heat.

Shred To cut into thin pieces using large holes of a grater.

Simmer To cook just below the boiling point so that tiny bubbles form on bottom and sides of pan.

Sliver To cut into long thin pieces.

Steam To cook food, covered, over a small amount of boiling water.

Stir To mix ingredients with a circular motion, using a spoon.

Stir-fry To quickly sauté meat or vegetables while stirring constantly in a hot wok or frypan.

Toast To brown by baking (eg. bread).

Toss To tumble ingredients lightly with a lifting motion.

Whip To beat rapidly with wire whisk or beater to incorporate air and make mixture light and fluffy.

Weight/Volume Equivalents Guide

To help you shop for your recipe ingredients, use this guide. Remember, the conversions listed are approximate only.

Vegetable	Weight	Volume	Approx. Purchase Size
Asparagus	1lb (500g)	4 cups (1L) 1-inch (2.5 cm) pieces	16 to 20 medium spears
Beans, Green or Wax	1 lb (500g)	3 cups (750 mL) 1-inch (2.5 cm) pieces	n/a
Beets	1 lb (500g)	2 cups (500mL) diced	3 medium
Broccoli	1 lb (500g)	3 $\frac{1}{2}$ cups (875 mL) florets	1 medium bunch
Cabbage	2 lb (1kg)	8 cups (2L) shredded	1 medium head
Carrots	1 lb (500g)	3 cups (750mL) sliced or shredded	5 to 6 medium
Cauliflower	2 lb (1kg)	4 cups (1 L) florets	1 large head
Celery	$\frac{1}{4}$ lb (125g)	1 cup (250mL) sliced	2 stalks
Cucumber, Field	$\frac{1}{2}$ lb (250g)	2 cups (500mL) sliced or diced	1 medium
Cucumber, Pickling	1 lb (500g)	2-$\frac{1}{2}$ cups (625 mL) sliced	5 medium
Eggplant	1 lb (500g)	2-$\frac{1}{2}$ cups (625 mL) diced	1 medium
Green Onions	$\frac{1}{4}$ lb (125g)	$\frac{1}{2}$ cup (125mL) 1 bunch, sliced	8 medium
Lettuce, Boston	$\frac{1}{2}$ lb (250g)	6 cups (1.5 L) bite size pieces	1 small head

WEIGHT/VOLUME EQUIVALENTS GUIDE

Vegetable	Weight	Volume	Approx. Purchase Size
Lettuce, Iceberg	1¼ lb (625g)	6 cups (1.5 L) shredded 10 cups (2.5 L) bite-size pieces	1 small head
Mushrooms	½ lb (250g)	3 cups (750mL)sliced	24 medium
Onions, Cooking	1 lb (500g)	2 cups (500mL) chopped 3 cups (750mL) sliced	5 medium
Onions, Spanish	1¼ lb (625g)	3 cups (750mL) chopped 4 cups (1L) sliced	1 large
Potatoes, Baking	1 lb (500g)	4 cups (1L) sliced or diced 1¾ cups (425mL) mashed	3 medium
Potatoes, New	1 lb (500g)	4 cups (1L) sliced or diced	7 medium
Peppers, Green, Red,Sweet or Yellow	1 lb (500g)	2 cups (500ml) chopped	2 large
Radishes	¼ lb (125g)	1 cup (250mL) sliced	12 medium
Rutabaga	2 lb (1kg)	4 cups (1L) diced	1 large
Spinach	10oz (284 g)	6 cups (1.5 L) torn 1½ cups (375mL) cooked	1 bag
Tomatoes	1 lb (500g)	3 cups (750mL) chopped	3 medium
Turnip, White	1 lb (500g)	3 cups (750 mL) cubed	4 medium
Zucchini	1 lb (500g)	3 cups (750mL) sliced	3 to 4 medium

WEIGHT / VOLUME EQUIVALENTS GUIDE

Fruit	Weight	Volume	Approx. Purchase Amount
Apples	1 lb (500g)	3 cups (750mL) chopped 4 cups (1 L) sliced	3 medium
Apricots	1 lb (500g)	$2\frac{1}{4}$ cups (550mL) chopped	8 to 12 medium
Bananas	1 lb (500g)	2 cups (500mL) sliced 1-$\frac{1}{3}$ (325 mL) mashed	3 medium
Black Currants	1 lb (500g)	3 cups (750mL) whole	$1\frac{1}{2}$ pints
Blueberries	1 lb (500g)	3 cups (750mL) whole $2\frac{1}{2}$ cups (625 mL) lightly crushed	$\frac{1}{2}$ pints
Cherries	1 lb (500g)	$2\frac{1}{2}$ cups (625mL) whole	n/a
Cranberries	12 oz (340g)	3 cups (750mL) whole	1 bag
Gooseberries	1 lb (500g)	4 cups (1 L) whole	2 pints
Grapes	1 lb (500g)	$2\frac{1}{2}$ cups (625 mL) whole	1 medium bunch
Lemons	6 oz (175g)	$\frac{1}{3}$ cup (75ml) juice	1 medium
Limes	3 oz (75g)	3 tbsp (50mL) juice	1 medium
Melons (Honeydew)	4 lb (2kg)	4 cups (1 L) balls	1 medium
Melons (Cantaloupe)	2 lb (1 kg)	2 cups (500mL) balls	1 medium

WEIGHT / VOLUME EQUIVALENTS GUIDE

Fruit	Weight	Volume	Approx. Purchase Amount
Oranges	$\frac{1}{2}$ lb (250g)	$\frac{1}{2}$ cup (125mL) juice	1 medium
Peaches	1 lb (500g)	$1\frac{1}{2}$ cups (375 mL) chopped 2 cups (500mL) sliced	4 medium
Pears	1 lb (500g)	$2\frac{1}{2}$ cups (625 mL) chopped 3 cups (750mL) sliced	4 medium
Pineapple	2 lb (1 kg)	3 cups (750 mL) 1-inch (2.5cm) cubes	1 medium
Plums	1 lb (500g)	$2\frac{1}{2}$ cups (625 mL) pitted and chopped	8 to 12 medium
Raspberries	1 lb (500g)	4 cups (1L) whole 2 cups (500mL) lightly crushed	2 pints
Rhubarb	1 lb (500g)	3 cups (750mL) 1-inch (2.5 cm) pieces	4 to 8 stalks
Strawberries	1 lb (500g)	4 cups (1L whole) 2 cups (500mL) lightly crushed	3 pints

Adapted from Canadian Living Magazine

EMERGENCY SUBSTITUTION GUIDE

It's always best to use the exact ingredients called for in a recipe. But if you don't have a particular ingredient on hand look below for a substitute that will give satisfactory results. We recommend that you avoid making more than one substitution in a recipe.

Ingredient	Substitution
Anchovy fillet, 1	½ tsp. (2mL) anchovy paste.
Baking powder, 1 tsp. (5mL)	¼ tsp. (1mL) baking soda plus ½ tsp. (2mL) cream of tartar.
Broth, 1 cup (250mL) regular strength chicken **or** beef	Dissolve 1 chicken **or** beef bouillon cube in 1 cup (250mL) hot water.
Buttermilk or Sour Milk, 1 cup (250mL)	Mix 1 tbsp.(15mL) white vinegar **or** lemon juice with 1 cup (250mL) Milk, allow to stand 5 minutes.
Cake flour, 1 cup (250mL)	1 cup (250mL) all-purpose flour minus 2 tbsp. (30mL); **or** all purpose flour sifted 3 times, then measured to make 1 cup (250mL).
Chili hot red, dried, whole	½ to 1 tsp. (2 to 5 mL) crushed red pepper.
Chili oil, ¼ tsp. (1mL)	¼ tsp. (1 mL) salad oil plus pinch of cayenne.
Chinese five spice, 1 tsp. (5mL)	¼ tsp. (1 mL) **each** crushed anise seeds, ground cinnamon, ground cloves and ground ginger.
Chocolate, 1 square (28g) unsweetened	3 tbsp. (45mL) unsweetened cocoa plus 1 tbsp. (15mL) butter, melted.
Coconut Milk, 1 cup (250mL)	1 cup (250mL) whipping cream plus ½ tsp. (2mL) **each** coconut extract and granulated sugar.
Corn starch 1 tbsp. (15mL) (used for thickening)	2 tbsp. (30mL) all purpose flour.
Corn syrup, (250mL)	Mix 1 cup (250 mL) granulated sugar with ¼ cup (50mL) liquid*. (Do not use substitution for making candy)

EMERGENCY SUBSTITUTION GUIDE

Ingredient	Substitution
Egg yolks, 2 (used for thickening in custards)	1 whole egg.
Fines herbs,1 tsp. (5 mL)	$\frac{1}{4}$ tsp. (1 mL) **each** dried thyme leaves, oregano leaves, sage leaves, and rosemary.
Ginger,$\frac{1}{2}$ tsp. (2 mL) grated fresh	$\frac{1}{4}$ tsp. (1mL) ground ginger.
Herbs, fresh, 1 tsp. (5mL)	$\frac{1}{2}$ tsp. (2mL) of the same herb, dried.
Honey,1 cup (250mL)	1 $\frac{1}{2}$ cups (375mL) granulated sugar plus $\frac{1}{4}$ cup (50mL) liquid*.
Italian herb seasoning, 1 tsp. (5 mL)	$\frac{1}{4}$ tsp. (1 mL) **each** dried thyme leaves, marjoram leaves, oregano leaves and basil.
Ketchup, or tomato-based chili sauce 1 cup (250mL)	1 can (7 $\frac{1}{2}$ ozs/213 mL) tomato sauce plus $\frac{1}{2}$ cup (125mL) granulated sugar and 2 tbsp. (30mL) white vinegar.
Mustard, dry, 1 tsp. (5 mL) (in wet mixtures)	1 tbsp. (15ml) prepared mustard.
Onion, $\frac{1}{4}$ cup (50ml) fresh, minced	1 tbsp. (15mL) instant minced onion (let stand in liquid as directed).
Parsley, fresh 2 tbsp. (30 mL) minced	1 tbsp. (15mL) dried parsley flakes.
Sesame oil, 1 tbsp. (15mL)	1 $\frac{1}{2}$ tsp. (7 mL) sesame seeds sautéed in $\frac{1}{2}$ tsp. (2 mL) vegetable oil.
Tomatoes, 1 can (19ozs/540 mL)	2 $\frac{1}{2}$ cups (525 mL) chopped, peeled fresh tomatoes, simmered about 10 minutes.

*Use the same liquid called for in the recipe. Equivalence is based on how the product functions in recipe, not sweetness.

XXIV

NOTES

RECIPES

XXVI

NOTES

APPETIZER COQUILLES

½ cup	butter	125 mL
4 cups	sliced, fresh mushrooms	1 L
1 pound	fresh or frozen scallops	500 g
1 cup	dry white wine	250 mL
¾ teaspoon	salt	3 mL
¼ teaspoon	ground thyme	1 mL
1	bay leaf	1
3 tablespoons	flour	50 mL
1¼ cups	milk	300 mL
¾ cup	fine, dry, bread crumbs	175 mL

Melt 2 tablespoons (25 mL) butter in a saucepan. Sauté mushrooms until tender. Cut scallops in half; add to mushrooms with wine, salt, thyme and bay leaf. Bring to a boil. Cover and simmer 8 minutes or until scallops are tender. Drain, reserving 1 cup (250 mL) broth. Discard bay leaf. Melt 3 tablespoons (50 mL) of remaining butter in a saucepan; blend in flour. Gradually stir in milk and reserved broth. Cook over medium heat, sitrring constantly, until mixture just comes to a boil and thickens. Add mushrooms and scallops. Spoon into 8 shallow individual baking dishes or shells. Melt remaining 3 tablespoons (50 mL) butter. Add bread crumbs and sprinkle on top of each dish. Bake in preheated 400°F (200°C) oven 10 minutes or until browned and bubbly.
    Makes 8 servings.

Avocado appetizer

3	tablespoons	butter	45 mL
2	tablespoons	flour	30 mL
1	cup	milk	250 mL
pinch		pepper	pinch
¼	cup	chopped chives	50 mL
¼	cup	chopped parsley	50 mL
pinch		onion salt	pinch
pinch		celery salt	pinch
pinch		garlic powder	pinch
1	cup	shredded Mozzarella cheese	250 mL
1	(6-ounce)	can crab meat, drained	1 (200 g)
2		ripe avocados, split in half	2

Melt butter in a saucepan over medium heat; add flour, stirring constantly. Gradually add milk, stirring constantly. When thickened add pepper, chives, parsley and seasonings. Combine thoroughly. Add cheese, stir over heat until melted, do not boil. Remove from heat and add drained crab meat which has been flaked.

Fill avocado halves.

Bake at 350°F (180°C) for 15 minutes.

Makes 4 servings as an appetizer.

Mexican Cheese Dip,
page 11

BLUE CHEESE DIP

1	cup	crumbled blue cheese	250 mL
1	tablespoon	cooking oil	15 mL
¾	cup	milk	175 mL
1	cup	chopped pecans or walnuts	250 mL

LOW-CAL ZESTY DIP

1	cup	buttermilk	250 mL
1	pound	dry cottage cheese	500 g
¼	cup	chopped green onions	50 mL
2		large garlic cloves, minced	2
½	teaspoon	salt	2 mL
		pepper, freshly ground	
⅛	teaspoon	cayenne pepper	0.5 mL

For each of the above dips, blend all ingredients in a blender or food processor until smooth. Chill until ready to serve. Note: If a thinner dip is desired, add additional milk. Makes about 2½ cups.

CHEESY SHRIMP TOASTS

1 cup	grated Canadian Cheddar cheese	250 mL
2 tablespoons	mayonnaise	30 mL
¼ cup	milk	50 mL
1 (4 ounce)	can shrimp drained and chopped	1 (113 g)
¼ cup	finely chopped green pepper	50 mL
2 tablespoons	finely chopped green onion	30 mL
¼ teaspoon	dried basil	1 mL
	salt, pepper to taste	
8 slices	thinly sliced bread	8
	softened butter	
	pimento stuffed olives	

Combine cheese and mayonnaise in a bowl. Gradually whisk in milk until mixture is well blended. Stir in shrimp, green pepper, green onion, basil, salt and pepper. Remove crusts from bread. Cut diagonally into 4 triangles. Spread with butter. Top with a spoonful of cheese mixture. Place on lightly greased cookie sheet. Bake in 350°F (180°C) oven for about 10 minutes until edges of toast are lightly brown. Garnish with olive slices.
Makes 32.

COQUILLES ST. JACQUES

3 tablespoons	butter, melted	45 mL
1 tablespoon	lemon juice	15 mL
4 cups	fresh mushrooms, sliced	1 L
1 tablespoon	green onion, sliced	15 mL
½ cup	white wine	125 mL
¼ teaspoon	thyme	1 mL
1	bay leaf	1
¼ teaspoon	salt	1 mL
dash	pepper	dash
1 pound	scallops	500 g

SAUCE

3 tablespoons	butter, melted	45 mL
3 tablespoons	flour	45 mL
½ cup	broth from scallops	125 mL
1 cup	18% cream	250 mL

Sauté mushrooms and onion in 3 tablespoons (45 mL) butter and lemon juice for 5 minutes. Set aside. In a small pot, combine wine, thyme, bay leaf, salt, pepper and scallops. (If scallops are large, cut in half across the grain.) Bring to a boil, reduce heat, and simmer for 5 minutes. Discard bay leaf, and drain, reserving ½ cup (125 mL) of the broth.

To make sauce, combine 3 tablespoons (45 mL) melted butter with flour. Gradually add broth and cream, stirring constantly. Cook over medium heat until smooth and thickened. Stir in scallops and sautéed mushrooms. Place in scallop shells and heat under broiler until hot and bubbly.

Makes enough to fill 8 shells.

Serves 8 as an appetizer or 4 as a main course.

CRAB ON MELON APPETIZER

2	tablespoons	butter	30 mL
2	tablespoons	flour	30 mL
1	cup	milk	250 mL
2	tablespoons	chopped fresh chives	30 mL
2	tablespoons	chopped fresh parsley	30 mL
⅛	teaspoon	cayenne	0.5 mL
⅛	teaspoon	celery salt	0.5 mL
⅛	teaspoon	garlic powder	0.5 mL
1	cup	shredded Mozzarella cheese	250 mL
6	ounces	canned crabmeat, drained and flaked	200 g
1		melon or cantaloupe, peeled and sliced	1

In a medium saucepan, melt butter; add flour and stir to make a smooth paste. Gradually add milk, stirring constantly; cook until thick and bubbly. Add chives, parsley and seasonings; mix thoroughly. Add cheese, stirring just until melted. Remove from heat and add crabmeat. Spoon over thin slices of melon.
Makes 6 servings.

DIPPITY-DO CHEESE FONDUE

¾ cup	milk	175 mL
3 cups	cubed process Swiss cheese slices	750 mL
	bread sticks, apple wedges, raw vegetables or bread cubes	

Place ½ cup (125 mL) milk and cheese cubes in a saucepan. Cook over low heat, stirring constantly, until cheese is melted and smoothly combined. Gradually stir in remaining milk and continue stirring until fondue is heated through. Pour sauce into a small pot and keep warm over a candle flame. Serve with assorted dippers.
Makes 2⅓ cups/575 mL.

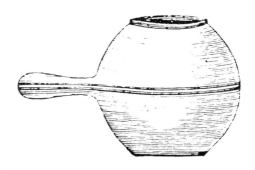

Gouda Dunk

2 tablespoons	butter	30 mL
2	green onions, finely chopped	2
2 tablespoons	flour	30 mL
2 teaspoons	chicken bouillon mix	10 mL
¼ teaspoon	Worcestershire sauce	1 mL
1½ cups	milk	375 mL
2½ cups	grated Gouda cheese	625 mL
1 teaspoon	chopped parsley	5 mL
	assorted fresh vegetables	
	cubed bread	

Melt butter in a medium saucepan. Add green onions. Cook until softened. Blend in flour, bouillon mix and Worcestershire sauce. Gradually stir in milk. Cook and stir over medium heat until mixture comes to a boil and thickens. Reduce heat, add cheese and stir until melted. Pour into a small pot over a candle warmer. Serve with vegetables and bread for dipping.

Makes 2 cups (500 mL).

HONEY GARLIC APPETIZER MEAT BALLS

2 pounds	lean ground beef	1 kg
1 cup	fine dry bread crumbs	250 mL
¾ cup	milk	175 mL
½ cup	finely-chopped onion	125 mL
2	eggs	2
2 teaspoons	salt	10 mL
1 tablespoon	butter	15 mL
4	cloves garlic, crushed	4
¾ cup	ketchup	175 mL
½ cup	liquid honey	125 mL
¼ cup	soy sauce	50 mL

Turn meat into a bowl and break up with a fork. Add and mix in bread crumbs, milk, onion, eggs and salt. Shape mixture into 1-inch (2.5 cm) balls. Place in a single layer in a 10 x 15 x ¾-inch (2 L) jelly roll pan. Bake in preheated 500°F (260°C) oven 12 to 15 minutes. Drain well. Melt butter in a saucepan; sauté garlic until tender. Add ketchup, honey and soy sauce. Bring to a boil. Reduce heat; cover and simmer 5 minutes. Add meatballs to sauce. Return to boil and simmer uncovered 5 to 10 minutes, stirring occasionally, or until sauce thickens slightly and glazes meatballs. Serve in a chafing dish with toothpicks.
Makes about 5½ dozen appetizers.

Hot Cheese Cocktail Dip

2 tablespoons	butter	30 mL
2 tablespoons	flour	30 mL
2 teaspoons	chicken bouillon mix	10 mL
pinch	dry mustard	pinch
1½ cups	milk	375 mL
1 cup	shredded old Canadian Cheddar cheese	250 mL
	radishes, celery sticks, cauliflowerets, carrot sticks, assorted crackers, etc.	

Melt butter in a small saucepan. Blend in flour, bouillon mix and dry mustard. Gradually stir in milk. Cook over medium heat, stirring constantly, until mixture just comes to a boil and thickens. Remove from heat. Add cheese and stir until melted. Pour into small pot over candle warmer. Serve with vegetables or crackers for dipping.

Makes 2 cups/500 mL.

MEXICAN CHEESE DIP

2 tablespoons	butter	30 mL
1	medium onion, chopped	1
1 cup	finely chopped tomatoes	250 mL
1½ tablespoons	flour	25 mL
1 cup	milk	250 mL
1 cup	cubed Canadian	250 mL
	Cheddar or Brick cheese	
1-2 tablespoons	chopped jalapeno chilies	15-30 mL
	or small hot peppers	

Sauté onion in butter; add tomatoes. Cook over medium heat about 5 minutes. Blend together flour and milk; pour over onions and tomatoes. Stir over low heat until thickened; add cheese and stir until melted. Add chilies. Serve hot.
Makes 2 cups.

MUSHROOM CROUSTADE APPETIZERS

⅓ cup	butter	75 mL
2½ cups	finely-chopped mushrooms	625 mL
¼ cup	finely-chopped green onion	50 mL
2 tablespoons	flour	30 mL
½ teaspoon	salt	2 mL
1 cup	milk	250 mL
1 tablespoon	finely-chopped parsley	15 mL
½ teaspoon	lemon juice	2 mL
48	Toast Cups	48

Melt butter in a saucepan. Sauté mushrooms and onions until tender and mushroom liquid has evaporated. Blend in flour and salt. Gradually stir in milk. Cook over medium heat, stirring constantly, until mixture just comes to a boil and thickens. Remove from heat and stir in parsley and lemon juice. Cover and refrigerate until serving time. Divide mixture evenly among Toast Cups. Bake 24 at a time in preheated 350°F (180°C) oven 15 minutes. Garnish with parsley to serve.

Makes 48 appetizers.

TOAST CUPS

¼ cup	butter, melted	50 mL
48	slices fresh bread	48

Brush melted butter in twenty-four 2-inch (5 cm) muffin cups. Cut a 3-inch (7.5 cm) round from each bread slice. Press bread rounds carefully into muffin cups. Brush each lightly with remaining melted butter. Bake in preheated 400°F (200°C) oven 12 to 15 minutes or until crisp and lightly browned. Remove from pans and repeat with remaining bread rounds.

SALMON MINI QUICHE

	pastry for 9-inch (23 cm) double pie shell	
1½ cups	grated Swiss cheese	375 mL
1 (7.5 ounce)	can salmon, drained, flaked	1 (213 g)
1 tablespoon	minced onion	15 mL
1 tablespoon	minced celery	15 mL
1 tablespoon	minced parsley	15 mL
	few drops Tabasco	
1 tablespoon	flour	15 mL
½ teaspoon	salt	2 mL
3	eggs, beaten	3
1 cup	table cream	250 mL
1 tablespoon	grated Parmesan cheese	15 mL

Roll out pastry and cut into twelve 5 inch (12 cm) circles. Set each circle loosely into 3 inch (7 cm) muffin cups. Chill. Toss together Swiss cheese, salmon, onion, celery, parsley, Tabasco, flour and salt. Spoon into pastry shells. Combine eggs and cream. Pour over cheese mixture in shells. Sprinkle each with Parmesan cheese. Bake in preheated 375°F (190°C) oven 25 to 30 minutes until set. Makes 1 dozen.

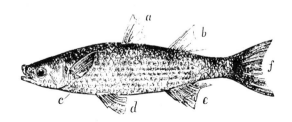

VERSATILE CHICKEN AND HAM PATÉ

2	tablespoons	butter	30 mL
3		green onions, finely chopped	3
1	pound	ground pork	500 g
1	pound	deboned chicken, chopped	500 g
6	ounces	cooked ham, chopped	200 g
1	tablespoon	finely chopped parsley	15 mL
1	teaspoon	salt	5 mL
¼	teaspoon	freshly ground pepper	1 mL
¼	teaspoon	allspice	1 mL
½	teaspoon	thyme	2 mL
2		eggs, beaten	2
1	cup	milk	250 mL
1	cup	fresh breadcrumbs	250 mL
3		hard boiled eggs (optional)	3

Melt butter in a small frypan; add green onions; cook over medium heat 2-3 minutes. In a large bowl, combine softened onions with remaining ingredients, except for the hard boiled eggs; mix well. Pack half the meat mixture into a 9″ x 5″ (2 L) loaf pan; place peeled hard boiled eggs down the centre, if desired; fill with the remaining meat mixture. Cover with foil; set in a large roasting pan containing hot water to reach halfway up the loaf pan. Bake in a 350°F (180°C) oven for 1½ hours.

To serve hot: Drain off excess fat, let rest in pan 15 minutes. Unmold and serve with tomato or cheese sauce.

To serve cold: Drain off excess fat, cool in pan, unmold and wrap well. Refrigerate. Flavour improves after 2-3 days well-wrapped in refrigerator. Serve with chili sauce or mustard-flavoured mayonnaise.

Makes 1 large loaf.

BANANA FRAPPE

1	cup	cold milk	250 mL
½	cup	yogurt	125 mL
2		medium ripe bananas	2
1	teaspoon	vanilla extract	5 mL
1		egg	1
2		whole walnuts, chopped	2

Place ingredients in a blender; blend until smooth and creamy.

Makes 1-2 servings.

BLENDER BREAKFASTS

BANANA ORANGE

2½ cups	milk	625 mL
2	eggs	2
1	medium banana, quartered	1
2 tablespoons	liquid honey	30 mL
⅓ cup	frozen orange juice	75 mL
	concentrate, thawed	

Makes 4 cups/1 L.

STRAWBERRY FROST

2 cups	milk	500 mL
2	eggs	2
1 cup	sliced frozen strawberries,	250 mL
	thawed	

Makes 4 cups/1 L.

PINEAPPLE GRAPEFRUIT WHIZ

2 cups	milk	500 mL
2	eggs	2
½ cup	crushed pineapple	125 mL
⅓ cup	frozen grapefruit juice	75 mL
	concentrate, thawed	
2 tablespoons	liquid honey	30 mL

Makes 3 cups/750 mL.

Combine ingredients for each drink in blender container. Cover and blend at high speed until smooth.

BLENDER BREAKFASTS

BREAKFAST-IN-A-GLASS

1	cup	milk	250 mL
2	tablespoons	frozen orange juice concentrate	30 mL
1		egg	1
		nutmeg, to taste	
¾	cup	orange yogurt	175 mL

Garnish with kiwi fruit. Makes 1 serving.

PINEAPPLE REFRESHER

1	cup	milk	250 mL
¼	cup	coconut cream	50 mL
1		egg	1
⅓	cup	crushed pineapple	75 mL
½	cup	fresh strawberries (optional)	125 mL

Makes 1 serving.

BANANA BUTTERMILK

1	cup	buttermilk	250 mL
1		banana, peeled	1
¼	cup	wheat germ	50 mL
2	teaspoons	honey or maple syrup	10 mL
⅓	cup	drained canned fruit	75 mL

Makes 1 serving.

Combine ingredients for each drink in a blender container. Cover and blend at high speed until smooth.

BLENDER MILKSHAKES

ORANGE BUTTERMILK COOLER

1½ cups	buttermilk, chilled	375 mL
1½ cups	orange juice, chilled	375 mL
1 tablespoon	lemon juice	15 mL
⅓ cup	sugar	75 mL

Makes 4 cups/1 L.

TIN ROOF

2 cups	milk	500 mL
2 cups	chocolate ice cream	500 mL
¼ cup	chocolate sundae sauce	50 mL
2 tablespoons	peanut butter	30 mL

Makes 4 cups/1 L.

CREAMY APRICOT

2 cups	milk	500 mL
2 cups	vanilla ice cream	500 mL
1 (14-ounce)	can apricots, drained	1 (398 mL)

Makes 4 cups/1 L.

BANANABERRY

2 cups	milk	500 mL
1 cup	raspberry sherbet	250 mL
2	bananas	2

Makes 3 cups/750 mL.

Combine ingredients for each drink in blender container. Cover and blend at high speed until smooth.

CANADIANA SHAKE

2	cups	cold milk	500 mL
¼	cup	maple syrup	50 mL
1	pint	maple walnut ice cream	0.5 L

Spoon ice cream into blender. Add remaining ingredients. Cover; blend until smooth and frothy.
Makes 4 servings.

PEACHES 'N' CREAM SHAKE

1	cup	cold milk	250 mL
1	cup	sliced peaches	250 mL
1	pint	peach or vanilla ice cream	0.5 L
2	tablespoons	frozen lemonade concentrate, thawed	30 mL

Combine peaches and milk in blender. Cover and blend at high speed until smooth. Spoon in ice cream, add lemonade concentrate. Cover and blend until smooth and frothy.
Makes 4 servings.

STRAWBERRY VELVET SHAKE

3 cups	milk	750 mL
1¼ cups	light cream	300 mL
3	eggs	3
¼ cup	sugar	50 mL
¼ teaspoon	salt	1 mL
1 teaspoon	vanilla	5 mL
3 cups	strawberry ice cream	750 mL

Scald milk and cream in top of double boiler. Beat eggs slightly; blend in sugar and salt. Stir in a little of the hot milk mixture, and return to top of double boiler. Cook over simmering water, stirring constantly, until custard thickens and will coat a metal spoon. Stir in vanilla. Chill. To serve, cut ice cream into large chunks. Place ice cream and custard in blender container. Cover and blend at low speed until thick and smooth. Pour into tall glasses.
Makes 6-7 cups/1.2-1.3 L.

CHOCOLATE PEANUTTY SHAKE

2 cups	chocolate milk	500 mL
¼ cup	peanut butter	50 mL
½ cup	vanilla ice cream	125 mL

Place ingredients in blender jar. Cover and blend until smooth, and frothy.

Makes about 3 cups/750 mL.

How to make the perfect milkshake

BASIC MILKSHAKE STEPS

1. Put two or more scoops of ice cream in a blender
2. Add ¾ cup of cold milk
3. Add flavouring
4. Blend until smooth
5. Pour into large glass

FLAVOURS

Vanilla – add 1 teaspoon vanilla
Chocolate – add 2 tablespoons chocolate syrup
Berry – add ⅓ cup berries, fresh or frozen
Fruit – add ⅓ cup fruit, fresh or canned

MILKSHAKES

GREAT GRAPE SHAKE-UP

1 cup	vanilla ice cream	250 mL
1 cup	milk	250 mL
¼ cup	frozen grape juice concentrate, thawed	50 mL

Makes about 2¼ cups/550 mL.

PEANUT BUTTER 'N' HONEY SHAKE

1 cup	vanilla ice cream	250 mL
1 cup	milk	250 mL
2 tablespoons	liquid honey	30 mL
2 tablespoons	peanut butter	30 mL

Makes about 2¼ cups/550 mL.

Spoon ice cream into blender container. Add remaining ingredients. Cover and blend at high speed until smooth and frothy.

STRAWBERRY BANANA SHAKE

1	medium banana, sliced	1
1 cup	cold chocolate milk	250 mL
2 cups	strawberry ice cream	500 mL
	vanilla ice cream (optional)	

Combine banana, chocolate milk and half strawberry ice cream in blender container. Cover and blend until smooth. Add remaining strawberry ice cream. Blend until desired consistency. Pour into tall chilled glasses. Top each with a scoop of vanilla ice cream.

Makes 3 cups/750 mL.

CHOCOLATE ORANGE BLOSSOM

| 2 cups | chocolate milk | 500 mL |
| 2 tablespoons | frozen orange juice concentrate | 30 mL |

Place ingredients in blender. Whirl until frothy. Makes about 2 cups/500 mL.

DESSERT SIPPERS

GRASSHOPPER

¾ cup	milk	175 mL
⅓ cup	mint liqueur	75 mL
2 cups	vanilla ice cream	500 mL
	grated chocolate	

Combine milk and mint liqueur in blender container. Spoon ice cream into container. Cover and blend at high speed until smoothly combined. Pour into chilled stemmed glasses, garnish with grated chocolate and serve with straws for sipping.
Makes about 3 cups/750 mL.

CRICKET

¾ cup	chocolate milk	175 mL
⅓ cup	chocolate or coffee liqueur	75 mL
2 cups	chocolate or coffee ice cream	500 mL
	grated chocolate	

Combine milk and chocolate liqueur in blender container. Spoon ice cream into container. Cover and blend at high speed until smoothly combined. Pour into chilled stemmed glasses, garnish with grated chocolate and serve with straws for sipping.
Makes about 3 cups/750 mL.

EGGNOG ROYALE

4 cups	coffee ice cream	1 L
8 cups	eggnog	2 L
4 cups	milk	1 L
3 cups	cold strong coffee	750 mL
	whiskey (optional to taste)	
2 cups	whipping cream	500 mL
	dash grated nutmeg	

The day before serving scoop ice cream into 10-12 balls; refreeze on cookie sheet. Just before serving, combine eggnog, milk, coffee and optional whiskey in large chilled punch bowl. Beat cream to form soft peaks and fold into eggnog mixture. Float ice cream balls on top and sprinkle with nutmeg. Serve at once.
Makes about 35 – 4 ounce/125 mL servings.

OPEN HOUSE COFFEE NOG

4 cups	cold milk	1 L
3 tablespoons	instant coffee crystals	45 mL
6	eggs	6
¾ cup	sugar	175 mL
1 cup	light rum	250 mL
1 cup	whipping cream	250 mL
2 cups	coffee or vanilla ice cream	500 mL

Heat 1 cup (250 mL) of the milk and stir in the coffee; chill. Beat eggs until thick and lemon coloured; gradually beat in sugar. Stir in coffee mixture, rum and remaining 3 cups (750 mL) milk. Whip cream until softly stiff; fold into milk mixture. Spoon ice cream on top of mixture in a punch bowl or spoon over individual servings. Serve immediately.
Makes about 11 cups/2.75 L.

FIRESIDE EGGNOG ALEXANDER

⅔ cup	chocolate liqueur	150 mL
⅓ cup	brandy	75 mL
1 L	dairy eggnog	1 L
	ground nutmeg	

Stir chocolate liqueur and brandy into eggnog. Chill well before serving over crushed ice. Sprinkle with nutmeg.
Makes about 5 cups/1.25 L.

FRESH LEMON COOLER

1½ cups	vanilla ice cream	375 mL
2 cups	milk	500 mL
¾ cup	freshly-squeezed lemon juice	175 mL
½ cup	sugar	125 mL
	yellow food colouring	
	(optional)	

Spoon ice cream into blender container. Add milk, lemon juice and sugar. Cover and blend at high speed until smooth. Tint a pale yellow if desired.
Makes about 5 cups/1.25 L.

LUSCIOUS LIME FROST

1½ cups	lime sherbet	375 mL
2 cups	milk	500 mL
½ cup	frozen limeade concentrate,	125 mL
	thawed	
	scoops of lime sherbet	
	mint leaves	

Spoon sherbet into blender container. Add milk and limeade concentrate. Cover and blend at high speed until smooth. Top each serving with a scoop of sherbet; garnish with mint leaves.
Makes about 5 cups/1.25 L.

FRUIT FLIP

1 cup	milk	250 mL
½ cup	natural yogurt	125 mL
1 (14 ounce)	can apricots, drained	1 (398 mL)
4	large ice cubes	4

Place ingredients in a blender jar and whirl until smooth. For variation use other favourite fruits, canned peaches, pineapple, mandarine oranges or fresh berries.
Makes 2 servings.

HAWAIIAN PUNCH

2 cups	vanilla ice cream	500 mL
3 cups	pineapple juice	750 mL
1 cup	orange juice	250 mL
4 cups	milk	1 L

Soften ice cream in a large mixing bowl. While beating ice cream, gradually add fruit juices then milk. Pour into chilled thermos containers.

Makes 10 cups/2.5 L.

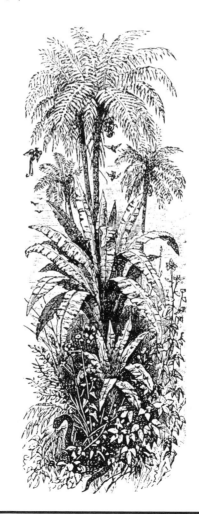

ITALIAN CAPPUCCINO COFFEE

1 cup	hot milk	250 mL
1 cup	hot espresso or strong	250 mL
	coffee	
2 teaspoons	sugar	10 mL
pinch	ground cinnamon	pinch
	grated semi-sweet chocolate	

Combine hot milk, hot coffee, sugar and cinnamon in a blender container. Cover and blend at high speed for 10 seconds or until frothy. Pour into cups and sprinkle with chocolate. Serve immediately.
Makes about 2 cups/500 mL.

CONTINENTAL HOT CHOCOLATE

2 cups	hot milk	500 mL
⅓ cup	chocolate syrup	75 mL
¼ cup	chocolate liqueur	50 mL
¼ teaspoon	almond extract	1 mL
½ cup	whipping cream	125 mL
	ground nutmeg	

Combine hot milk, syrup, liqueur and almond extract. Whip cream until softly stiff. Pour milk mixture into cups or mugs. Top with whipped cream and sprinkle with nutmeg.
Makes about 3 cups/750 mL.

P-NUTTY WARM UP

½ cup	liquid honey	125 mL
⅓ cup	smooth peanut butter	75 mL
4 cups	milk	1 L
	nutmeg (optional)	

In a saucepan smoothly combine honey and peanut butter. Gradually stir in milk. Cook over medium heat, stirring constantly until mixture is hot. Pour into mugs and sprinkle with nutmeg if desired.
Makes about 5 cups/1.25 L.

HOT BANANA COCOA

1 tablespoon	unsweetened cocoa	15 mL
1 tablespoon	sugar	15 mL
1 tablespoon	cold milk	15 mL
2 tablespoons	puréed banana	30 mL
1 cup	hot milk	250 mL
	ground cinnamon	

Combine cocoa and sugar in a mug. Blend in 1 tablespoon (15 mL) cold milk. Add banana purée. Stir in 1 cup (250 mL) hot milk; sprinkle with cinnamon. Serve immediately.
Makes about 1 cup/250 mL.

CAFE AU LAIT

1	cup	milk	250 mL
1	teaspoon	instant coffee crystals	5 mL
		cinnamon stick	

Heat milk but do not boil. Add coffee crystals; stir well. Serve with cinnamon stick.
Makes 1 serving.

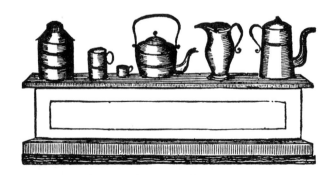

JOGGERS' NOG

1 cup	milk	250 mL
1 tablespoon	peanut butter	15 mL
1 tablespoon	honey (optional)	15 mL
½	banana	½
	dash of grated nutmeg	

Place all ingredients except nutmeg, in a blender jar; whirl until smooth. Serve with a sprinkling of nutmeg.

Makes 1 serving.

LIME REFRESHER

3	cups	cold milk	750 mL
1	pint	vanilla ice cream	0.5 L
6	ounces	limeade concentrate	200 g
4	teaspoons	lime sherbet	20 mL
		mint leaves	

Blend ice cream and lime concentrate until smooth. Add milk and combine. Pour mixture into four tall glasses. Add scoop of sherbet and garnish with mint leaf.
Makes 4 servings.

RASPBERRY FROSTY

2	cups	milk, divided	500 mL
¾	cup	raspberries, fresh or frozen	175 mL
2	tablespoons	sugar	30 mL

Freeze 1 cup milk in ice cube tray. Place frozen milk cubes in blender with fruit, sugar and 1 cup milk. Blend for approximately 1 minute. Note: Other fruits, fresh or frozen may be substituted.
Makes 2-3 servings.

Milk julep

2 cups	milk	500 mL
1 cup	chocolate mint ice cream	250 mL
	fresh mint	

Combine milk and ice cream in blender. Whirl until smooth. Garnish with fresh mint.

Makes 3 cups/750 mL.

MOCHA FLOAT

3	cups	milk	750 mL
1	tablespoon	instant coffee	15 mL
⅓	cup	chocolate syrup	75 mL
1	tablespoon	sugar	15 mL
4	scoops	vanilla ice cream	4

Combine milk, coffee, chocolate syrup and sugar in blender. Mix thoroughly. Pour into glasses and top with ice cream.

Makes 4 servings.

YOGURT FLIP

1	cup	milk	250 mL
½	cup	fruit-flavoured yogurt	125 mL
1		egg	1
1	tablespoon	honey	15 mL

Whirl all together in blender. Serve in chilled glasses. Top with fruit or coconut, if desired.

Makes 3 servings.

Purple Cow

| ¾ cup | frozen grape juice concentrate | 175 mL |
| 1½ cups | milk | 375 mL |

Combine ingredients in blender container. Cover and blend until smooth.

Makes 2 ¼ cups/550 mL.

PEACHY BANANA FLIP

1	banana	1
3	ripe medium peaches (or 5 drained canned peach halves)	3
1½ cups	milk	375 mL
1 cup	unflavoured yogurt	250 mL
½ teaspoon	vanilla	2 mL
1 tablespoon	sugar	15 mL
3	cracked ice cubes	3

Peel and slice banana. Peel peaches, remove pit and cut in chunks. Combine fruit in blender container with remaining ingredients. Cover and blend until smooth and fluffy. Serve at once. Makes about 5 cups/1.25 L.

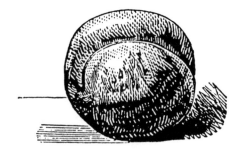

RASPBERRY COOLER

2 cups	buttermilk	500 mL
¾ cup	raspberry yogurt	175 mL
10 ounces	frozen raspberries, partially thawed	300 g

Combine ingredients in blender container. Cover and blend until smooth. Serve at once.

Makes 4 cups/1 L.

RASPBERRY FRAPPE

2 cups	fresh or frozen raspberries	500 mL
½ cup	lemon juice	125 mL
¼ cup	sugar	50 mL
1½ cups	cold milk	375 mL
1 cup	soft vanilla ice cream	250 mL
	crushed ice	
	lime slices	

Combine raspberries, lemon juice and sugar in a blender container. Cover and blend. Strain to remove seeds. Mix well with milk and ice cream. Fill chilled glasses with crushed ice and add raspberry mixture. Decorate each glass with a lime slice and serve with straws.

Makes about 6 servings.

SUNRISE STARTER

1	cup	milk	250 mL
1		egg	1
1	tablespoon	maple syrup	15 mL
1	teaspoon	instant coffee crystals	5 mL

Combine ingredients in blender until smooth.
Makes 1 large serving.

NOTES

A #1 SEAFOOD CHOWDER

6	slices bacon	6
2 (5 ounce)	cans baby clams (optional)	2 (142 g)
	water	
2	large potatoes, diced	2
1 (5 ounce)	can lobster meat (optional)	1 (142 g)
1 (5 ounce)	can crab meat	1 (142 g)
4 ounces	fresh mushrooms, finely chopped	125 g
2	large celery stalks, finely chopped	2
5	green onions, finely chopped	5
¼ teaspoon	soy sauce	1 mL
½ teaspoon	Worcestershire sauce (optional)	2 mL
3 cups	milk	750 mL
1 cup	10% cream	250 mL
2 tablespoons	butter	30 mL

Fry bacon until very crisp. When done, set on paper towel to absorb fat, and crush to make fine bacon bits. Set aside for garnishing the chowder when serving. Retain 2 tablespoons (30 mL) of the bacon fat in frypan for sautéing the vegetables later. While the bacon is cooking, measure the juice from the clams and add enough water to make 2 cups (500 mL). Add potatoes, and bring to a boil. When potatoes begin to boil, add baby clams, lobster and crab, and simmer. Sauté the mushrooms, celery and green onions in bacon fat for 5 minutes until vegetables are soft. Add soy sauce and Worcestershire sauce. In a double boiler or heavy saucepan, heat milk, cream and butter very slowly until hot. When ready, add milk mixture and sautéed vegetables to the simmering potatoes and seafood. Pour into individual bowls and garnish with bacon bits.

Makes 6 servings.

Autumn Vegetable Soup

2 tablespoons	butter	30 mL
2	medium onions, coarsely chopped	2
2	cloves garlic, minced	2
3	medium carrots, coarsely chopped	3
2	ribs celery, coarsely chopped	2
1	small zucchini, coarsely chopped	1
2	medium potatoes, peeled and diced	2
2	medium tomatoes, peeled and chopped	2
3 tablespoons	chopped fresh parsley (or 1 teaspoon (5 mL) dried)	45 mL
2 cups	chicken stock	500 mL
¼ teaspoon	dried thyme	1 mL
1 teaspoon	salt (or more to taste)	5 mL
¼ teaspoon	pepper	1 mL
2 cups	milk	500 mL

Melt butter in a large saucepan. Add onions and garlic. Cook until fragrant but do not brown. Add carrots, celery, zucchini, potatoes, tomatoes, half of the parsley (reserve remaining for garnish), stock, thyme, salt and pepper. Bring to a boil, cover, reduce heat and cook 25 minutes or until vegetables are very tender. Purée soup in blender, food processor or food mill. Return to heat. Add milk. Cook until heated thoroughly. Do not allow soup to boil. Taste and add seasoning if necessary. Garnish each serving with reserved parsley.

Makes 6 servings.

BORSCHT

5 cups	water	1.25 L
1 pound	pork spareribs, cut into pieces	500 g
2 cups	sliced mushrooms	500 mL
6	medium beets, peeled and grated	6
1	medium potato, diced	1
1	medium onion, diced	1
½ cup	tomato juice	125 mL
1-2 tablespoons	lemon juice	15-30 mL
2 tablespoons	flour	30 mL
2 cups	milk	500 mL
2 teaspoons	salt	10 mL
¼ teaspoon	pepper	1 mL
	fresh dill, chopped	

In a large pot, bring water and ribs to a boil. Reduce heat and simmer uncovered for 30 minutes. Add vegetables, tomato and lemon juice; simmer an additional 20 minutes. Blend flour with milk until smooth and pour into soup. Stir occasionally over low heat until thickened; season with salt and pepper. Garnish with dill.

Note: 6 cups of canned or frozen beets may be used.

Makes 6 to 8 servings.

Bean and Bacon Potage

2	slices chopped bacon	2
⅓ cup	chopped onion	75 mL
2 teaspoons	chicken bouillon mix	10 mL
2 cups	milk	500 mL
1 (19-ounce)	can baked beans	1 (540 mL)
1 cup	shredded Canadian Cheddar cheese	250 mL
¾ teaspoon	Worcestershire sauce	3 mL
	chopped parsley	

Cook bacon in a saucepan until crisp. Drain well; set aside. Sauté onion in bacon drippings until tender. Blend in bouillon mix. Stir in milk. Add beans, cheese and Worcestershire sauce. Cook over medium heat, stirring constantly, until cheese melts and soup is heated through. Do not boil. Transfer mixture to a blender container. Add reserved bacon. Cover and blend until smooth. Garnish with parsley to serve.

Makes 5 cups/1.25 L.

BROCCOLI SOUP

2 tablespoons	butter	30 mL
1	onion, chopped	1
1 pound	broccoli, trimmed and chopped	500 g
1 cup	chicken stock	250 mL
1½ cups	milk	375 mL
½ teaspoon	salt	2 mL
¼ teaspoon	freshly ground pepper	1 mL
½ teaspoon	curry powder, or to taste	2 mL
½ cup	sour cream	125 mL
	lemon juice	

Melt butter in a heavy saucepan; add onion; cook gently until softened. Add broccoli pieces and stock; simmer, covered, about 15 minutes. Remove; blend or purée vegetables until smooth. Return to pot and stir in milk, salt, pepper and curry powder; heat through but do not boil. Stir in sour cream and a squeeze of lemon juice; adjust seasonings before serving. Serve hot or chilled.

Makes 4 servings.

Variations: Stir a few tablespoons of crumbled blue cheese into hot soup before serving. Toss a few small sprigs of broccoli into lightly salted, boiling water for a few minutes. Drain and use to garnish each bowl of soup.

CANADIANA CHEDDAR CHEESE SOUP

3 tablespoons	butter	45 mL
¼ cup	finely-chopped onion	50 mL
¼ cup	finely-grated carrot	50 mL
3 tablespoons	flour	45 mL
1 tablespoon	chicken bouillon mix	15 mL
½ teaspoon	paprika	2 mL
½ teaspoon	dry mustard	2 mL
2 cups	milk	500 mL
2 cups	water	500 mL
2 cups	shredded old Canadian Cheddar cheese	500 mL
	toasted croutons	

Melt butter in a saucepan. Sauté onion and carrot until tender. Blend in flour, bouillon mix, paprika and mustard. Gradually stir in milk and water. Cook over medium heat, stirring constantly, until mixture just comes to a boil and thickens. Remove from heat. Add cheese and stir until melted. Garnish with toasted croutons to serve.

Makes about 5 cups/1.25 L.

CAULIFLOWER BISQUE WITH CHEESY CROUTONS

1		medium cauliflower	1
2	tablespoons	butter	30 mL
1		onion, chopped	1
1		potato, peeled and diced	1
1	cup	water	250 mL
½	teaspoon	salt	2 mL
2	cups	milk	500 mL
½	teaspoon	freshly ground pepper	2 mL
¼	teaspoon	grated nutmeg	1 mL

Trim cauliflower into florets. Drop into a pot of lightly salted boiling water and cook two minutes. Drain and set aside. In a large saucepan melt butter, add onion and potato and cook over medium heat for a few minutes. Add blanched cauliflower, water and salt. Cover and simmer 10 minutes. Blend or purée vegetables until smooth. Return to pot, stir in milk, pepper, nutmeg and additional salt to taste. Heat through but do not boil. Serve very hot with cheesy croutons.
Makes 4 servings.

Variation: Try with other vegetable combinations. Replace cauliflower with 4 cups (1 L) chopped carrots, squash, broccoli or celery.

CHEESY CROUTONS

Toss 1 cup (250 mL) bread cubes in a frying pan with 2 tablespoons (30 mL) melted butter. Spread in a single layer on a baking sheet, sprinkle with 1 tablespoon (15 mL) grated Parmesan cheese and bake for 15 minutes in a 300°F (150°C) oven until cubes are crisp and lightly browned.

CHEESE CAULIFLOWER SOUP

2 cups	potato, chopped	500 mL
2 cups	cauliflower, chopped	500 mL
1 cup	carrots, chopped	250 mL
1 cup	onion, chopped	250 mL
1	large clove garlic	1
4 cups	chicken stock or water	1 L
1 ½ teaspoons	salt (if using water)	7 mL
¾ cup	milk	175 mL
1 ½ cups	sharp Cheddar cheese, grated	375 mL
¼ teaspoon	dill weed	1 mL
¼ teaspoon	dry mustard	1 mL
dash	black pepper	dash
1 ½ cups	cooked cauliflower, chopped	375 mL
¾ cup	buttermilk	175 mL

Place chopped potatoes, cauliflower, carrots and onions in a large pot. Add garlic, salt (if being used) and stock. Bring to a boil; cover and simmer for 15 minutes. Let cool for 10 minutes, then whirl in blender, 2 cups (500 mL) at a time, until smooth. Return to the pot and add milk, cheese, dill weed, mustard, and black pepper. Heat gently, until cheese is melted. Just before serving, add the cooked cauliflower and buttermilk. Reheat to serving temperature and serve immediately.
Makes 10 servings.

CHEESE ZUCCHINI SOUP

4	strips bacon	4
½ cup	finely chopped onion	125 mL
¼ cup	chopped green pepper	50 mL
2½ cups	zucchini, cut in	625 mL
	¼ inch/6 mm slices	
1 tablespoon	chopped pimento	15 mL
1 cup	water	250 mL
½ teaspoon	salt	2 mL
¼ cup	butter	50 mL
¼ cup	all-purpose flour	50 mL
1 teaspoon	salt	5 mL
¼ teaspoon	pepper	1 mL
2½ cups	milk	625 mL
½ teaspoon	Worcestershire sauce	2 mL
1 cup	shredded Canadian	250 mL
	Cheddar cheese	250 mL

Cook bacon until crisp in skillet; set aside for garnish. Sauté onion and green pepper in bacon fat until tender. Add zucchini, pimento, water and salt. Cover; bring to a boil, turn down to simmer and cook about 5 minutes or until zucchini is tender. Meanwhile prepare cheese soup base. Melt butter in a 3-quart (3 L) saucepan; blend in flour, salt and pepper. Remove from heat; stir in milk and Worcestershire sauce. Heat to boiling, stirring constantly. Boil and stir 1 minute. Remove from heat; stir in cheese until melted. Return to low heat to finish melting if necessary. (Do not boil.) Add vegetables with liquid to soup base. Heat to serving temperature. Garnish with crumbled bacon.

Makes about 6 cups/1.5 L.

Note: Additional milk may be added to achieve desired consistency of soup.

CHEESY BROCCOLI SOUP

3 tablespoons	butter, melted	45 mL
½ cup	celery, chopped	125 mL
½ cup	onion, chopped	125 mL
4 cups	fresh broccoli, chopped	1 L
2 ½ cups	water	625 mL
2	packets instant chicken bouillon powder	2
¼ cup	flour	50 mL
3 cups	milk	750 mL
dash	pepper	dash
½ cup	Swiss cheese, grated	125 mL

In a large saucepan, sauté celery and onion in butter for 5 minutes until vegetables are soft but not brown. Add broccoli, water, and instant bouillon to the vegetables. Bring to a boil, reduce heat, cover and simmer for 10 minutes or until vegetables are tender. Combine flour, milk and pepper. Stir into vegetable mixture. Cook over medium heat, stirring constantly until mixture just comes to a boil, and is thickened. Pour into individual serving bowls and top with grated cheese.

Makes 6 servings.

CHEESY VEGETABLE SOUP

¼	cup	butter	50 mL
¾	cup	sliced celery	175 mL
½	cup	thinly sliced carrot	125 mL
¼	cup	chopped onion	50 mL
¼	cup	all-purpose flour	50 mL
1	(10-ounce)	can chicken broth	1 (284 mL)
3	cups	milk	750 mL
3	cups	shredded Canadian Cheddar cheese, divided	750 mL
6		slices Italian bread	6
		butter	
		chopped parsley	

For soup, melt butter in 2 quart (2 L) saucepan; sauté celery, carrot and onion until tender. Stir in flour. Cook until smooth, stirring constantly. Gradually stir in broth. Boil and stir for 1 minute. Stir in milk. Heat to simmering point, stirring constantly. Remove from heat and stir in 2 cups (500 mL) cheese until melted. Return to low heat to finish melting cheese if necessary. (Do not boil.) Meanwhile toast bread on both sides; butter one side.

Sprinkle buttered side with remaining cheese and parsley.

Broil until cheese melts. Top each serving of soup with a piece of hot toast. Serve immediately.

Makes about 6 cups/1.5 L.

CHILLED VICHYSSOISE SUPREME

3 tablespoons	butter	45 mL
1 cup	chopped onion	250 mL
4 cups	thinly-sliced pared potatoes	1 L
3½ cups	water	875 mL
4 teaspoons	chicken bouillon mix	20 mL
½ teaspoon	salt	2 mL
¼ teaspoon	pepper	1 mL
2¼ cups	milk	550 mL
	chopped parsley	

Melt butter in a large saucepan. Sauté onion until tender. Add potatoes, water, bouillon mix, salt and pepper. Bring to boil. Reduce heat, cover and simmer 25 to 30 minutes or until potatoes are tender. Pour half of mixture into blender container. Cover and blend until smooth. Pour into a large bowl. Purée remaining mixture and add to bowl. Stir milk into soup. Cover and chill. Serve well chilled. Garnish with parsley.

Makes about 8 cups/2 L.

CHUNKY CHICKEN CHOWDER

2½-3 pound		cut up broiler fryer chicken	1.2 to 1.5 kg
2		small onions, chopped	2
1½	cups	boiling water	375 mL
1	cup	sliced celery	250 mL
1½	teaspoon	salt	7 mL
pinch		pepper	pinch
¼	teaspoon	thyme	1 mL
3½	cups	milk	875 mL
3	tablespoons	flour	45 mL

Combine chicken and onions in a large saucepan. Add boiling water; cover and simmer 1 hour or until tender. Remove chicken. Add celery, salt, pepper and thyme to broth; cover and simmer 10 minutes or until celery is tender. Meanwhile remove bones and skin from chicken. Cut meat into 1" (2.5 cm) strips. Combine milk and flour. Add to celery mixture, stirring constantly until thickened and mixture comes to a boil. Add chicken. Heat thoroughly.
Makes 6-6½ cups/1.5 L.

CLAM CHOWDER WITH BACON AND CROUTONS

¼ cup	butter	50 mL
2	onions, finely chopped	2
¼ cup	all purpose flour	50 mL
4 cups	milk	1 L
3	medium potatoes, peeled and diced	3
1 teaspoon	dried thyme	5 mL
¼ teaspoon	salt (or more to taste)	1 mL
¼ teaspoon	pepper	1 mL
2 (5 ounce)	tins clams with juices	2 (142 g)

GARNISH

6	slices bacon, cooked crisp and chopped	6
½ cup	bread croutons	125 mL
3	green onions, chopped	3

Melt butter in a large saucepan or Dutch oven. Add onions. Cook until fragrant but do not brown. Add flour. Cook over medium low heat, stirring, for 5 minutes. Cool slightly. Whisk in milk. Bring to a boil. Add potatoes, thyme, salt and pepper. Reduce heat, cover and simmer until potatoes are tender (about 20 minutes). Add clams and juices and heat thoroughly. Do not boil. Taste and add seasoning, if necessary. Serve garnished with green onions, bacon bits and croutons.

Makes 6 servings.

CLAM SOUP

1 cup	milk	250 mL
1 cup	10% cream	250 mL
1 (5 ounce)	can baby clams, with juice	1 (142 g)
2 tablespoons	butter	30 mL
2 tablespoons	sliced green onion, tops only	30 mL
½ teaspoon	salt	2 mL
¼ teaspoon	pepper	1 mL

In saucepan, combine milk, cream, clam juice, butter, green onion tops, salt and pepper. Place pan on very low heat, and slowly heat to melt butter, and infuse the flavour of the green onion. Do NOT simmer. When the mixture is piping hot, add the clams and reheat slowly to serving temperature.
Makes 2 servings.

Cold Pink Beet Soup

1 cup	milk	250 mL
½ cup	sour cream	125 mL
2 tablespoons	white vinegar	30 mL
½ teaspoon	salt	2 mL
1 teaspoon	sugar	5 mL
3	eggs, hard-cooked and sliced	3
1 (14 ounce)	can beets with juice	1 (398 mL)
½	medium cucumber, finely chopped	½
1¼ cups	cold water	300 mL

Combine milk, sour cream, vinegar, salt, sugar, eggs, beets with juice, and cucumber in a blender. Purée until smooth. Add water. Refrigerate at least 4 hours before serving.
Makes 6 servings.

COLD BEET BORSCHT WITH CUCUMBERS

4 cups	buttermilk	1 L
1 cup	sour cream	250 mL
1 (14 ounce)	can diced beets with juice	1 (398 mL)
1½ cups	cucumber, peeled, seeded and diced	375 mL
2 teaspoons	dill weed	10 mL
2 tablespoons	green onions, minced	30 mL
1 tablespoon	red wine vinegar	15 mL
¼ cup	sugar	50 mL
1 teaspoon	prepared mustard	5 mL
4	chopped hard-cooked eggs	4

Whisk buttermilk and sour cream together in a large bowl. Add beets, cucumber, dill weed and onions. In a small bowl, blend vinegar, sugar and mustard. Stir into beet mixture. Add chopped eggs. Cover and chill at least 4 hours or overnight.

Makes 6 servings.

CORN 'N' BEEF SOUP

¼ cup	butter	125 mL
1	small onion, chopped	1
2 cups	diced potatoes	500 mL
1¼ cup	water	300 mL
1 tablespoon	all purpose flour	15 mL
3 cups	milk	750 mL
1 (3 ounce)	package smoked sliced beef, chopped	1 (75 g)
1 teaspoon	beef stock base	5 mL
1 (14 ounce)	can kernel corn	1 (398 mL)
½ teaspoon	celery seed	2 mL
1 cup	sour cream	250 mL
	chopped parsley	
	salt	
	freshly ground pepper	

Melt butter in large saucepan. Add onion and potato and cook for two minutes. Add water; cover, bring to a boil. Lower heat and simmer until potatoes are tender. Stir in flour and cook for 1 minute. Gradually stir in milk. Stir in beef, beef stock base, corn and seasonings. Heat through for flavours to blend. Gently stir in sour cream. Reheat and serve garnished with chopped parsley. Makes 8 servings.

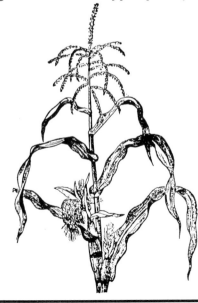

CORN CHOWDER

1	tablespoon	butter	15 mL
3		slices bacon	3
1		large onion, chopped	1
4		large potatoes, peeled and diced	4
3	cups	milk	750 mL
1	cup	creamed corn, fresh or canned	250 mL
2	cups	corn kernels	500 mL
1	teaspoon	salt	5 mL
1	teaspoon	finely chopped parsley	5 mL

Heat butter in a large heavy pan. Add bacon and onion and cook until tender. Add potatoes and cook over medium heat 5 mins. Stir in 2 cups (500 mL) milk. Bring just to a boil. Cover and simmer until potatoes are tender. Gently stir in the creamed corn, whole kernels and remaining milk. Heat through. Season. Serve with croutons and sliced cooked sausage.
Makes 4-6 servings.

CREAM OF TOMATO AND LEEK SOUP

3 tablespoons	butter	45 mL
3	leeks, white part only, sliced	3
2	cloves garlic, minced	2
1	carrot, chopped	1
1	rib celery, chopped	1
3 tablespoons	all purpose flour	45 mL
8	ripe tomatoes,	8
	peeled and chopped	
2 ½ cups	milk	625 mL
3 tablespoons	tomato paste	45 mL
1	bay leaf	1
½ teaspoon	dried thyme	2 mL
	or 1 teaspoon (5 mL) fresh	
1 ½ teaspoons	salt (or to taste)	7 mL
¼ teaspoon	black pepper	1 mL
3	green onions, sliced	3
½ cup	sour cream – optional	125 mL

Melt butter in a large saucepan. Add leeks, garlic and cook without browning until tender. Add carrots and celery; cook 5 minutes. Add flour and cook, without browning, 5 minutes. Add tomatoes, bring to a boil. Add milk, tomato paste, bay leaf, thyme, salt, pepper and slowly bring to a boil, stirring occasionally. Lower heat, cover, simmer gently 30 minutes. Remove bay leaf. Purée, heat thoroughly and season to taste. If soup is thicker than you like add additional milk. Serve sprinkled with green onions and a spoonful of sour cream.

Note: If tomatoes are not in season use one 28 ounce (796 mL) tin tomatoes with the juices. Do not worry if soup looks slightly curdled while cooking – it will come together when puréed.

Makes 6 servings.

CREAM OF ASPARAGUS SOUP

1	pound	fresh asparagus	500 g
2	cups	chicken broth or water	500 mL
1		small onion, finely chopped	1
3	tablespoons	butter	45 mL
3	tablespoons	flour	45 mL
2	cups	milk	500 mL
		salt and pepper to taste	

Discard the white part of asparagus stalks; rinse well in cold water. Cut into 1″ (3 cm) pieces. Place in saucepan with water or chicken stock and onion; cover and bring to a boil. Cook until asparagus is tender, about 10 to 15 minutes. Reserve a few tips for garnish. Purée soup in blender or food processor. Melt butter in saucepan; stir in flour and cook until smooth and bubbly. Add milk and seasonings; cook, stirring constantly until sauce thickens and comes to a boil. Add asparagus purée. Adjust seasoning to taste. Serve hot or cold, garnished with reserved asparagus. Note: If too thick when served cold, thin with additional milk.

Makes 6 servings.

CREAM OF RUTABAGA SOUP

4	slices bacon	4
2	medium onions, chopped	2
2	stalks celery, chopped	2
2	cloves garlic, minced	2
2 cups	rutabaga, diced	500 mL
1	medium potato, diced	1
3 cups	chicken stock	750 mL
1 tablespoon	sugar	15 mL
¼ teaspoon	white pepper	1 mL
1 cup	milk	250 mL

Cook bacon until crisp. Remove from pan. Crumble and set aside. Sauté onions, celery, and garlic in bacon fat for 5 minutes. Add rutabaga, potato, chicken stock, sugar and pepper; cover and simmer for 30 minutes. Blend in batches, in blender or food processor. Return to saucepan and stir in milk. Heat gently to serving temperature. Do not boil. Garnish with crumbled bacon.
Makes 6 servings.

Mushroom and Leek Soup,
page 94

CREAMED PUMPKIN SOUP

2 tablespoons	butter	30 mL
¼ cup	whole wheat flour	50 mL
2 cups	milk	500 mL
1	packet instant chicken bouillon powder	1
1 cup	hot water	250 mL
1 (14 ounce)	can cooked pumpkin	1 (398 mL)
1 teaspoon	very finely grated ginger	5 mL
½ teaspoon	salt	2 mL
dash	cayenne	dash
1 tablespoon	honey	15 mL

Melt butter in an 8-cup (2 L) saucepan. Add the flour, stirring until the flour absorbs all the fat. Gradually add 1 cup (250 mL) milk and continue to stir the mixture over a medium heat until it is smooth and thickened. Dissolve the instant bouillon in hot water. To the flour mixture, add the second cup (250 mL) of milk, bouillon, pumpkin, ginger, salt, cayenne and honey. Reheat to serving temperature.

Makes 6 servings.

CREAMY BROCCOLI POTAGE

4 cups	chopped fresh broccoli	1 L
½ cup	chopped celery	125 mL
½ cup	chopped onion	125 mL
2½ cups	water	625 mL
3 tablespoons	butter	45 mL
2 teaspoons	chicken bouillon mix	10 mL
2 teaspoons	salt	10 mL
¼ cup	flour	50 mL
3 cups	milk	750 mL

Place broccoli, celery, onion, water, butter, bouillon mix and salt in a large saucepan. Bring to a boil. Reduce heat; cover and simmer 10 minutes or until vegetables are tender. Smoothly combine flour and milk. Stir in vegetable mixture. Cook over medium heat, stirring constantly, until mixture just comes to a boil and thickens. Ladle into bowls to serve.

Makes about 7 cups/1.75 L.

CREAMY ONION SOUP

3 cups	onions, thinly sliced and separated	750 mL
2 tablespoons	butter	30 mL
2 tablespoons	flour	30 mL
5 cups	milk	1.25 L
1 teaspoon	salt	5 mL
dash	pepper	dash
dash	nutmeg	dash

Cook onions in butter about 10 minutes until soft. Sprinkle with flour, and stir until blended. Gradually add milk, and stir until combined well. Heat, stirring constantly, until just under the boil. Cover, and simmer 20 minutes, stirring occasionally. Add salt, pepper and nutmeg, stirring until blended. Serve with crusty French bread.

Makes 8 servings.

CREAMY ITALIAN MINESTRONE

2 (10-ounce)	cans beef broth	2 (284 mL)
1 (12-ounce)	package frozen chopped spinach	1 (340 g)
1½ cups	sliced quartered zucchini	375 mL
1 (10-ounce)	can tomato soup	1 (284 mL)
1 (5¼-ounce)	can tomato paste	1 (156 mL)
4 cups	milk	1 L
1¼ teaspoons	salt	6 mL
1 teaspoon	Italian seasoning	5 mL
¼ teaspoon	garlic powder	1 mL
1½ cups	macaroni, cooked and drained	375 mL
	grated Canadian Parmesan cheese	

Place beef broth and spinach in a large saucepan. Cook over low heat until spinach is thawed. Add zucchini. Bring to a boil. Reduce heat. Cover and simmer 10 minutes. Smoothly combine tomato soup and tomato paste. Stir into pan along with milk, salt, Italian seasoning and garlic powder. Cook over medium heat, stirring constantly, until mixture just comes to a boil. Add macaroni just before serving. Ladle into bowls and sprinkle with Parmesan cheese to serve.

Makes 12 cups/3 L.

CREOLE BISQUE

2 tablespoons	butter	30 mL
2 tablespoons	flour	30 mL
1 teaspoon	salt	5 mL
1 (19 ounce)	can stewed tomatoes with onions and peppers	1 (540 mL)
½ pound	frozen fish fillets, cut into 1″ (2.5 cm) pieces	250 g
½ pound	zucchini, sliced ¼″ (0.5 cm) thick	250 g
2 cups	milk	500 mL

In a large saucepan, melt butter. Stir in flour and salt, and blend until smooth. Slowly add 1 cup (250 mL) of milk, blending until smooth. Add the tomatoes, fish and zucchini. Bring to boil, reduce heat, cover and simmer 5-10 minutes, or until fish is cooked and zucchini is tender. Stir in remainder of milk. Heat to serving temperature.

Makes 6 servings.

CREAMY MUSHROOM SOUP

3 cups	thinly-sliced, fresh mushrooms	750 mL
¼ cup	chopped onion	50 mL
2 teaspoons	chicken bouillon mix	10 mL
2 cups	water	500 mL
¼ cup	butter	50 mL
¼ cup	flour	50 mL
1½ teaspoons	salt	7 mL
pinch	poultry seasoning	pinch
3 cups	milk	750 mL

Combine mushrooms, onion, bouillon mix and water in a medium saucepan. Bring to a boil. Reduce heat; cover and simmer 15 minutes. Melt butter in a large saucepan. Blend in flour, salt and poultry seasoning. Gradually stir in milk. Cook over medium heat, stirring constantly, until mixture just comes to a boil and thickens. Stir in undrained mushroom mixture. Serve hot.

Makes about 6 cups/1.5 L.

DELICATE CURRIED CRAB SOUP

3 tablespoons	butter	45 mL
2 tablespoons	flour	· 30 mL
½ teaspoon	curry powder	2 mL
4 cups	milk	1 L
1 (5 ounce)	can crab meat	1 (142 g)
¼ cup	dry white wine	50 mL
1 cup	sour cream	250 mL
dash	salt	dash
	snipped chives	

Melt butter in saucepan. Blend in flour and curry powder. Add milk gradually, blending well and stirring constantly, until thickened and bubbly. Add crab meat and wine, combining well. Cool. Blend about ½ cup (125 mL) of the soup into the sour cream until smooth. Then stir sour cream mixture into the rest of the soup. Chill several hours or overnight. If soup is too thick, thin with milk until the desired consistency is reached. Serve cold, garnished with chives.

Makes 6 servings.

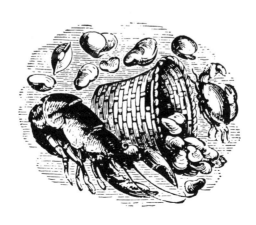

CURRIED AVOCADO SOUP

1	large onion, diced	1
1	stalk celery, diced	1
2 tablespoons	butter, melted	30 mL
4 tablespoons	flour	60 mL
2 cups	chicken stock	500 mL
1 tablespoon	lime juice	15 mL
1 teaspoon	fresh tarragon	5 mL
2 teaspoons	vinegar	10 mL
2 teaspoons	horseradish	10 mL
1	clove garlic, crushed	1
$\frac{1}{2}$ teaspoon	curry powder	2 mL
1 teaspoon	salt	5 mL
$\frac{1}{8}$ teaspoon	pepper	0.5 mL
dash	allspice	dash
1	avocado	1
2 cups	milk	500 mL

Sauté onion and celery in butter until soft. Add flour and blend well. Stir in chicken stock, and heat, stirring constantly, until thickened. Add lime juice, tarragon, vinegar, horseradish, garlic, curry powder, salt, pepper and allspice. Peel and cube avocado. Mash with fork, or in blender. Remove soup mixture from heat, and blend with avocado. Slowly, add milk, stirring constantly. Chill thoroughly.

Makes 4 servings.

EASY SALMON BISQUE

1 (14-ounce)	can New England style clam chowder	1 (398 mL)
1 (10-ounce)	can cream of celery soup	1 (284 mL)
1 (14-ounce)	can stewed tomatoes	1 (398 mL)
1 (7¾-ounce)	can salmon	1 (220 g)
2½ cups	milk	625 mL
¾ teaspoon	seasoned salt	3 mL

Combine clam chowder and celery soup in a saucepan. Add and break up tomatoes. Remove skin and bones from salmon, if desired and break up into large pieces. Add to pan. Stir in milk and salt. Cook over medium heat, stirring occasionally, until heated through. Do not boil.

Makes 8 cups/2 L.

FRENCH ONION SOUP AU LAIT

¼ cup	butter	50 mL
7 cups	sliced onion	1.75 L
2 tablespoons	flour	30 mL
3 (10-ounce)	cans beef broth	3 (284 mL)
3 cups	milk	750 mL
¼ cup	butter, melted	50 mL
1	small clove garlic, minced	1
6	slices French bread cut 1-inch (2.5 cm) thick	6
1½ cups	shredded Canadian Swiss cheese	375 mL

Melt ¼ cup (50 mL) butter in a large saucepan. Sauté onion until golden. Blend in flour. Gradually stir in beef broth. Cook over medium heat, stirring constantly, until mixture just comes to a boil and thickens. Reduce heat, cover and simmer 30 to 40 minutes. Stir in milk. Reheat to serving temperature; do not boil. Combine ¼ cup (50 mL) melted butter and garlic. Brush both sides of bread slices with butter mixture. Place on shallow baking pan and toast in preheated 325°F (160°C) oven 10 minutes; turn each slice and toast 5 minutes longer or until lightly browned. To serve, ladle about 1⅓ cups (325 mL) of soup into ovenproof soup bowls. Top each with 1 slice of toast and ¼ cup (50 mL) cheese. Return to oven 10 minutes or until cheese is melted.

Makes 6 servings.

FRESH CAULIFLOWER AND HAM CHOWDER

2 cups	frozen hash brown potatoes	500 mL
2 cups	sliced cauliflowerets	500 mL
½ cup	chopped onion	125 mL
1 cup	water	250 mL
1 tablespoon	chicken bouillon mix	15 mL
2 tablespoons	flour	30 mL
3 cups	milk	750 mL
2 cups	diced cooked ham	500 mL

Combine potatoes, cauliflower and onion in a large saucepan. Add water and bouillon mix. Bring to a boil. Reduce heat; cover and simmer 10 to 12 minutes. Blend in flour. Gradually stir in milk. Cook over medium heat, stirring constantly, until mixture just comes to a boil and thickens. Stir in ham and heat through.

Makes about 7 cups/1.75 L.

GARDEN TURKEY SOUP

1	turkey carcass	1
	leftover turkey gravy	
10 cups	water	2.5 L
1	onion, quartered	1
2	celery stalks	2
3	sprigs parsley	3
1	bay leaf	1
1 teaspoon	thyme	5 mL
1 tablespoon	salt	15 mL
½ teaspoon	pepper	2 mL
2	potatoes, peeled and diced	2
4 tablespoons	butter, melted	60 mL
1	onion, finely chopped	1
2	celery stalks, sliced	2
2	carrots, coarsely grated	2
¼ cup	flour	50 mL
2 cups	milk	500 mL
3 tablespoons	parsley, chopped	45 mL
½ teaspoon	marjoram	2 mL
½ teaspoon	dill	2 mL

Break carcass into pieces small enough to fit into a large pot. Add any leftover gravy and water. Add onion, celery, parsley, bay leaf, thyme, salt and pepper. Bring to boil and simmer 1½ hours. Remove carcass, and save any meat, setting it aside. Strain broth and reserve. Discard vegetables and carcass. Cover the potatoes with some of the broth, and boil until tender. Purée in the blender or strain through a sieve. Meanwhile, sauté chopped onion, sliced celery, and grated carrot in butter for 5 minutes, until tender. Add flour and stir to blend. Gradually add the milk, stirring constantly until thick. Add the puréed potatoes and remaining broth. Add any turkey meat, parsley, marjoram and dill. Simmer gently 10 minutes. Makes 12 servings.

GOLDEN CARROT SOUP

⅓ cup	butter	75 mL
½ cup	chopped onion	125 mL
2 cups	thinly-sliced carrots	500 mL
1 tablespoon	chicken bouillon mix	15 mL
3 cups	water	750 mL
¼ cup	uncooked long grain rice	50 mL
½ teaspoon	salt	2 mL
2 cups	milk	500 mL

Melt butter in a large saucepan. Sauté onion until tender. Add carrots, bouillon mix, water, rice and salt. Bring to a boil. Reduce heat; cover and simmer until carrots are tender and rice is cooked. Pour half the mixture into blender container. Cover and blend until smooth. Repeat with remaining mixture. Return all to saucepan. Stir in milk and heat through.
Makes about 6 cups/1.5 L.

GREEN SPLIT PEA SOUP

1½ cups	green split peas	375 mL
4½ cups	water	1.25 L
2 teaspoons	salt	10 mL
3 cups	finely-chopped cooked ham	750 mL
½ cup	finely-chopped onion	125 mL
½ cup	grated carrot	125 mL
2 cups	milk	500 mL
	cayenne	
	nutmeg	

Wash peas; drain. Add water and salt to peas. Bring to a boil; boil 2 minutes. Remove from heat. Cover and let stand 1 hour. Do not drain water. Add ham, onion and carrot to peas. Return to a boil. Cover and simmer about 1½ hours or until peas are tender. Stir in milk, cayenne and nutmeg to taste. Heat through and serve.

Makes about 9 cups/2.25 L.

HARVEST CHOWDER

2	tablespoons	butter	30 mL
½	cup	finely chopped onion	125 mL
4	cups	chopped vegetables (peas, zucchini, beans, carrots, potatoes, turnip, celery, etc.)	1 L
2	cups	drained and chopped canned tomatoes	500 mL
2	cups	water or stock	500 mL
1		bay leaf	1
1	tablespoon	chopped fresh parsley	15 mL
½	teaspoon	pepper flakes	2 mL
1	teaspoon	salt	5 mL
4	cups	milk	1 L
1	cup	cooked macaroni	250 mL

In a large heavy pot toss onions in butter until lightly browned. Add chopped vegetables and toss over heat for 2 minutes; add tomatoes, stock or water, herbs and seasonings. Cover and simmer until vegetables are just tender. Stir in milk and pasta; heat through. Serve hot, garnished with garlic-flavoured croutons, and fresh, chopped parsley. Accompany with grated Parmesan cheese and fresh crusty bread.
Makes 6 servings.

Hearty ham bone chowder

1 pound	dried navy or pea beans	500 g
	water	
2½ pound	ham shank with meat	2.25 kg
1 cup	chopped onion	250 mL
1 (19-ounce)	can stewed tomatoes	1 (540 mL)
2 cups	milk	500 mL
1 teaspoon	salt	5 mL

Wash beans; drain well. Place in a large saucepan. Cover with water. Heat to boiling. Boil 2 minutes. Remove from heat. Cover and let stand 1 hour. Drain beans and measure liquid; add water to make 6 cups (1.5 L). Return beans and liquid to pan. Add ham shank and onion. Bring to boil. Reduce heat; cover and simmer 2 hours or until beans are tender. Remove ham shank from broth. Strain beans from broth. Reserve 3 cups (750 mL) whole beans. Mash remaining beans and return to broth. Remove meat from bone and cut into small pieces. Add and break up tomatoes, reserved whole beans, meat, milk and salt to broth. Heat through.

Makes about 14 cups/3.5 L.

HEARTY CLAM SOUP

2	slices bacon, diced	2
½ cup	chopped onion	125 mL
1 cup	diced raw potatoes	250 mL
1 teaspoon	salt	5 mL
¼ teaspoon	celery salt	1 mL
½ cup	boiling water	125 mL
4 cups	milk	1 L
3 tablespoons	flour	45 mL
2 (5-ounce)	cans baby clams	2 (142 g)

Cook bacon until crisp in a large saucepan. Add onion and sauté until tender. Add potatoes, salt, celery salt and water. Cover and simmer over low heat 10 minutes or until potatoes are cooked. Combine milk and flour. Stir into potato mixture. Cook over medium heat, stirring constantly until thick and mixture comes to a boil. Add undrained clams and heat through. Do not boil. Serve hot.

Makes 6-7 cups/1.75 L.

HERBED FRESH TOMATO BISQUE

3 cups	chopped, peeled tomatoes	750 mL
2 cups	water	500 mL
½ cup	chopped onion	125 mL
1 (5½-ounce)	can tomato paste	1 (156 mL)
2 teaspoons	sugar	10 mL
2 teaspoons	salt	10 mL
½ to 1 teaspoon	basil leaves	2 to 5 mL
¼ to ½ teaspoon	ground thyme	1 to 2 mL
pinch	pepper	pinch
2 tablespoons	butter	30 mL
2 tablespoons	flour	30 mL
1 cup	milk	250 mL

In a medium saucepan combine tomatoes, water, onion, tomato paste, sugar, salt, basil, thyme and pepper. Cook over medium heat, stirring occasionally, until mixture comes to a boil. Reduce heat, cover and simmer 10 minutes. In a large saucepan melt butter. Blend in flour. Gradually stir in milk. Cook over medium heat, stirring constantly, until mixture just comes to a boil and thickens. Gradually stir in tomato mixture. Reheat if necessary. Do not boil.

Makes about 6 cups/1.5 L.

HUNGARIAN GOULASH SOUP

1 pound	lean stewing beef	500 g
2 cups	sliced onion	500 mL
1 (10-ounce)	can beef broth	1 (284 mL)
1 tablespoon	paprika	15 mL
1 teaspoon	salt	5 mL
1 teaspoon	caraway seeds	5 mL
2 cups	1-inch/2.5 cm carrot sticks	500 mL
3 tablespoons	flour	45 mL
3 cups	milk	750 mL
	chopped parsley	

Cut meat into ½-inch (12 mm) pieces. Combine meat, onion, beef broth, paprika, salt and caraway seeds in a saucepan. Bring to a boil. Reduce heat; cover and simmer 45 minutes. Add carrots and continue cooking 15 minutes or until meat and carrots are tender. Smoothly combine flour and milk. Stir into meat mixture. Cook over medium heat, stirring constantly, until mixture just comes to a boil and thickens. Garnish with parsley if desired.

Makes about 6 cups/1.5 L.

ICED CUCUMBER SOUP

2	large cucumbers	2
½ teaspoon	salt	2 mL
1½ cups	plain yogurt	375 mL
1 cup	milk	250 mL
2 teaspoons	finely grated onion	10 mL
½ teaspoon	dill weed	2 mL

Pare cucumbers; split lengthwise. Scoop out seeds. Grate cucumbers to make about 1½ cups (375 mL). Place grated cucumbers in a sieve; sprinkle with salt, and let stand for 15 minutes. Combine yogurt, milk, onion and dill, blending thoroughly. Drain cucumbers. Stir into yogurt mixture. Cover and chill. Garnish with a slice of cucumber or snipped fresh dill weed.
Makes 4 cups/1 L.

ICED SPINACH SOUP

¼ cup	butter	50 mL
¼ cup	finely chopped onion	50 mL
¼ cup	all purpose flour	50 mL
1 teaspoon	salt	5 mL
½ teaspoon	dry mustard	2 mL
¼ teaspoon	ground nutmeg	1 mL
1¼ cups	chicken stock	300 mL
1 (10 ounce)	package frozen chopped spinach, thawed	1 (300 g)
½ cup	grated carrots	125 mL
2½ cups	milk	625 mL

Melt butter in a medium saucepan. Add onions and cook until softened. Blend in flour, salt, mustard and nutmeg. Gradually whisk in chicken stock. Bring to a boil stirring constantly. Add spinach and carrots. Cook over medium heat stirring occasionally until carrots are tender and spinach thawed. Purée in blender or food processor until smooth. Stir in milk. Cover and chill. This soup is also good served hot.

Makes 6 servings.

KETTLE OF FISH CHOWDER

1 pound	cod or halibut fillets	500 g
3	slices bacon, chopped	3
⅓ cup	chopped onion	75 mL
⅓ cup	chopped celery	75 mL
2 cups	diced potatoes	500 mL
⅔ cup	water	150 mL
2 teaspoons	salt	10 mL
3 cups	milk	750 mL
1½ cups	frozen peas	375 mL
1½ teaspoons	Worcestershire sauce	7 mL
3 tablespoons	flour	45 mL
1¼ cups	light cream	300 mL
	chopped parsley or paprika	

Cut fish into bite-size pieces. Cook bacon in a saucepan until crisp. Drain, reserving drippings; set bacon aside. Sauté onion and celery in reserved drippings until tender. Add potatoes, fish, water and salt. Bring to a boil. Reduce heat; cover and simmer 10 to 15 minutes or until potatoes and fish are tender. Stir in milk, peas, crisp bacon and Worcestershire sauce. Smoothly combine flour and cream; add to mixture. Cook over medium heat, stirring constantly, until mixture just comes to a boil and thickens. Sprinkle with chopped parsley or paprika to serve.

Makes about 7½ cups/1.75 L.

LETTUCE SOUP

1		small romaine lettuce	1
1	tablespoon	butter	15 mL
1		carrot, grated	1
4		green onions, finely chopped	4
1	cup	chicken stock	250 mL
		lemon juice	
		pinch marjoram	
		salt and pepper	
1	cup	milk	250 mL
2		egg yolks	2

Clean and dry lettuce, remove tough stalks and shred leaves finely. Melt butter in a heavy saucepan. Add lettuce, carrot and green onions. Toss lightly over medium heat. Stir in stock and marjoram. Cover and simmer for 10 minutes. Combine milk and egg yolks; stir slowly into hot mixture. Warm through but do not boil. Add lemon juice and seasoning to taste. Serve immediately.

Makes 4 servings.

LOBSTER BISQUE

3 tablespoons	butter, melted	45 mL
3 tablespoons	flour	45 mL
dash	pepper	dash
¼ teaspoon	celery salt	1 mL
2 cups	milk	500 mL
1 cup	chicken stock	250 mL
2 teaspoons	instant onion flakes	10 mL
10 ounces	cooked lobster, canned or frozen	300 g
1 teaspoon	paprika	5 mL
¼ cup	10% cream	50 mL
1 tablespoon	sherry	15 mL

Combine butter, flour, pepper and celery salt. Gradually add milk and chicken stock, and cook over medium heat, stirring constantly, until smooth and thickened. Add onion flakes, lobster, and paprika. Heat gently for 10 minutes, stirring occasionally. Add cream and sherry and reheat to serving temperature.
Makes 4 servings.

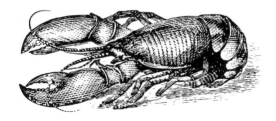

LO CAL CREAM OF ZUCCHINI SOUP

1	tablespoon	butter	15 mL
1		onion, chopped	1
1		carrot,chopped	1
2	cups	sliced zucchini	500 mL
2	cups	chicken broth	500 mL
1	teaspoon	tarragon	5 mL
½	teaspoon	salt	2 mL
¼	teaspoon	garlic powder	1 mL
2	cups	milk	500 mL

Sauté onion, carrot and zucchini in butter. Add chicken broth, seasonings and milk; cook for 10 minutes, or until vegetables are tender. Purée in blender or food processor until smooth. Serve garnished with paprika or thin zucchini slices.
Makes 4 diet servings.

MARTHA WASHINGTON'S CREAM OF CRAB SOUP

2	eggs, hard-cooked	2
2 tablespoons	butter	30 mL
1 tablespoon	grated lemon rind	15 mL
1 tablespoon	flour	15 mL
1 teaspoon	Worcestershire sauce	5 mL
⅛ teaspoon	mace	0.5 mL
2 (5 ounce)	cans crab meat, drained	2 (142 g)
1 tablespoon	butter, melted	15 mL
3	fresh mushrooms	3
3	celery stalks, finely chopped	3
1	small green onion, top and bulb, chopped	1
3 cups	milk, scalded	750 mL
2 cups	10% cream	500 mL
dash	pepper	dash

Make a paste of the eggs, 2 tablespoons (30 mL) butter, lemon rind, flour, Worcestershire sauce, and mace. Mix with the crab meat. In 1 tablespoon (15 mL) butter, sauté the mushrooms, celery, and green onion for about 5 minutes until soft. In a double boiler, or very heavy saucepan, combine the scalded milk, cream, and pepper. Add the crab meat paste and vegetables, and heat slowly until the mixture thickens.

Makes 6 servings.

MELON AND POTATO SOUP

2 cups	diced cantaloupe	500 mL
3	medium potatoes, boiled, peeled and diced	3
2 cups	milk	500 mL
4 tablespoons	butter	60 mL
½ teaspoon	salt	2 mL
¼ teaspoon	pepper	1 mL
4	egg yolks	4
¼ cup	sour cream	50 mL
1 tablespoon	milk	15 mL

Place half the melon in a blender with ½ cup (125 mL) of milk. Blend to a purée. Repeat with remaining melon, potatoes, and ½ cup (125 mL) more of milk. Melt butter, add the purée, and simmer for two minutes. Add remaining 1 cup (250 mL) more milk, salt and pepper, and continue cooking for 5 minutes over low heat. Beat the egg yolks until creamy, and stir gradually into the soup. Remove the saucepan from the heat as soon as the egg yolks have been added. Thin sour cream with 1 tablespoon (15 mL) milk, and swirl a garnish into each bowl. Serve immediately.

Makes 4 servings.

MUSHROOM AND LEEK SOUP

½ cup	butter	125 mL
2	bunches leeks	2
½ pound	mushrooms, chopped	250 g
¼ cup	flour	50 mL
1 teaspoon	salt	5 mL
dash	cayenne pepper	dash
1 cup	chicken broth	250 mL
3 cups	milk	750 mL
1 tablespoon	dry sherry (or lemon juice)	15 mL

Wash leeks very well; slice and use white part only. In ¼ cup (50 mL) butter, sauté leeks until tender but not brown. Remove and set aside. In remaining butter, sauté mushrooms until soft – about 10 minutes. Blend in flour, salt and cayenne. Gradually stir in broth and milk. Cook, stirring, until mixture thickens and comes to a boil. Add leeks, sherry, salt and pepper to taste. Simmer for 10 minutes. Serve with thin slices of lemon and a sprinkling of parsley, if desired.

Makes 6 servings.

ONION SOUP AU LAIT

1 tablespoon	butter	15 mL
1	medium onion, sliced	1
3-4	garlic cloves, minced	3-4
3 cups	milk	750 mL
1 teaspoon	salt	5 mL
	cayenne pepper to taste	
	pinch of nutmeg	
½ cup	grated Parmesan cheese	125 mL
4	slices French bread, toasted	4
1 cup	grated Mozzarella cheese	250 mL

In a saucepan, sauté onion and garlic in butter slowly until soft. Add milk; simmer for 20 minutes. Add seasonings and 1 tablespoon (15 mL) Parmesan cheese. Place toasted French bread slices in bottom of 4 individual ovenproof soup bowls; sprinkle Mozzarella cheese and remaining Parmesan over toast; pour in soup. Heat in oven or under broiler until cheese is melted and golden brown.

Makes 4 servings.

OYSTER STEW

¼ cup	onion, finely chopped	50 mL
¼ cup	butter, melted	50 mL
2 tablespoons	flour	30 mL
1 teaspoon	salt	5 mL
dash	pepper	dash
4 (5 ounce)	tins oysters, with liquid	4 (142 g)
3 cups	milk	750 mL

Sauté onion in butter until soft but not brown, about 5 minutes. Stir in flour, salt and pepper until smooth. Add oysters with liquid; simmer until edges of oysters begin to curl. Stir in milk gradually. Bring to serving temperature over low heat, stirring constantly.

Makes 6 servings.

POTAGE JARDINIERE

2	medium onions, chopped	2
3 tablespoons	butter	45 mL
4	large potatoes, diced	4
1 tablespoon	chopped chives (optional)	15 mL
3 tablespoons	chopped fresh parsley	45 mL
3 stalks	celery, including leaves, chopped	3
1 or 2	large carrots, sliced	1 or 2
2 teaspoons	salt	10 mL
¼ teaspoon	paprika	2 mL
1½ cups	boiling water	375 mL
4 tablespoons	butter	60 mL
2 tablespoons	flour	30 mL
1 teaspoon	salt	5 mL
¼ teaspoon	pepper	1 mL
4 cups	milk	1 L
2	chicken bouillon cubes	2

In a large saucepan or stockpot, sauté onion in 3 tablespoons (45 mL) butter until golden. Add potatoes, chives, parsley, celery, carrots, salt, paprika and boiling water; bring to boil and simmer until vegetables are tender, about 20 minutes. Purée mixture by pressing through a sieve or using a blender. Melt 4 tablespoons (60 mL) butter in the same saucepan. Blend in flour and let mixture bubble. Remove from heat and add milk and bouillon cubes. Cook over medium heat, stirring constantly, until thick and smooth. Season with salt and pepper. Add vegetable purée and heat through. To serve cold, thin slightly with more milk, chill and garnish with chopped chives.
Makes eight 8 oz. servings (250 mL each).

POTATO-BACON HARVEST SOUP

8	strips bacon, cut up	8
1 cup	chopped onion	250 mL
2 cups	cubed potatoes	500 mL
1 cup	water	250 mL
½ teaspoon	salt	2 mL
pinch	pepper	pinch
1 (10-ounce)	can cream of chicken soup	1 (284 mL)
1 cup	sour cream	250 mL
1¾ cups	milk	425 mL
2 tablespoons	chopped parsley	30 mL

Cook bacon until crisp in a 3-quart (3 L) saucepan. Add onion; sauté 3 minutes. Pour off drippings. Add potatoes, water, salt, and pepper; bring to a boil. Cover; simmer 10-15 minutes or until potatoes are tender. Gradually stir in soup, sour cream, milk and parsley. Bring to serving temperature over low heat, stirring occasionally. Do not boil.

Makes about 7 cups/1.75L.

PROTEIN POWER CHEDDAR SOUP

2 tablespoons	butter, melted	30 mL
2 tablespoons	onion, finely chopped	30 mL
⅓ cup	flour	75 mL
1¼ teaspoons	dry mustard	6 mL
¼ teaspoon	paprika	1 mL
2 teaspoons	Worcestershire sauce	10 mL
6 cups	milk	1.5 L
3	packets instant chicken bouillon powder	3
1½ cups	celery, thinly sliced	375 mL
2½ cups	Cheddar cheese, grated	625 mL

In a large pot, sauté onion in butter for about 5 minutes, until soft. Blend in flour, mustard, paprika and Worcestershire sauce. Gradually add milk, stirring constantly, and continue to cook over medium heat until smooth and thickened. Add instant bouillon powder and celery. Bring to a boil, and stir 1 minute. Remove from heat. Add cheese and stir until melted. Serve topped with finely chopped green pepper or cooked, crumbled bacon.

Makes 8 servings.

SWANKY FRANKY CHOWDER

2 tablespoons	butter	30 mL
1 tablespoon	flour	15 mL
1 teaspoon	chili powder (optional)	5 mL
½ teaspoon	prepared mustard	2 mL
2 cups	milk	500 mL
1 (19-ounce)	can baked beans	1 (540 mL)
¼ cup	ketchup	50 mL
2 tablespoons	molasses	30 mL
4	wieners, thinly sliced	4
	crushed corn chips	

Melt butter in a medium saucepan. Blend in flour, chili powder and mustard. Gradually stir in milk. Cook over medium heat, stirring constantly, until mixture just comes to a boil and thickens. Stir in baked beans, ketchup, molasses and wieners. Heat through. Serve hot, topped with crushed corn chips.
Makes 5 cups/1.25 L.

WATERCRESS VICHYSSOISE

2 tablespoons	butter, melted	30 mL
4	leeks, white part only, finely sliced	4
1	medium onion, sliced	1
2	bunches watercress, finely chopped	2
5	medium potatoes, peeled and sliced	5
4 cups	chicken stock	1 L
2 teaspoons	salt	10 mL
1/8 teaspoon	white pepper	0.5 mL
2 cups	milk	500 mL
1 cup	18% cream	250 mL

Cook leeks and onion in butter until they just begin to turn golden. Add watercress, potatoes, chicken stock, salt and pepper. Simmer for 40 minutes. Rub mixture through a sieve, or whirl in blender or food processor. Add the milk and cream. Chill thoroughly. Makes 8 servings.

ZUCCHINI BUTTERMILK BISQUE

¼ cup	butter, melted	50 mL
1	medium onion, chopped	1
1	clove garlic, minced	1
2 pounds	zucchini, coarsely grated	1 kg
1 teaspoon	curry powder	5 mL
2 cups	chicken broth	500 mL
dash	pepper	dash
3 cups	buttermilk	750 mL
	sour cream or yogurt for garnish	

Sauté onion and garlic in butter until soft, about 5 minutes. Add zucchini, and cook, covered, until soft, about 10 minutes. Add curry powder, chicken broth and pepper, stirring until well combined. Purée mixture in small batches in a blender. Transfer to a bowl and add buttermilk. Chill well. Garnish with a dollop of sour cream or yogurt.

Makes 6 servings.

AMERICAN STYLE ENCHILADAS

3	eggs, beaten	3
1½ cups	milk	375 mL
1 cup	all purpose flour	250 mL
1½ teaspoons	salt	7 mL
2 tablespoons	oil	30 mL
½ pound	ground beef	250 g
½ pound	ground pork	250 g
1	medium onion, chopped	1
½ cup	chopped green pepper	125 mL
2	cloves garlic, minced	2
1 tablespoon	chili powder	15 mL
plus 1 teaspoon		5 mL
1 (14 ounce)	can meatless spaghetti sauce	1 (398 mL)
1 (7½ ounce)	can tomato sauce	1 (213 mL)
1 cup	grated Canadian cheddar cheese	250 mL

Combine eggs and milk in a medium sized bowl. Whisk in flour and ½ teaspoon (2 mL) salt. Beat until smooth. Heat a 6″ (15 cm) skillet until hot. Lightly grease. Pour in ¼ cup (50 mL) batter. Swirl to cover bottom of pan. Cook briefly. Turn. Lightly brown. Stack while making remaining pancakes. Set aside. Heat oil in a large frying pan. Add beef and pork and cook until well browned. Drain off excess fat. Add onion, green pepper and garlic. Cook until heated through. Stir in chili powder, salt and 4 tablespoons (60 mL) tomato sauce. Simmer uncovered 10 minutes. Cool. To assemble: spoon ¼ cup (50 mL) meat mixture on each pancake. Fold envelope style and place in a greased 9″x 13″x 2″ (3.5 L) pan. Combine spaghetti sauce with remaining tomato sauce and pour on top. Sprinkle with cheddar cheese. Bake in 350°F (180°C) oven for 30 minutes until hot and bubbly.
Makes 12 enchiladas.

BARBEQUED BEEF LOAF

2 pounds	lean ground beef	1 kg
1½ cups	soft bread crumbs	375 mL
1 cup	milk	250 mL
2	eggs, slightly beaten	2
1 (42 g)	envelope onion soup mix	1 (42 g)
¼ cup	ketchup	50 mL
1 tablespoon	brown sugar	15 mL
1 tablespoon	prepared mustard	15 mL
1 teaspoon	Worcestershire sauce	5 mL

Turn meat into a bowl and break up with a fork. Add and mix in bread crumbs, milk, eggs and onion soup mix. Press lightly into a 9x5x3-inch (1.5 L) loaf pan. Bake in preheated 350°F (180°C) oven 1 hour. Drain off pan drippings. Combine ketchup, sugar, mustard and Worcestershire sauce. Spread over top of meat loaf. Return to oven and bake an additional 30 minutes. Cut in slices to serve.

Makes 6 to 8 servings.

BEEF CASSEROLE WITH PUFFED TOPPING

2 cups	cubed cooked beef	500 mL
¼ cup	butter	50 mL
3 tablespoons	flour	45 mL
2 cups	milk	500 mL
1	beef bouillon cube	1
1 tablespoon	finely chopped green onion	15 mL
1 tablespoon	finely chopped parsley	15 mL
½ cup	sliced mushrooms	125 mL
	salt, pepper to taste	
¼ pound	Canadian Cheddar cheese	125 g
2	eggs, separated	2

Preheat oven to 375°F (190°C).

Melt butter in a large heavy frying pan. Stir in flour and blend with butter. Cook 1 minute. Gradually whisk in milk. Add bouillon cube. Stir over medium heat until nicely thickened. Add mushrooms, green onion and parsley. Heat through. Transfer to an 8-inch square (2 L) greased baking pan. Cut cheese into thin strips. Beat yolks until thick and lemony in colour. Beat whites until stiff. Arrange cheese slices on top of beef. Fold egg yolks into whites and spoon over cheese and meat. Bake approximately 20 minutes, until puffed and golden.

Makes 4 servings.

BEEF LIVER WITH TOAST POINTS

⅓ cup	all-purpose flour	75 mL
1 pound	beef liver, sliced in ½" (12 mm) thick pieces	500 g
½ cup	butter	125 mL
½ cup	chopped onion	125 mL
¼ cup	chopped green pepper	50 mL
¼ cup	all-purpose flour	50 mL
1 teaspoon	salt	5 mL
½ teaspoon	thyme	2 mL
½ teaspoon	rosemary	2 mL
pinch	pepper	pinch
1 cup	milk	250 mL
1 cup	broth	250 mL
1 (10 ounce)	can sliced mushrooms, drained	1 (284 mL)
6	slices white bread, cut into 2 triangles, each	6

Dredge beef liver in ⅓ cup (75 mL) flour. In a large skillet melt ¼ cup (50 mL) butter, add liver and sauté until golden brown and fully cooked, about 15 minutes. Keep warm until sauce is prepared.

Melt remaining butter in a saucepan, add onion and green pepper. Cook until tender, about 5 minutes. Stir in flour and seasonings until smooth. Remove from heat, stir in milk and broth. Heat to boiling, boil and stir 1 minute. Stir in mushrooms.

To serve, place toast points on each serving plate. Divide liver evenly over toast. Spoon sauce over and serve immediately.

Makes 6 servings.

BEEF STUFFED ACORN SQUASH

2	medium acorn squash	2
¾ pound	lean ground beef	375 g
¼ cup	chopped onion	50 mL
¼ cup	chopped celery	50 mL
3 tablespoons	flour	45 mL
½ teaspoon	salt	2 mL
¼ teaspoon	ground sage	1 mL
1 cup	milk	250 mL
¾ cup	cooked rice	175 mL
¼ cup	shredded Canadian Colby or Cheddar cheese	125 mL

Cut each squash in half; discard seeds. Place cut side down in a 3-quart (3 L) rectangular baking pan. Bake in preheated 350°F (180°C) oven 50 to 55 minutes or until tender. Sauté beef, onion and celery in a large frypan until meat is browned; drain off any fat. Blend in flour, salt and sage. Stir in milk. Cook over medium heat, stirring constantly, until mixture just comes to a boil and thickens. Stir in cooked rice. Turn squash cut side up in pan; fill squash cavities with equal amounts of meat mixture. Return to oven and bake an additional 30 minutes. Top with cheese; bake 3 minutes longer. Makes 4 servings.

BEEF AND RICE CRISP

1 pound	lean ground beef	500 g
1 cup	coarsely-chopped onion	250 mL
½ cup	coarsely-chopped green pepper	125 mL
2 tablespoons	ketchup	30 mL
¼ teaspoon	salt	1 mL
½ teaspoon	dry mustard	2 mL
1½ cups	cooked rice	375 mL
1 cup	milk	250 mL
1 (10-ounce)	can cream of mushroom soup	1 (284 mL)
1½ cups	shredded Canadian Cheddar cheese	375 mL
1 teaspoon	Worcestershire sauce	5 mL
2 cups	corn flakes, coarsely crushed	500 mL
3 tablespoons	butter, melted	45 mL

Sauté beef, onion and green pepper in a large frypan until meat is browned; drain off any fat. Blend in ketchup, salt and dry mustard. Turn meat mixture into a 2-quart (2 L) round casserole; spread rice on top. Gradually stir milk into mushroom soup. Stir in cheese and Worcestershire sauce. Pour over meat and rice layers. Combine crushed corn flakes and butter; sprinkle evenly over casserole. Bake in preheated 375°F (190°C) oven 35 to 40 minutes or until hot and bubbly. Makes 6 servings.

BUTTERMILK LIVER

2 tablespoons	butter	30 mL
½ cup	bacon, diced	125 mL
1-2	large onions, sliced	1-2
1 pound	beef liver, thinly sliced	500 g
½ teaspoon	thyme	2 mL
1½ cups	buttermilk	375 mL
	salt	
	freshly ground pepper	
3 tablespoons	chopped fresh parsley	45 mL
½ cup	dry breadcrumbs	125 mL

Melt one tablespoon butter in a large skillet. Add bacon pieces and cook over medium heat until soft. Remove with slotted spoon and set aside. Add onion to pan and cook until soft. Remove to a buttered 4 cup (1 L) baking dish. Trim liver and cut into serving size pieces. Add to skillet and brown, 1-2 minutes on each side. Arrange liver over onions, season well with salt, pepper and thyme and add buttermilk. Toss breadcrumbs with remaining butter, bacon bits and parsley and sprinkle over top. Bake in preheated 350°F (180°C) oven covered for 15 minutes. Remove covering and continue baking until liver is tender and top is lightly browned. Serve over hot buttered rice.

Makes 4 servings.

CABBAGE CARAWAY BEEF BAKE

1 tablespoon	vegetable oil	15 mL
1	small onion, chopped	1
1	garlic clove, minced	1
1 pound	ground beef	500 g
½	medium cabbage, slivered	½
¾ cup	butter	175 mL
½ cup	flour	125 mL
4 cups	milk	1 L
1 teaspoon	caraway seeds	5 mL
¼ teaspoon	grated nutmeg	1 mL
1 teaspoon	salt	5 mL
	pinch freshly ground pepper	
½ cup	dry breadcrumbs	125 mL

Heat oil in large frying pan. Add onion and garlic and saute until tender. Stir in ground beef and cook until lightly browned. Drain off excess fat and set aside. Steam cabbage 5 minutes. Drain and set aside. Melt ½ cup (125 mL) butter in a saucepan. Stir in flour and cook over medium heat for two minutes. Remove from heat, quickly blend in milk. Return to heat and cook, stirring constantly until mixture comes to a boil and is smooth and thick. Season to taste with caraway, nutmeg, salt and pepper. Combine meat with sauce and shredded cabbage and turn mixture into a 2 quart (2L) baking dish. Melt remaining butter and toss with breadcrumbs; sprinkle crumbs over meat mixture. Bake in a preheated 350°F (180°C) oven for about 30 minutes until sauce is bubbling and top is golden.
Makes 4 servings.

CABBAGE AND LIVER FINGERS

1 pound	beef liver, thinly sliced	500 g
2 tablespoons	vegetable oil	30 mL
2 cups	finely shredded cabbage	500 mL
1/4 cup	butter	50 mL
1/2	small onion, minced	1/2
1/4 cup	flour	50 mL
2 cups	milk	500 mL
	pinch mace	
1 teaspoon	dry mustard	5 mL
	salt	
	freshly ground pepper	
1/2 cup	seasoned breadcrumbs	125 mL

Trim liver and cut into finger sized strips. Heat oil in large frying pan and add liver strips a few at a time to brown lightly on both sides. Season, remove and set aside. Steam cabbage in 1" (2.5 cm) boiling water for 5 minutes. Drain and set aside. Melt butter in saucepan, add onion and cook until soft. Stir in flour and cook for 2 minutes. Remove from heat; whisk in milk. Return to heat, stirring constantly, while mixture comes to a boil and thickens. Season with mace, mustard, salt and pepper. Add cabbage to sauce and turn into a buttered 4 cup (1L) baking dish. Top with liver strips and sprinkle with seasoned breadcrumbs. Bake uncovered in preheated 350°F (180°C) oven until heated through, about 20 minutes.
Makes 4-6 servings.

CABBAGE ROLLS

1 large	head of cabbage	1
2 cups	milk	500 mL
½ cup	water	125 mL
½ cup	long grain rice	125 mL
1¼ pounds	ground beef	625 g
1 small	onion, minced	1
1 teaspoon	salt	5 mL
¼ teaspoon	pepper	1 mL
1 (14 ounce)	can tomato sauce	1 (398 mL)
2 tablespoons	brown sugar	30 mL

Cook cabbage in simmering water just enough to soften leaves. Separate leaves. Trim thick stem on each leaf with knife. Heat 1 cup (250 mL) milk and water in a medium saucepan. Add rice. Cover and cook until done. Combine beef with remaining milk, onion, salt, pepper and cooked rice. Mix well. Spoon 2 tablespoons (30 mL) meat mixture on each cabbage leaf. Roll up envelope style. Place in a 9"x 13"x 2" (3.5 L) baking pan. Pour tomato sauce over. Sprinkle with brown sugar. Cover and cook in 325°F (160°C) oven for 1 hour. Uncover and continue cooking 30 minutes more until rolls are tender.

Makes 18 cabbage rolls; 6 servings.

CHEESEBURGER PIE

1 cup	variety baking biscuit mix	250 mL
¼ cup	milk	50 mL
1 tablespoon	vegetable oil	15 mL
1	medium onion, chopped	1
1	garlic clove, minced	1
1 pound	ground beef	500 g
2 tablespoons	variety baking biscuit mix	30 mL
1 tablespoon	Worcestershire sauce	15 mL
	salt	
	freshly ground pepper	
2	medium tomatoes	2
2	eggs, beaten	2
1 cup	shredded Canadian Cheddar cheese	250 mL
½ teaspoon	oregano	2 mL

Preheat oven to 375°F (190°C).

Combine baking mix and milk to make a soft dough. Knead gently 5 times on a lightly floured surface and roll out to fit into a 9″ (1 L) pie plate. Trim and flute edges and set aside. Heat oil in a large skillet and sauté onions and garlic until soft. Add beef and brown lightly. Drain off excess fat. Stir 2 tablespoons (30 mL) baking mix, Worcestershire sauce and seasoning into meat and turn into pastry lined pie plate. Arrange tomato slices on top. Combine eggs, cheese and oregano and spoon over tomatoes. Bake about 30 minutes. Serve in wedges, with chili sauce.

Makes 6 servings.

CHEESEBURGER MEAT BALLS

1	pound	lean ground beef	500 g
¼	cup	fine dry bread crumbs	50 mL
1		single serving package mushroom soup mix	1
½	cup	milk	125 mL
2	tablespoons	butter	30 mL
2	tablespoons	all-purpose flour	30 mL
¼	teaspoon	dry mustard	1 mL
¼	teaspoon	salt	1 mL
1¼	cups	milk	300 mL
2	cups	shredded Canadian Cheddar cheese	500 mL

Combine beef, crumbs, mushroom soup mix and ½ cup (125 mL) milk; shape into 1-inch (2.5 cm) balls. Place in single layer in large shallow pan or heavy jelly roll pan. Bake at 500°F (260°C) for 8-10 minutes; drain well. Melt butter in saucepan. Blend in flour, dry mustard and salt. Gradually stir in 1¼ cups (300 mL) of milk. Cook and stir over medium heat until mixture comes to a boil. Remove from heat. Add cheese; stir until melted. Pour over meatballs. Serve with rice or spaghetti.
  Makes 4 servings.

CHEESEBURGER CASSEROLE

1 (11-ounce)	package frozen peas and carrots	1 (312 g)
1 pound	lean ground beef	500 g
½ cup	chopped onion	125 mL
¼ cup	butter	50 mL
¼ cup	flour	50 mL
½ teaspoon	salt	2 mL
1½ cups	milk	375 mL
1½ cups	shredded Canadian Cheddar cheese	375 mL
½ teaspoon	Worcestershire sauce	2 mL
2 cups	hot mashed potatoes*	500 mL

Cook vegetables according to package directions. Drain well; set aside. Combine beef and onions in a saucepan; sauté until meat is browned and onions are tender. Drain; discard drippings. Add meat mixture to peas and carrots. Melt butter in a saucepan. Blend in flour and salt. Gradually stir in milk. Cook over medium heat, stirring constantly, until mixture just comes to a boil and thickens. Remove from heat. Add 1 cup (250 mL) of the cheese; stir until melted. Stir in Worcestershire sauce. Add meat and cooked vegetables; mix well. Turn into a 1½-quart (1.5 L) rectangular baking dish and bake in preheated 375°F (190°C) oven 20 to 25 minutes. Spoon potatoes around edge of casserole. Top with remaining cheese. Return to oven and bake an additional 5 minutes or until cheese melts.

Makes 4 to 6 servings.

* Make up your favourite recipe of mashed potatoes or use instant.

CORNED BEEF CASSEROLE

2 cups	shell macaroni	500 mL
1 (12 ounce)	can corned beef	1 (340 g)
¼ pound	grated Mozzarella cheese	125 g
1 (10 ounce)	can cream of chicken soup	1 (227 g)
1 cup	milk	250 mL
½ cup	chopped onions	125 mL
¼ teaspoon	ground pepper	1 mL
½ teaspoon	dried basil	2 mL
½ cup	breadcrumbs	125 mL
1 tablespoon	melted butter	15 mL
	parsley	
	olive slices	

Cook macaroni in boiling salted water until tender. Drain. Combine remaining ingredients except butter and breadcrumbs. Grease a 3 quart (3 L) shallow baking dish. Alternate layers of macaroni with corned beef. Combine butter and crumbs. Sprinkle on top. Bake in 375°F (190°C) oven for 1 hour until nicely browned. Garnish with parsley and olive slices if desired.

Makes 6-8 servings.

CORNED BEEF AND CABBAGE BAKE

12 cups	shredded cabbage	3 L
1 cup	water	250 mL
1 teaspoon	salt	5 mL
¼ cup	butter	50 mL
¼ cup	flour	50 mL
1 teaspoon	seasoned salt	5 mL
¼ teaspoon	dry mustard	1 mL
2 cups	milk	500 mL
1½ cups	shredded Canadian Colby cheese	375 mL
1 (12-ounce)	can corned beef	1 (340 g)

Combine cabbage, water and 1 teaspoon (5 mL) salt in a saucepan. Bring to boil. Reduce heat; cover and simmer until tender. Drain well. Melt butter in a large saucepan. Blend in flour, 1 teaspoon (5 mL) seasoned salt and dry mustard. Gradually stir in milk. Cook over medium heat, stirring constantly, until mixture just comes to a boil and thickens. Remove from heat. Add cheese and stir until melted. Cut corned beef into 12 slices. Alternately layer the cabbage, corned beef and sauce in a 2½-quart (2.5 L) casserole. Bake in preheated 350°F (180°C) oven 30 to 35 minutes or until hot and bubbly.
    Makes 6 servings.

CREAMY VEAL RAGOUT

1 pound	stewing veal	500 g
1 tablespoon	cooking oil	15 mL
2½ cups	water	625 mL
½ cup	chopped onion	125 mL
½ cup	chopped celery	125 mL
1¾ teaspoons	salt	10 mL
¾ teaspoon	ground thyme	3 mL
1½ cups	cubed potatoes	375 mL
1½ cups	cubed carrots	375 mL
1½ cups	frozen peas	375 mL
¼ cup	flour	50 mL
1 cup	milk	250 mL

Cut veal into 1-inch (2.5 cm) cubes. Brown in hot oil in a large saucepan. Add water, onion, celery, salt and thyme. Bring to a boil. Reduce heat; cover and simmer 30 minutes. Add potatoes and carrots. Simmer an additional 20 minutes. Add peas. Smoothly combine flour and milk. Add to saucepan. Cook over medium heat, stirring constantly, until mixture just comes to a boil and thickens. Serve hot.

Makes 4 servings.

CURRIED BEEF LOAF

3 slices	home style bread	3
1 cup	milk	250 mL
1½ pounds	ground beef	750 g
½ pound	ground pork	250 g
1	onion, finely chopped	1
1	garlic clove, minced	1
1 tablespoon	curry powder	15 mL
1 teaspoon	salt	5 mL
2 tablespoons	chopped fresh parsley	30 mL
2	eggs, beaten	2
	freshly ground pepper	

Remove crusts from bread; tear in pieces and let soak in milk. Toss together in a large bowl with remaining ingredients. Press lightly into a 9"x 5"x 3" (1.5 L) loaf pan. Cover with foil and bake in preheated 350°F (180°C) oven about 1½ hours. Drain off pan juices. Slice to serve. If to be served cold, leave loaf overnight in refrigerator.

Makes 6-8 servings.

CURRIED BEEF BAKE

3 tablespoons	butter	45 mL
2	onions, chopped	2
1	garlic clove, minced	1
2	slices, white bread	2
1¼ cups	milk	300 mL
2	eggs	2
1½ pounds	ground beef	750 g
2 tablespoons	regular curry powder	30 mL
1 tablespoon	sugar	15 mL
2 tablespoons	lemon juice	30 mL
1 teaspoon	salt	5 mL
¼ teaspoon	ground pepper	1 mL
½ cup	sliced almonds	125 mL
½ cup	raisins	125 mL

Cook onion and garlic in butter until soft. Soak bread in milk. In a large bowl combine the ground meat with 1 beaten egg, the cooked onions and garlic and the remaining ingredients. Squeeze bread, reserving milk and add to the meat mixture. Pile mixture into a well buttered 9″ square (2.5L) pan. Beat milk with remaining egg. Pour over meat. Bake for 1 hour in 350°F (180°C) oven until meat is cooked and top is lightly browned. Drain off excess fat before serving.

Makes 4 servings.

CURRIED LIVER STRIPS

1 pound	liver, cut in strips	500 g
⅓ cup	flour	75 mL
⅓ cup	butter	75 mL
1 cup	finely-chopped onion	250 mL
1 to 2 teaspoons	curry powder	5 to 10 mL
½ teaspoon	salt	2 mL
2 cups	milk	500 mL
1 (10-ounce)	can sliced mushrooms, drained	1 (284 mL)
	hot cooked rice	

Coat liver lightly with flour; set aside. Reserve leftover flour. Melt some of the butter in a large frypan. Sauté liver strips a few at a time until browned, adding more butter to pan as needed. Remove meat from pan. Add onion and sauté until tender. Blend in reserved flour, curry powder and salt. Gradually stir in milk. Return meat to pan; add mushrooms. Cook over medium heat, stirring constantly, until mixture just comes to a boil and thickens. Serve over hot cooked rice.

Makes 4 to 6 servings.

GREEK BEEF AND MACARONI CASSEROLE

2 cups	elbow macaroni	500 g
1 tablespoon	oil	15 mL
1 pound	ground beef	500 g
½ cup	chopped onion	125 mL
1 (5 ½ ounce)	can tomato paste	1 (156 mL)
½ teaspoon	salt	2 mL
3 tablespoons	flour	45 mL
½ cup	mayonnaise	125 mL
3 cups	milk	750 mL
3	eggs	3
½ cup	grated Parmesan cheese	125 mL
¼ teaspoon	grated nutmeg	1 mL

Cook macaroni in lightly salted water until tender. Drain. Heat oil in a large frying pan. Add beef and lightly brown. Add onion. Cook until softened. Drain off excess fat. Blend in tomato paste, salt. Set aside. Whisk flour with mayonnaise in a small saucepan. Gradually blend in milk. Cook stirring constantly over medium heat until nicely thickened. Remove from heat. Beat eggs with cheese and nutmeg. Gradually add to milk sauce. Mix macaroni and sauce together. Lightly grease a 9-inch square (2.5 L) baking dish. Spoon half the macaroni mixture into dish. Cover with meat mixture. Top with remaining macaroni. Bake in 325°F (160°C) oven until nicely browned, approximately 45 minutes. Let cool 15 minutes before serving.

Makes 6 servings.

HAMBURGER HOT CAKES

1 tablespoon	oil	15 mL
½ pound	ground beef	250 g
¾ cup	chopped onion	175 mL
¾ cup	chopped celery	175 mL
¼ cup	chopped green pepper	50 mL
1 (10 ounce)	can tomato soup	1 (284 mL)
1 teaspoon	salt	5 mL
1 teaspoon	Worcestershire sauce	5 mL
½ teaspoon	celery seed	2 mL
¼ teaspoon	ground pepper	1 mL
2 cups	all purpose flour	500 mL
4 teaspoons	baking powder	20 mL
1	egg	1
1¼ cups	milk	300 mL

Heat oil in a large frying pan. Add beef and cook until browned. Stir in onion, celery, green pepper. Cook until just tender. Drain off excess fat. Stir in soup, salt, celery seed and pepper. Simmer 5 minutes. Remove from heat. Combine flour and baking powder in a large mixing bowl. Whisk egg with milk in a small bowl. Add to dry ingredients and mix quickly just to moisten. Add meat mixture. Blend well. Heat a large lightly greased griddle or frying pan. Spoon ¼ cup (50 mL) batter onto griddle for each hot cake. Brown on both sides (5-8 minutes) to cook through. Serve as is or with a mushroom sauce.

Makes 15 hot cakes; 6 servings.

MUSHROOM SAUCE

Heat 1 10 ounce (284 mL) can cream of mushroom soup in a small saucepan. Add milk to desired consistency for sauce. Heat through.

HEARTY BEEF GOULASH

1	tablespoon	oil	15 mL
2		slices bacon, chopped	2
1		garlic clove, minced	1
2		carrots, sliced	2
2		medium onions, sliced	2
1		green pepper, sliced	1
¼	pound	whole small mushrooms	100 g
1½	pounds	stewing beef, cubed	750 g
¼	cup	flour	50 mL
3	tablespoons	paprika	45 mL
1	cup	beef stock	250 mL
1		bay leaf	1
1	cup	milk	250 mL
1	tablespoon	flour	15 mL
		salt	
		freshly ground pepper	

Heat oil in heavy frypan; add bacon and cook until lightly browned; remove with a slotted spoon to a large pot. Add garlic, carrots, onions, green pepper and mushrooms to frypan and stir over medium heat for five minutes; remove and add to pot.

Toss beef chunks in flour; brown well in frypan, adding more oil if necessary; add to vegetables. Mix paprika well into meat and vegetable mixture; stir in stock and bay leaf; bring slowly to a boil.

Lower heat; simmer, covered, until meat is tender, about 1½ hours. Add milk, mixed with remaining flour. Cook, stirring over medium heat until sauce is thickened. Season to taste.

Makes 4 servings.

LIKEABLE LIVER AND RICE CASSEROLE

½ cup	long grain rice	125 mL
1 ½ cups	milk	375 mL
½ cup	water	125 mL
½ teaspoon	salt	2 mL
1	bay leaf	1
¾ pound	beef liver, thinly sliced	350 g
2 tablespoons	butter	30 mL
2	onions, chopped	2
1 teaspoon	basil	5 mL
	salt, freshly ground pepper	
2	eggs, beaten	2
½ cup	grated Canadian Cheddar cheese (optional)	125 mL

Combine rice, milk, water, salt and bay leaf in top of a double boiler. Set over medium heat and cook, stirring occasionally until rice is tender and liquid is absorbed, about 45 minutes. Set aside; remove bay leaf. Trim liver and cut into 1″ (5 cm) pieces. Melt butter in large fry pan. Saute liver pieces a few at a time until lightly browned, remove meat from pan. Add more butter if necessary and saute onions until tender. Add herbs and seasoning to taste. Combine liver, seasoned onions, and cooked rice with eggs. Spoon mixture into a greased 4 cup (1L) casserole and top with grated cheese if desired. Bake in preheated 350°F (180°C) oven 30 minutes before serving.

Makes 4 servings.

LIVER WITH BACON 'N' ONION SAUCE

6	slices bacon, chopped	6
1 cup	chopped onion	250 mL
2 tablespoons	flour	30 mL
2 teaspoons	chicken bouillon mix	10 mL
1½ cups	milk	375 mL
1½ pounds	sliced beef liver	750 g

Cook bacon in a saucepan until crisp. Drain, reserving 3 tablespoons (45 mL) drippings; set bacon aside. Sauté onion in reserved drippings until tender. Blend in flour and bouillon mix. Gradually stir in milk. Cook over medium heat, stirring constantly, until mixture just comes to a boil and thickens. Keep warm. Pan fry or broil liver to desired degree of doneness. To serve pour sauce over individual servings of liver and top with crisp bacon.

Makes 6 servings.

MEXICAN BEAN BEEF CASSEROLE

1 tablespoon	butter	15 mL
3	bacon strips, diced	3
1	medium onion, chopped	1
1	garlic clove, minced	1
½ pound	ground beef	250 g
½ teaspoon	ground cumin	2 mL
1 (14 ounce)	can corn niblets	1 (398 mL)
1 (14 ounce)	can kidney beans	1 (398 mL)
1 (14 ounce)	can lima beans	1 (398 mL)
1	green pepper, chopped	1
	tabasco, to taste	
	pinch cayenne	
¾ cup	cornmeal	175 mL
2 cups	milk	500 mL
3	eggs, separated	3
1 teaspoon	baking powder	5 mL
¼ cup	grated Canadian Cheddar cheese	50 mL
	salt	
	freshly ground pepper	
	paprika	

In a large dutch oven, melt butter and sauté bacon until lightly browned. Add onion and garlic and cook until soft. Stir in ground beef and brown lightly. Drain away excess fat. Sprinkle on cumin. Add corn, beans, green pepper and season to taste. Let simmer for a few minutes. In a large saucepan add cornmeal to cold milk and bring to a boil, stirring constantly. Reduce heat and continue stirring until mixture becomes smooth and thick. Remove from heat and leave 5 minutes. Beat in egg yolks, baking powder, cheese and seasonings. Beat egg whites until stiff and fold into corn mixture. Lightly grease inside dutch oven above bean layer and pour in corn mixture. Sprinkle top with paprika. Bake in a preheated 350°F (180°C) oven for about one hour or until topping is cooked and lightly browned.
Makes 4-6 servings.

Mexican Casserole

3 cups	elbow macaroni	750 mL
2 tablespoons	finely chopped parsley	30 mL
6 tablespoons	grated Parmesan cheese	90 mL
2 tablespoons	butter	30 mL
½ teaspoon	salt	2 mL
1 tablespoon	oil	15 mL
1 pound	ground beef	500 g
1	medium onion, chopped	1
1	clove garlic, chopped	1
1 teaspoon	chili powder	5 mL
½ teaspoon	dried oregano	2 mL
1 (5½ ounce)	can tomato paste	1 (156 mL)
3	eggs, beaten	3
1½ cups	milk	375 mL

Cook macaroni in lightly salted water until tender. Drain and combine with parsley, 2 tablespoons (30 mL) Parmesan cheese, butter and salt. Set aside. Heat oil in a large heavy frying pan. Add beef and lightly brown. Add onion and garlic. Cook to soften. Drain off excess fat. Stir in chili powder, oregano and tomato paste. Cook 5 minutes. To assemble, spoon half the macaroni into a lightly greased 9-inch square (2.5L) baking dish. Top with meat mixture, then remaining macaroni. Beat eggs and milk together. Pour over macaroni. Sprinkle with remaining Parmesan. Bake in 350°F (180°C) oven for 45 minutes until hot and nicely browned.
    Makes 6 servings.

MY OWN MINI MEAT LOAF

1½ pounds	lean ground beef	750 g
1	egg, slightly beaten	1
⅔ cup	milk	150 mL
1 cup	fresh bread crumbs	250 mL
1 teaspoon	salt	5 mL
1 teaspoon	Worcestershire sauce	5 mL

Turn meat into a bowl and break up with a fork. Add and mix in egg, milk, bread crumbs, salt and Worcestershire sauce. Press meat mixture into 9 large muffin cups; level off tops. Bake in preheated 350°F (180°C) oven 35 to 40 minutes. Remove from pans. Pour EASY MUSHROOM SAUCE over individual servings.

Makes 9 mini meat loaves.

EASY MUSHROOM SAUCE

1 (10-ounce)	can cream of mushroom soup	1 (284 mL)
⅔ cup	milk	150 mL

Place mushroom soup in a small saucepan. Gradually stir in milk. Cook over medium heat, stirring constantly until heated through.

Makes 1¾ cups/425 mL.

PEPPERY BEEF STROGANOFF

1½ pounds	sirloin steak (at least 1″/2.5 cm thick)	750 g
1 tablespoon	ground black pepper	15 mL
⅓ cup	vegetable oil	75 mL
⅓ cup	butter	75 mL
2	onions, sliced	2
2	cloves garlic, minced	2
1 pound	mushrooms, sliced	500 g
3 tablespoons	brandy, rye or beef stock	45 mL
2 tablespoons	all-purpose flour	30 mL
1½ cups	milk, hot	375 mL
1 teaspoon	Worcestershire sauce	5 mL
1 tablespoon	tomato paste	15 mL
2	dill pickles, cut into matchsticks	2
½ cup	sour cream	125 mL
	Tabasco sauce, salt and pepper to taste	
1 pound	fettucine noodles	500 g
3 tablespoons	butter	45 mL
3 tablespoons	chopped fresh parsley	45 mL

Slice steak into thin strips. Pat meat dry and dust with pepper. Heat half the oil in a large skillet; cook half the meat until nicely browned. Reserve. Repeat with remaining meat. Discard excess oil. Return pan to heat; melt half the butter. Add onions and garlic; cook until tender. Add mushrooms and brandy; cook until mushrooms are tender and any liquid in the pan has evaporated. Add to reserved meat. Return pan to heat; add remaining butter. Stir in flour; cook 5 minutes without browning. Whisk in milk; Worcestershire sauce, tomato paste and bring to a boil. Cook 5 minutes. Add pickles, sour cream and meat. Heat gently. Season to taste. Cook fettucine in boiling salted water until tender. Toss with butter; season to taste. Serve with beef; sprinkle with parsley.
Serves 4 to 6.

Asparagus Cheddar Quiche,
page 144

Saucy Short Ribs

3 pounds	lean beef short ribs	1.5 kg
	cooking oil	
4 cups	diced onion	1 L
1 (10-ounce)	can beef broth	1 (284 mL)
2½ teaspoons	seasoned salt	12 mL
1 teaspoon	Worcestershire sauce	5 mL
¼ teaspoon	garlic powder	1 mL
¼ teaspoon	pepper	1 mL
⅓ cup	flour	75 mL
1 cup	milk	250 mL
	hot cooked noodles	

Trim excess fat from ribs and cut into serving size pieces. Heat a small amount of oil in a large saucepan. Brown meat, a few pieces at a time, adding more oil to pan as needed. Drain off any remaining fat. Return ribs to pan. Add onions, beef broth, salt, Worcestershire sauce, garlic powder and pepper. Bring to a boil. Reduce heat; cover and simmer 1½ hours or until meat is tender. Skim off any accumulated fat. Smoothly combine flour and milk. Gradually stir into meat mixture. Cook over medium heat, stirring constantly, until mixture just comes to a boil and thickens. Serve over noodles.

Makes 6 servings.

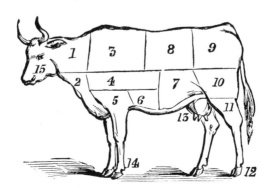

SAUERBRATEN SAUCED BEEF LOAF

1½ pounds	lean ground beef	750 g
1 cup	fresh rye bread crumbs	250 mL
⅔ cup	milk	150 mL
¼ cup	finely-chopped onion	50 mL
1 teaspoon	salt	5 mL
¼ teaspoon	caraway seed	1 mL
1	egg, slightly beaten	1
1½ cups	water	375 mL
2 teaspoons	beef bouillon mix	10 mL
¼ cup	firmly-packed brown sugar	50 mL
¼ cup	raisins	50 mL
3 tablespoons	lemon juice	45 mL
½ cup	coarse ginger snap crumbs	125 mL

Combine beef, bread crumbs, milk, onion, salt, caraway seed and egg. Shape mixture into a round loaf; set aside. In a 10-inch (25 cm) frypan combine water, bouillon mix, sugar, raisins, lemon juice and ginger snap crumbs. Cook over medium heat, stirring constantly, until mixture comes to a boil and thickens. Add beef loaf. Spoon sauce over loaf. Heat to boiling. Cover and simmer about 1 hour, basting occasionally. Remove loaf from pan. Slice and serve with sauce.
    Makes 6 servings.

Savoury Veal Casserole

6 tablespoons	butter	90 mL
2	medium onions, chopped	2
1½ pounds	ground veal	750 g
1 tablespoon	tomato paste	15 mL
½ teaspoon	salt	2 mL
¼ teaspoon	pepper	1 mL
3 tablespoons	finely chopped parsley	45 mL
½ cup	beef stock	125 mL
¼ cup	dry sherry	50 mL
	oil	
4	medium eggplant	4
6 tablespoons	breadcrumbs	90 mL
½ cup	grated Parmesan cheese	125 mL
2 tablespoons	flour	30 mL
1¾ cups	milk	425 mL
1	egg yolk	1

Melt 2 tablespoons (30 mL) butter in a large frying pan. Cook onions until softened. Add veal and cook until just browned. Remove excess fat. Add tomato paste, salt, pepper, parsley, beef stock, 2 tablespoons (30 mL) of sherry. Simmer 15 minutes covered. Stir in 3 tablespoons (45 mL) breadcrumbs. Cut eggplants into ⅜″ (1 cm) slices. Fry in oil until lightly browned. Set aside. Prepare sauce. Melt remaining butter in saucepan. Whisk in flour. Cook briefly. Gradually add milk and cook until lightly thickened. Stir in remaining sherry. Remove from heat, add sauce slowly to beaten egg yolk.

To assemble: grease a 9-inch square (2.5 L) casserole dish. Dust with 1 tablespoon (15 mL) breadcrumbs. Line bottom of dish with one third of the eggplant slices; layer with half the meat mixture. Sprinkle generously with 2 tablespoons (30 mL) Parmesan cheese. Repeat layers. Top with another layer of eggplant and pour sauce over all. Sprinkle with remaining Parmesan cheese and breadcrumbs and dot with butter. Bake in 350°F (180°C) oven for 45 minutes until golden brown.

Makes 4-6 servings.

SCALLOPED HAMBURGER CASSEROLE

1 tablespoon	butter	15 mL
1 ½ pounds	ground beef	750 g
2	medium onions, chopped	2
1 teaspoon	salt	5 mL
¼ teaspoon	pepper	1 mL
½ teaspoon	oregano	2 mL
3 tablespoons	all purpose flour	45 mL
2 ½ cups	milk	625 mL
¼ cup	finely chopped parsley	50 mL
4 cups	thinly sliced peeled potatoes	1 L
2 cups	grated Canadian cheddar cheese	500 mL

Melt butter in a large frying pan. Add beef and onion. Cook until meat is lightly browned. Drain off all but 2 tablespoons fat (30 mL). Stir in salt, pepper, oregano and flour. Cook briefly. Blend in milk and cook and stir over medium heat until nicely thickened. Add parsley. Arrange alternate layers of potatoes and beef mixture in greased 9 inch square (2.5 L) casserole dish. Cover and bake in 350°F (180°C) oven for 1 ½ hours, or until potatoes are tender. Uncover. Sprinkle with cheese. Reduce heat to 300°F (150°C) and continue cooking, 20 minutes. Remove from oven and let rest 20 minutes before serving.

Makes 6 servings.

SHEPHERD'S PIE

1 tablespoon	oil	15 mL
1 pound	ground beef	500 g
2	medium onions, chopped	2
1	clove garlic, chopped	1
1 teaspoon	salt	5 mL
¼ teaspoon	pepper	1 mL
¼ teaspoon	dried thyme	1 mL
3 tablespoons	all purpose flour	45 mL
1 (10 ounce)	can beef broth	1 (284 g)
1 teaspoon	Worcestershire sauce	5 mL
2	carrots, diced	2
4	large potatoes	4
1 tablespoon	butter	15 mL
1	egg, beaten	1
½ cup	milk	125 mL
½ cup	grated Canadian cheddar cheese	125 mL
2 tablespoons	grated Parmesan cheese	30 mL
2 tablespoons	buttered bread crumbs	30 mL

Cook ground beef, onions and garlic in oil until beef is lightly browned. Add salt, pepper, thyme and flour. Blend well. Gradually stir in beef broth. Cook until nicely thickened. Stir in Worcestershire sauce and carrots. Cover and simmer over low heat for 20 minutes. Spoon into a buttered 9-inch square (2.5 L) casserole dish. Peel and cook potatoes until tender. Mash with butter. Blend ½ beaten egg with milk, and cheddar cheese. Beat into mashed potatoes. Spread on top of meat mixture. Brush with reserved egg. Sprinkle with Parmesan and breadcrumbs. Bake in 400°F (200°C) oven for 15 minutes or until golden brown.

Makes 6 servings.

SPAGHETTI MEAT BAKE

½ pound	lean ground beef	250 g
1 (14-ounce)	can spaghetti sauce	1 (398 mL)
1 (200 g)	package spaghetti	1 (200 g)
1 tablespoon	butter	15 mL
1 tablespoon	flour	15 mL
¼ teaspoon	salt	1 mL
1 cup	milk	250 mL
½ cup	shredded Canadian Colby cheese	125 mL
2 tablespoons	grated Canadian Parmesan cheese	30 mL

Brown beef in a saucepan; drain off drippings. Return meat to pan. Add spaghetti sauce. Bring to a boil. Reduce heat; simmer, uncovered, 8 to 10 minutes. Break spaghetti into thirds. Cook according to package directions; drain well. Melt butter in a saucepan. Blend in flour and salt. Gradually stir in milk. Cook over medium heat, stirring constantly, until mixture just comes to a boil and thickens. Remove from heat. Add half the Colby and Parmesan cheeses and stir until melted. Combine cooked spaghetti and meat sauce. Spread half the spaghetti mixture in bottom of a 1½-quart (1.5 L) shallow rectangular baking dish. Spoon half the cheese sauce over spaghetti. Repeat layers. Sprinkle remaining Colby and Parmesan cheeses on top. Bake in preheated 325°F (160°C) oven 20 minutes or until hot and bubbly.
Makes 4 to 6 servings.

SPICY BEEF CASSEROLE

2 tablespoons	oil	30 mL
1	medium onion, chopped	1
1	red pepper, chopped	1
2	garlic cloves, chopped	2
1 pound	ground beef	500 g
2 teaspoons	ground chili powder	10 mL
1 cup	elbow macaroni	250 mL
2 cups	frozen whole kernel corn	500 mL
1 (14 ounce)	can tomatoes,	1 (384 mL)
	drained and chopped	
1 teaspoon	salt	5 mL
1 cup	buttermilk	250 mL
1 cup	grated Mozarella cheese	250 mL
1 cup	grated Canadian Cheddar cheese	250 mL

Preheat oven to 425°F (220°C). Cook onion, red pepper and garlic in oil until softened. Add ground beef and cook until lightly browned. Drain off excess fat. Mix beef and vegetables with chili powder, macaroni, corn, tomatoes, salt and buttermilk. Turn into a 9″ square (2.5 L) lightly greased baking dish. Top with grated cheeses. Lower heat to 375°F (190°C) and bake for one hour until hot and bubbly and macaroni is tender.

Makes 6 servings.

SWEDISH MEAT BALLS

1 pound	lean ground beef	500 g
1⅓ cups	milk	325 mL
½ cup	fine, dry, bread crumbs	125 mL
¼ cup	finely-chopped onion	50 mL
1	egg	1
1 teaspoon	salt	5 mL
¼ cup	butter	50 mL
¼ cup	flour	50 mL
1 (10-ounce)	can beef broth	1 (284 mL)
1 teaspoon	dill weed	5 mL
	hot cooked noodles	

Turn meat into a bowl and break up with a fork. Add and mix in ⅓ cup (75 mL) of the milk, bread crumbs, onion, egg and salt. Shape mixture into 1-inch (2.5 cm) balls. Place in a single layer in a 9x13x2-inch (3.5 L) rectangular baking pan. Bake in preheated 500°F (260°C) oven 8 to 10 minutes. Drain well. Melt butter in a saucepan. Blend in flour. Gradually stir in remaining milk and beef broth. Add dill weed. Cook over medium heat, stirring constantly, until mixture just comes to a boil and thickens. Add meatballs to sauce and heat through. Serve over noodles.

Makes 4 or 5 servings.

SWEET AND SOUR PINEAPPLE MEAT BALLS

2 pounds	lean ground beef	1 kg
1 cup	fine, dry, bread crumbs	250 mL
¾ cup	milk	175 mL
½ cup	finely-chopped onion	125 mL
2	eggs	2
2 teaspoons	salt	10 mL
2 (19-ounce)	cans pineapple chunks	2 (540 mL)
1 cup	thinly-sliced green pepper	250 mL
½ cup	soy sauce	125 mL
½ cup	cider vinegar	125 mL
½ cup	firmly-packed brown sugar	125 mL
⅓ cup	corn starch	75 mL
1 cup	water	250 mL
	hot cooked rice	

Turn meat into a bowl and break up with a fork. Add and mix in bread crumbs, milk, onion, eggs and salt. Shape mixture into 1-inch (2.5 cm) balls. Place in a single layer in a 10x15x¾-inch (2 L) jelly roll pan. Bake in preheated 500°F (260°C) oven 12 to 15 minutes. Drain well. In a large saucepan combine pineapple chunks and syrup, green pepper, soy sauce, vinegar and sugar. Smoothly combine corn starch and water. Stir into saucepan. Cook over medium heat, stirring constantly, until mixture comes to a boil and thickens. Reduce heat and continue to cook an additional 2 minutes, stirring occasionally. Add meat balls to sauce and heat through. Serve over rice.

Makes 8 to 10 servings.

VEAL BURGERS WITH MUSHROOM SAUCE

1½ pounds	ground veal (or pork or beef)	750 g
1	egg	1
½ cup	fresh breadcrumbs	125 mL
1 teaspoon	salt	5 mL
¼ teaspoon	freshly ground black pepper	1 mL
½ teaspoon	dried tarragon	2 mL
½ teaspoon	grated lemon peel	2 mL
6 tablespoons	butter – divided	90 mL
1	onion, chopped	1
½ pound	mushrooms, sliced	250 g
⅓ cup	all purpose flour	75 mL
1 cup	chicken stock	250 mL
1 tablespoon	lemon juice	15 mL
2 cups	milk	500 mL
½ teaspoon	salt or more to taste	2 mL
¼ teaspoon	freshly ground black pepper	1 mL
1 teaspoon	dried tarragon	5 mL
¼ teaspoon	Tabasco sauce	1 mL
1 teaspoon	Worcestershire sauce	5 mL
3 tablespoon	chopped fresh parsley (if available)	45 mL

Combine veal with egg, breadcrumbs, salt, pepper, tarragon and lemon peel. Shape into 4 patties approximately ½" (1 cm) thick. Heat 3 tablespoons (45 mL) of the butter in a large skillet. Brown veal patties on each side. Remove from pan and reserve. Discard excess fat in pan. Return pan to heat. Heat remaining butter in pan. Add onions and mushrooms. Cook a few minutes. Sprinkle with flour. Cook 5 minutes, scraping bottom of pan if necessary. Cool slightly. Whisk in stock, lemon juice, milk, salt, pepper, tarragon, Tabasco, and Worcestershire sauces. Bring to a boil. Reduce heat, add the veal burgers. Simmer gently, covered, 15 minutes. Taste sauce and add seasoning if necessary.
Serve sprinkled with parsley.
Makes 4 servings.

VEAL SUPREME

¼ cup	butter	50 mL
1 pound	veal scallops, flattened	500 g
2 cups	sliced, fresh mushrooms	500 mL
2 tablespoons	flour	30 mL
1 teaspoon	chicken bouillon mix	5 mL
½ teaspoon	salt	2 mL
1 cup	milk	250 mL
3 tablespoons	lemon juice	45 mL
	chopped parsley	

Melt 2 tablespoons butter in frypan. Sauté veal until brown on both sides. Set meat aside. Add remaining butter to pan. Saute mushrooms until tender and liquid has evaporated. Blend in flour, bouillon mix and salt. Gradually stir in milk. Cook over medium heat, stirring constantly, until mixture just comes to a boil and thickens. Stir in lemon juice. Return meat to pan; heat through. Sprinkle with parsley to serve.
Makes 4 servings.

WIENER AND SPAGHETTI CASSEROLE

3 tablespoons	butter	45 mL
3 tablespoons	finely chopped green pepper	45 mL
2	green onions, chopped	2
3 tablespoons	flour	45 mL
2 ¼ cups	milk	550 mL
1 cup	grated Canadian Cheddar cheese	250 mL
1 teaspoon	salt	5 mL
¼ teaspoon	pepper	1 mL
6	wieners	6
2 cups	broken dried spaghetti	500 mL
1 tablespoon	finely chopped parsley	15 mL
	paprika	

Melt butter in a medium sauce pan. Add green pepper and onion. Cook until just softened. Blend in flour. Cook briefly. Gradually whisk in milk. Cook and stir over medium heat until nicely thickened. Stir in cheese, salt and pepper. Continue cooking until cheese has melted. Cut wieners in ½" (1 cm) pieces. Cook spaghetti in boiling water until tender. Drain and combine with cheese sauce, wieners, and parsley. Pour into lightly greased 8" square (2 L) baking dish. Bake in 350°F (180°C) oven for about 25 minutes until hot and bubbly. Sprinkle with paprika before serving.
     Makes 4 servings.
*This is delicious made with pepperoni instead of wieners.

ASPARAGUS RAREBIT

2 (12 ounce)	tins asparagus	2 (341 mL)
6 tablespoons	butter, melted	90 mL
6 tablespoons	flour	90 mL
3 cups	milk	750 mL
1 cup	Cheddar cheese, grated	250 mL
¾ teaspoon	dry mustard	3 mL
1½ teaspoons	Worcestershire sauce	7 mL
¾ teaspoon	salt	3 mL
dash	pepper	dash
6	English muffins, split and toasted	6

In a saucepan, combine butter and flour. Add milk, stirring constantly, and continue cooking until smooth and thickened. Add cheese, mustard, Worcestershire sauce, salt and pepper. Stir until cheese is melted. Meanwhile, heat asparagus in its own liquid until hot. Toast the English muffins. Arrange hot asparagus on English muffins. Pour sauce over all.

Makes 6 servings.

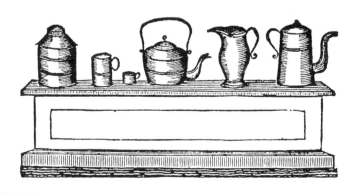

ASPARAGUS CHEDDAR QUICHE

1	partially baked 10" or 11" pastry shell	1
1 tablespoon	Dijon mustard	15 mL
2 tablespoons	butter	30 mL
1	onion, finely chopped	1
¼ cup	finely chopped fresh dill (or 1 teaspoon dried)	50 mL
1 pound	cooked asparagus, coarsely chopped	500 g
1½ cups	grated Old Cheddar cheese	375 mL
3	eggs	3
2	egg yolks (or 1 additional egg)	2
1½ cups	milk	375 mL
1 teaspoon	salt	5 mL
¼ teaspoon	pepper	1 mL
	nutmeg and cayenne to taste	

Preheat oven to 375°F (190°C).

Brush partially baked pastry shell with mustard. Reserve. Melt butter. Cook onion until tender but do not brown. Combine onion with dill, asparagus and cheese. Spoon evenly into pastry shell. Beat eggs with yolks or additional egg. Add milk, salt, pepper, nutmeg and cayenne. Pour mixture over asparagus and cheese. Bake 35 to 45 minutes or until just set. Cool 10 minutes before serving.

Makes 10"/26 cm quiche.

Asparagus Cheese Souffle Roll

⅓ cup	butter	75 mL
⅓ cup	flour	75 mL
½ teaspoon	salt	2 mL
1¼ cups	milk	300 mL
½ cup	grated Canadian Cheddar cheese	125 mL
7	eggs, separated	7
¼ teaspoon	cream of tartar	1 mL
2 tablespoons	butter	30 mL
1	small onion, finely chopped	1
⅔ cup	sliced almonds	150 mL
2 tablespoons	flour	30 mL
1 teaspoon	salt	5 mL
½ teaspoon	nutmeg	2 mL
	pepper, freshly ground	
1 pound	asparagus, cooked, drained and chopped	500 g
1½ cups	grated Swiss cheese	375 mL
1 cup	sour cream	250 mL
1 tablespoon	Dijon mustard	15 mL

Line a 15" x 10" (2 L) jelly roll pan with foil; grease well. Melt ⅓ cup (75 mL) butter in saucepan; stir in flour and set until smooth. Gradually add milk; bring to boil, stirring. Simmer until thick. Blend in Cheddar and lightly beaten egg yolks; remove from heat. Beat egg white with cream of tartar until soft peaks form. Fold one third whites into cheese mixture. Carefully fold in remaining whites. Turn into prepared pan. Bake at 350°F (180°C) for 15-20 minutes, until puffed and firm to the touch.

While soufflé roll is baking, sauté onion and almonds in 2 tablespoons (30 mL) butter. Sprinkle with 1 tablespoon (15 mL) flour; stir to blend. Add seasonings and asparagus. Stir for 5 minutes, or until most moisture evaporates. Remove from heat.

With metal spatula, loosen edges of soufflé roll; turn out on tea towel; remove foil gently. Spread asparagus mixture gently on top; sprinkle with cheese. Roll up jelly-roll fashion; place on serving dish. Let stand a few minutes in warm oven. Mix sour cream and mustard together, serve over soufflé roll.

Makes 8 servings.

Bacon, Cheese and Onion Tart

1		9" (23 cm) pie crust, partially baked	1
4		slices bacon, chopped	4
2		medium onions, sliced	2
½	teaspoon	thyme	2 mL
¼	teaspoon	nutmeg	1 mL
1	teaspoon	finely chopped parsley	5 mL
1	cup	grated Swiss cheese	250 mL
3		eggs	3
1	cup	milk	250 mL

Set oven to 350°F (180°C).

In a large frying pan sauté bacon until softened. Remove with a slotted spoon and place in prepared pie crust. Add onions to the same pan and sauté until tender. Season with thyme, nutmeg and parsley. Spread on top of bacon. Top with Swiss cheese. Beat eggs well. Blend with milk and pour over onion mixture. Bake for 35 minutes or until set.

Makes 4 servings.

BACON AND TOMATO STACK-UPS

1 tablespoon	butter	15 mL
1 tablespoon	flour	15 mL
1 teaspoon	chicken bouillon mix	5 mL
1 cup	milk	250 mL
1 cup	shredded Canadian Cheddar cheese	250 mL
12	slices cooked back bacon	12
12	slices tomato	12
6	split, toasted, English muffins	6

Melt butter in a small saucepan. Blend in flour and bouillon mix. Gradually stir in milk. Cook over medium heat, stirring constantly, until mixture just comes to a boil and thickens. Remove from heat. Add cheese; stir until melted. Place a slice of back bacon and a slice of tomato on each toasted English muffin half. To serve, place 2 muffin halves on a plate and pour about ¼ cup (50 mL) cheese sauce over each serving.

Makes 6 servings.

BAKED CHEDDAR STRATA

6	slices buttered bread, crusts removed	6
2 cups	milk	500 mL
2 cups	Cheddar cheese, grated	500 mL
4	eggs, well-beaten	4
½ teaspoon	salt	2 mL
¼ teaspoon	pepper	1 mL

Arrange the bread slices in bottom of well-greased 9"x 9" (23 cm x 23 cm) baking dish. Pour ½ cup (125 mL) milk over bread. Spread grated cheese evenly over the bread. Combine eggs with remaining milk, salt and pepper, and pour over the bread and cheese. Bake at 350°F (180°C) for 40 minutes.

Makes 6 servings.

BAKED BRUNCH EGGS

3 tablespoons	butter	45 mL
2 cups	sliced fresh mushrooms	500 mL
2 tablespoons	flour	30 mL
1 (10-ounce)	can cream of mushroom soup	1 (284 mL)
1 cup	milk	250 mL
8	eggs	8
7 or 8	slices French bread, cut ½-inch (12 mm) thick and buttered on both sides	7 or 8

Melt butter in a saucepan. Sauté mushrooms until tender and liquid has evaporated. Blend in flour, then mushroom soup. Gradually stir in milk. Cook over medium heat, stirring constantly, until mixture just comes to a boil and thickens. Spread 1½ cups (375 mL) of sauce in a 2-quart (2 L) shallow rectangular baking dish. Break 8 eggs carefully on top of sauce in dish, spacing evenly. Spoon remaining sauce around eggs. Cut each slice of buttered bread in half. Stand halved bread slices around outside edge of dish, crust side up. Bake in preheated 350°F (180°C) oven about 15 to 20 minutes.
Makes 4 servings.

BREAKFAST EGG CASSEROLE

2 cups	Cheddar cheese, grated	500 mL
2 tablespoons	butter, melted	30 mL
1 or 2	large Spanish onions, coarsely chopped	1 or 2
4	slices bacon	4
6	eggs, beaten	6
1½ cups	milk	375 mL
½ teaspoon	pepper	2 mL

Arrange cheese on the bottom of a well-greased 9"x 9" (23 cm x 23 cm) baking dish. Sauté onions in butter until soft, about 5 minutes. Spread over cheese layer. Cook bacon until done but not too crisp. Chop and sprinkle over onions. Combine eggs, milk and pepper. Pour over other ingredients in pan. Bake at 350°F (180°C) for 60-70 minutes until puffed, and set in the centre.
Makes 6 servings.

CANNELLONI

FILLING

1	(8 ounce)	package cannelloni noodles	1 (250 g)
¼	cup	butter	50 mL
⅓	cup	chopped onion	75 mL
1	(8 ounce)	package frozen spinach,	1 (250 g)
		cooked, drained and chopped	
3	cups	finely chopped ham	750 mL
½	cup	Parmesan cheese	125 mL
1	cup	Ricotta cheese or	250 mL
		cottage cheese	
1		egg	1

WHITE SAUCE

⅓	cup	butter	75 mL
⅓	cup	flour	75 mL
2	cups	milk	500 mL
1	teaspoon	salt	5 mL

TOMATO SAUCE

2	cups	tomato sauce	500 mL
½	teaspoon	basil	2 mL
dash		cayenne pepper	dash
2	tablespoons	Parmesan cheese	30 mL
¼	cup	chopped fresh parsley	50 mL

Cook cannelloni noodles according to package directions. Sauté onion in butter until soft. Stir in spinach; cook until moisture is gone. Remove from heat; add ham, cheeses, egg and seasonings to taste. Prepare a white sauce with butter, flour, milk and salt.

Prepare tomato sauce, season with basil and cayenne, to taste.

Assemble cannelloni as follows: Pour ½ cup (125 mL) tomato sauce in bottom of 9" x 13" (3.5 L) baking dish. Fill cooked cannelloni with ham filling. Place on top of tomato sauce. Pour white sauce over all; top with remaining tomato sauce. Sprinkle with Parmesan and parsley. Bake covered at 375°F (190°C) for 20-25 minutes or until bubbly.

Makes 8 servings.

CHEDDAR CHEESE SOUFFLE

3 tablespoons	butter	45 mL
¼ cup	flour	50 mL
½ teaspoon	salt	2 mL
1 cup	milk	250 mL
1½ cups	shredded Canadian Cheddar cheese	375 mL
4	eggs, separated	4

Melt butter in a medium saucepan. Blend in flour and salt. Gradually stir in milk. Cook over medium heat, stirring constantly, until mixture just comes to a boil and thickens. Remove from heat. Add cheese and stir until melted. Beat egg ·yolks well. Stir a small amount of hot mixture into yolks, return to saucepan and blend thoroughly. Beat egg whites until stiff but not dry. Stir large spoonful of beaten egg whites into cheese sauce. Fold in remaining egg whites. Turn into greased 6-cup (1.5 L) souffle dish. Bake in preheated 325°F (160°C) oven 55 minutes. Serve immediately.

Makes 6 servings.

CHEDDAR PINWHEEL BRUNCH

5	eggs, well beaten	5
1½ cups	shredded Canadian Cheddar cheese	375 mL
1 cup	milk	250 mL
1 tablespoon	finely-chopped green onion	15 mL
½ teaspoon	salt	2 mL
8	slices day-old bread	8
	soft butter	
6	slices cooked bacon, crumbled	6

Combine eggs, cheese, milk, green onion and salt. Pour into a 9-inch (1 L) pie plate. Trim crusts from bread; lightly butter. Cut each slice of bread in half diagonally. Stand bread triangles upright in egg mixture to form a circular pattern. Sprinkle bacon over bread. Bake in preheated 350°F (180°C) oven 40 to 45 minutes or until knife inserted near centre comes out clean. Cut and serve.

Makes 4 to 6 servings.

CHEESE BAKE

2	cups	cooked rice	500 mL
3	cups	shredded carrots	750 mL
2	cups	grated Cheddar cheese	500 mL
½	cup	milk	125 mL
2		eggs, beaten	2
2	tablespoons	finely chopped onion	30 mL
1½	teaspoons	salt	7 mL
¼	teaspoon	pepper	1 mL
½	cup	sliced mushrooms	125 mL
6		thick tomato slices	6

Combine rice, carrots, 1½ cups (375 mL) cheese, milk, beaten eggs, onion, seasonings and mushrooms. Pour into a greased 9″ (2.5 L) square baking dish. Top with tomato slices and sprinkle with remaining cheese. Bake at 350°F (180°C) for 1 hour.

Makes 6 servings.

CHEESE AND VEGETABLE STRATA

12	slices white bread, crusts removed	12
12	slices processed Swiss cheese	12
1 (10 ounce)	box frozen broccoli, chopped, thawed and drained	1 (284 g)
6	eggs, beaten	6
3 cups	milk	750 mL
2 tablespoons	instant minced onion	30 mL
½ teaspoon	salt	2 mL
dash	pepper	dash
¼ teaspoon	dry mustard	1 mL

Arrange slices of bread in bottom of 9"x 13" (23 cm x 33 cm) baking dish. Place a layer of cheese over the bread. Add a layer of broccoli. Combine eggs, milk, minced onion, salt, pepper and mustard. Pour over layers in baking dish. Refrigerate at least 6 hours or overnight. Bake at 325°F (160°C) for 55 minutes or until set. Makes 8 servings.

CHEESE ROULADE
WITH MUSHROOMS AND BACON

FILLING

8	slices bacon, chopped	8
½ cup	green onions, chopped	125 mL
½ pound	mushrooms, chopped	250 g
1 tablespoon	lemon juice	15 mL

For filling, cook bacon until crisp. Add green onions, mushrooms and lemon juice. Sauté for about 5 minutes until soft. Keep warm.

ROULADE

2 tablespoons	fine breadcrumbs	30 mL
2 tablespoons	butter	30 mL
2 tablespoons	flour	30 mL
1 cup	milk	250 mL
½ teaspoon	salt	2 mL
¼ teaspoon	dry mustard	1 mL
⅛ teaspoon	pepper	0.5 mL
8	eggs, separated	8
½ cup	Parmesan cheese, grated	125 mL

For roulade, grease well a 10"x 15"x ¾" (2 L) jelly roll pan. Line with wax paper, and grease the paper very well. Sprinkle with breadcrumbs. In a heavy pan, melt butter; stir in flour. Gradually add milk slowly and cook, stirring constantly, until smooth and thickened. Add salt, mustard and pepper. Remove from heat. Beat egg yolks until light-coloured, and whisk into mixture. Stir in half of the Parmesan cheese. Beat egg whites until stiff but not dry. Stir some of sauce mixture into whites. Then fold egg whites into sauce gently but thoroughly. Spread evenly in prepared pan. Bake at 400°F (200°C) for 15 minutes, or until firm to the touch.

continued on next page

CHEESE ROULADE
WITH MUSHROOMS AND BACON (continued)

CHEESE SAUCE

6 tablespoons	butter	90 mL
4 tablespoons	flour	60 mL
2 cups	milk	500 mL
1 teaspoon	salt	5 mL
¼ teaspoon	dry mustard	1 mL
¼ teaspoon	pepper	1 mL
2 cups	Cheddar cheese, grated	500 mL
¼ cup	18% cream	50 mL
	bacon curls for garnish	

For cheese sauce, melt butter in a heavy saucepan. Stir in flour. Whisk in milk slowly, stirring constantly over medium heat, until smooth and thickened. Add salt, mustard, pepper and cheese and stir until cheese is melted. Blend in cream, and keep hot without boiling.

When roulade is baked, invert the pan on a tea towel sprinkled with remainder of the grated Parmesan cheese. Lift off pan, and carefully peel off the paper. Drain any excess liquid from the warm filling. Add ¼ cup (50 mL) of the cheese sauce to the filling. Spread the roulade evenly with the filling and roll up starting at the short side. Place seam side downward on serving platter. Pour some of the hot sauce down the centre of roulade. Garnish with parsley or bacon curls. Serve rest of the sauce separately. Cut into slices to serve.

Makes 8 servings.

CHEESE BLINTZES

3	eggs, beaten	3
2	eggs, separated	2
1½ cups	all purpose flour	375 mL
1½ cups	milk	375 mL
6 tablespoons	sugar	90 mL
1 tablespoon	melted butter	15 mL
1 pound	cream cheese	500 g
2 tablespoons	sour cream	30 mL
1 teaspoon	grated lemon rind	5 mL
	butter	

In a large bowl combine whole eggs, egg whites, flour, milk, 2 tablespoons (30 mL) sugar. Add butter. Beat until smooth. Refrigerate batter for 30 minutes. In another bowl blend together cheese, sour cream, remaining sugar, egg yolks, lemon rind. Set aside.

To make blintze pancakes: lightly butter a 6 inch (15 cm) hot skillet. Pour ¼ cup (50 mL) batter into pan. Swirl to coat bottom. Pour off excess. Cook over medium heat until edges are dry and bottom of pancake is lightly browned. Repeat with remaining batter, stacking prepared pancakes between sheets of waxed paper. Spoon 2 tablespoons (30 mL) on bottom of half of browned side of each pancake. Fold in edges and roll up. When ready to serve cook gently in batter seam side down until nicely browned. Serve with Fresh Fruit Sauce.

Makes 12 blintzes.

FRESH FRUIT SAUCE:

1 cup	fresh fruit puree (peach, strawberry or raspberry)	250 mL
2 tablespoons	sugar	30 mL
1 tablespoon	lemon juice	15 mL
1 tablespoon	orange liqueur (optional)	15 mL

Combine ingredients. Blend well.

CHEESE VEGETABLE CASSEROLE

½	cup	chopped onion	125 mL
¼	cup	butter, melted	50 mL
¼	cup	all-purpose flour	50 mL
¼	teaspoon	dry mustard	1 mL
2	cups	milk	500 mL
1½	cups	shredded old Canadian Cheddar cheese	325 mL
1	(8 ounce)	package noodles, cooked, drained	1 (300 g)
1	(8 ounce)	package frozen peas and carrots, cooked, drained	1 (300 g)
4		hard-boiled eggs, sliced	4
½	cup	fresh bread cubes	125 mL
2	tablespoons	butter, melted	30 mL

Sauté onion in butter until tender; stir in flour and mustard. Cook over low heat until smooth, stirring constantly. Remove from heat. Gradually stir in milk. Bring to a boil over medium heat stirring constantly. Boil and stir 1 minute. Remove from heat; stir in cheeses until melted.

Place half the noodles in a buttered 2 quart (2 L) casserole; top with half the vegetables, half the egg slices and half the cheese sauce. Repeat layers. Top casserole with bread cubes tossed with butter.

Bake in preheated 350°F (180°C) oven 25-30 minutes or until heated through. Garnish with tomato wedges.

Makes 6 to 8 servings.

CHEESE 'N' CHILI CASSEROLE

1 tablespoon	flour	15 mL
2 teaspoons	instant minced onion	10 mL
¼ teaspoon	salt	1 mL
1 teaspoon	chili powder	5 mL
3½ cups	milk	875 mL
2 cups	dry macaroni	500 mL
1 cup	green pepper, chopped	250 mL
2 cups	Cheddar cheese, grated	500 mL
1 (19 ounce)	can kidney beans, drained	1 (540 mL)
8	thin fingers of Cheddar cheese	8

Combine flour, minced onion, salt and chili powder in a large saucepan. Gradually stir in the milk. Add macaroni and green pepper. Bring to a boil, stirring constantly. Reduce heat, cover and simmer for 10 minutes, stirring frequently. Stir in grated cheese until melted, then add 1 cup (250 mL) beans. Pour into a greased 6-cup (1.5 L) casserole. Place remaining beans on top. Top with cheese fingers, arranged spoke-fashion. Bake at 350°F (180°C) for 8-10 minutes, until cheese fingers are melted and casserole is heated through.

Makes 8 servings.

CHEESE SOUFFLE IN PEPPER CUPS

6	green peppers, tops removed and seeded	6
3 tablespoons	butter	45 mL
3 tablespoons	all-purpose flour	45 mL
1½ cups	milk	375 mL
1 cup	grated Canadian Cheddar cheese	250 mL
½ teaspoon	salt	2 mL
¼ teaspoon	pepper	1 mL
¼ teaspoon	garlic powder	1 mL
½ teaspoon	onion powder	2 mL
dash	hot pepper sauce	dash
3	eggs, separated	3
½ cup	crushed corn chips	125 mL

Prepare green peppers and stand them in greased baking pan. In a saucepan, melt butter. Stir in flour and cook over medium heat for 2 minutes. Gradually add milk, stirring constantly, and cook until thick and smooth. Blend in cheese and seasonings.

In a small bowl, beat egg yolks slightly. Add small amount of hot cheese sauce; mix well. Return to remaining sauce and continue to cook for 3 minutes over medium heat. In a mixing bowl, beat egg whites until stiff. Fold into cheese sauce. Spoon into pepper cups, sprinkle with corn chips and bake in 375°F (190°C) oven for 30 minutes or until soufflé is puffed and golden brown.

Makes 6 servings.

CHEESE SPINACH ROLL

ROLL

1 ½ pounds	fresh spinach	750 g
¾ cup	sharp Cheddar cheese, grated	175 mL
¼ cup	Parmesan cheese, finely grated	50 mL
½ cup	dry breadcrumbs	125 mL
2 tablespoons	onion, finely grated	30 mL
½ teaspoon	salt	2 mL
dash	pepper	dash
5	eggs, separated	5
6 tablespoons	butter, melted	90 mL

FILLING

4 tablespoons	butter	60 mL
4 tablespoons	flour	60 mL
2 cups	milk	500 mL
1 ½ teaspoons	prepared mustard	7 mL
½ teaspoon	salt	2 mL
¼ teaspoon	pepper	1 mL
4	eggs, hard-cooked and coarsely chopped	4
2 tablespoons	pimento, chopped	30 mL

Grease a 10"x 15" (25 cm x 38 cm) jelly roll pan. Line with wax paper and grease again. Wash spinach well and cook until just limp. Drain well. Chop finely. Combine with cheeses, breadcrumbs, onion, salt and pepper. Beat egg yolks until light-coloured. Stir in melted butter. Blend into spinach mixture. Beat egg whites until stiff, but not dry. Gently fold into spinach mixture. Pour into prepared pan. Spread evenly and bake at 375°F (190°C) for 15-18 minutes. Turn out onto a cloth towel. Let stand 10 minutes and then remove paper. Roll up jelly roll fashion with both towel and roll, starting from narrow side. Let cool. Refrigerate if not using within the hour.

To prepare filling, melt butter in a saucepan. Blend in flour. Add milk, stirring constantly over low heat until smooth and thickened. Add mustard, salt, pepper, eggs and pimento.

Unroll spinach roll, remove towel, spread filling, and reroll. Place on serving platter with seam side down. Can be served hot or cold.

Makes 6 servings.

CHEESY BISCUIT BAKE

⅓ cup	chopped onion	75	mL
1½ cups	fresh mushrooms, sliced	375	mL
⅓ cup	butter, melted	75	mL
4 tablespoons	flour	60	mL
½ teaspoon	salt	2	mL
1 teaspoon	dry mustard	5	mL
⅛ teaspoon	pepper	0.5	mL
2⅓ cups	milk	575	mL
½ cup	Cheddar cheese, grated	125	mL
1 (10 ounce)	box frozen peas, cooked and drained	1 (284	g)
4	eggs, hard-cooked and sliced	4	

TOPPING

2 cups	prepared biscuit mix	500	mL
1 cup	Cheddar cheese, grated	125	mL
⅔ cup	milk	175	mL

Sauté onion and mushrooms in butter. Blend in flour, salt, mustard and pepper. Gradually add 2⅓ cups (575 mL) milk, and cook, stirring constantly, until smooth and thickened. Add ½ cup (125 mL) cheese, and stir until melted. Fold in peas and eggs. Pour into greased 8-cup (2 L) baking dish. Bake at 425°F (220°C) for 10 minutes. Meanwhile, combine biscuit mix with half of cheese. Stir in ⅔ cup (175 mL) milk. Turn dough out on a lightly floured board. Pat into a 9" (23 cm) circle. Score dough into 8 wedges. Remove baking dish from oven. Top hot mixture with biscuit wedges. Return to oven and bake 10-12 minutes, or until biscuits are browned. Remove from oven. Sprinkle biscuits with remainder of cheese. Return to oven until cheese is melted.

Makes 8 servings.

CHEESY TOMATO PIE

2 cups	fresh bread crumbs	500 mL
¼ cup	Swiss cheese, grated	50 mL
2 tablespoons	butter, melted	30 mL
2	large tomatoes, sliced	2
2	green onions, chopped	2
1 teaspoon	salt	5 mL
½ teaspoon	sugar	2 mL
½ teaspoon	basil	2 mL
¼ teaspoon	pepper	1 mL
1 tablespoon	butter	15 mL
2	eggs, lightly beaten	2
1 cup	10% cream	250 mL
½ cup	Swiss cheese, grated	125 mL

Mix bread crumbs, ¼ cup (50 mL) cheese and melted butter together. Pat into bottom of a 9″ (23 cm) pie plate. Bake at 400°F (200°C) for 10 minutes, or until lightly browned. Let cool. Arrange tomato slices in layers on the crust. Sprinkle with green onions, salt, sugar, basil and pepper. Dot with butter. Beat eggs with cream and ½ cup (125 mL) cheese. Pour over tomatoes. Bake at 375°F (190°C) for 30-40 minutes, or until knife inserted in the centre comes out clean.

Makes 6 servings.

CHICKEN QUICHE AMANDINE

1	9" (1 L) unbaked, deep pie shell	1
½ cup	chopped cooked chicken	125 mL
3 tablespoons	sliced almonds	45 mL
1½ cups	shredded Swiss cheese	375 mL
3	eggs	3
1½ cups	milk	375 mL
½ teaspoon	salt	2 mL
¼ teaspoon	nutmeg	1 mL
pinch	pepper	pinch
2 tablespoons	grated Parmesan cheese	30 mL

Place chicken and almonds in pie shell. Sprinkle evenly with Swiss cheese. Beat eggs slightly; blend in milk, salt, nutmeg and pepper. Pour over cheese. Sprinkle with Parmesan cheese. Bake in preheated 375°F (190°C) oven 30-35 minutes or until a knife inserted near centre comes out clean. Let stand 10 minutes before serving.

Makes 6 servings.

CORNED BEEF OVEN OMELETTE

8	eggs	8
1 cup	milk	250 mL
3	(2 ounce/50 g) packages corned beef	3
1 cup	grated Mozzarella cheese	250 mL
¼ cup	grated onion	50 mL
¼ teaspoon	thyme	1 mL
	salt	
	freshly ground pepper	

Preheat oven to 325°F (160°C).

Beat eggs with milk. Chop corned beef finely and add to eggs with cheese, onion and seasonings. Pour into a greased 8" (1L) baking dish. Bake 45 minutes or until omelette is set and top is golden.

Makes 4 servings.

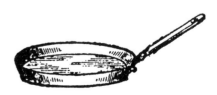

CRAB QUICHE

1	unbaked 9″ (23 cm) pie shell	1
2 cups	Swiss cheese, grated	500 mL
1 (5 ounce)	can crab meat, drained	1 (142 g)
3	green onions, chopped	3
2	eggs, beaten	2
¼ teaspoon	dry mustard	1 mL
1 cup	milk	250 mL

Place 1 cup (250 mL) cheese in pie shell. Distribute crab meat over cheese. Sprinkle with green onions. Combine eggs, mustard and milk. Pour into pie shell. Bake at 325°F (160°C) for 50 minutes or until set in centre. Sprinkle on remainder of cheese and return to oven until melted. Let stand for 10 minutes before serving.

Makes 6 servings.

DAIRY DELICIOUS COTTAGE MOLD

2		envelopes unflavoured gelatin	2
½	cup	cold water	125 mL
2	cups	cottage cheese	500 mL
1½	cups	milk	375 mL
½	cup	sour cream	125 mL
¼	cup	lemon juice	50 mL
2		green onions cut in 1″ (2.5 cm) pieces	2
1	teaspoon	salt	5 mL
¼	teaspoon	Tabasco sauce	1 mL
		marinated vegetables: asparagus, green beans or artichoke hearts	

In small saucepan, sprinkle gelatin over water; let stand for 10 minutes to soften. Dissolve over low heat, stirring constantly. Put cottage cheese, milk, sour cream, lemon juice, onions, salt and Tabasco in blender. Cover and blend at medium speed for 20 seconds then at high speed until smooth.

Add dissolved gelatin and blend at low speed for a few seconds. Pour into 5-cup (1.25 L) mold. Chill until set. Unmold and serve with marinated vegetables.

Makes 6 to 8 servings.

DEEP DISH TUNA QUICHE

PASTRY

1½ cups	whole wheat flour	375 mL
½ teaspoon	tarragon, crumbled	2 mL
¼ teaspoon	salt	1 mL
½ cup	shortening	125 mL
½ cup	mozzarella cheese, grated	125 mL
3 tablespoons	water	45 mL

FILLING

2 (7 ounce)	cans chunk tuna	2 (198 g)
3	eggs, beaten	3
1⅔ cups	milk	400 mL
1 cup	mozzarella cheese, grated	250 mL
1 teaspoon	tarragon	5 mL
⅓ cup	Parmesan cheese, grated	75 mL
¼ teaspoon	salt	1 mL
2 tablespoons	parsley, chopped	30 mL
1 (12 ounce)	can asparagus spears	1 (341 mL)

Mix flour, ½ teaspoon (2 mL) tarragon and ¼ teaspoon (1 mL) salt in a bowl. Cut in shortening and ½ cup (125 mL) mozzarella cheese until mixture is crumbly. Add water and stir to form dough. Press on the bottom and 2″ (5 cm) up the sides of an 8″ (20 cm) springform pan. Bake at 375°F (190°C) for 10 minutes. Drain tuna and arrange on crust. Combine eggs with milk, 1 cup (250 mL) mozzarella cheese, 1 teaspoon (5 mL) tarragon, all but 2 tablespoons (30 mL) of the Parmesan cheese, ¼ teaspoon (1 mL) salt and parsley. Pour over the tuna. Arrange asparagus spears decoratively across the top. Sprinkle with the reserved Parmesan cheese. Bake at 325°F (160°C) for 1¼ hours, or until set in the centre. Let stand 15 minutes before serving. Remove sides of pan. Cut in wedges.

Makes 6 servings.

Egg, Cheese and Olive Casserole

8	slices white bread, crusts removed	8
2 cups	Mozzarella cheese, grated	500 mL
2 cups	Cheddar cheese, grated	500 mL
⅔ cup	ripe olives, drained, pitted & sliced	150 mL
6	eggs	6
1 teaspoon	dry mustard	5 mL
4 cups	milk	1 L

Grease a 9" x 13" (23 cm x 33 cm) baking pan. Arrange layers of bread, cheeses and olives. Beat eggs; add mustard and milk. Pour over layers in baking pan. Refrigerate at least 6 hours, or overnight. Bake at 350°F (180°C) for 1 hour, or until puffed and golden.

Makes 6-8 servings.

EGGS BENEDICT AU GRATIN

2 tablespoons	butter	30 mL
2 tablespoons	all-purpose flour	30 mL
½ teaspoon	salt	2 mL
½ teaspoon	dry mustard	2 mL
pinch	pepper	pinch
1½ cups	milk	375 mL
1 cup	shredded Canadian Cheddar cheese	250 mL
1 teaspoon	Worcestershire sauce	5 mL
6	English muffins, split in half, toasted and buttered	6
6	slices bacon, cooked	6
6	eggs, poached	6

Melt butter in a 1-quart (1 L) saucepan; blend in flour, salt, mustard and pepper, until smooth. Remove from heat, stir in milk. Heat to boiling, stirring constantly. Boil and stir 1 minute. Remove from heat; stir in cheese until melted. If necessary, return to low heat to finish melting cheese. (Do not boil.) Add Worcestershire sauce. Arrange bacon on six muffin halves; top with poached eggs.

Spoon ¼ cup (50 mL) cheese sauce over eggs.

Serve with remaining muffin halves.

Makes 6 servings.

GOLDEN CHEESE AND RICE BAKE

½ cup	onion, finely chopped	125 mL
½ cup	green pepper, finely chopped	125 mL
2 tablespoons	butter, melted	30 mL
3 cups	cooked rice	750 mL
½ cup	parsley, finely chopped	125 mL
1 teaspoon	salt	5 mL
¼ teaspoon	pepper	1 mL
2 cups	old Cheddar cheese, grated	500 mL
3	eggs, beaten	3
1 cup	milk	250 mL

Sauté onions and green pepper in butter for 5 minutes until soft. Combine with rice, parsley, salt, pepper and cheese. Mix lightly. Combine eggs and milk, and stir into rice mixture. Pour into greased 6-cup (1.5 L) casserole, and bake at 350°F (180°C) for one hour.

Makes 6 servings.

Harvest lasagne

9	lasagne noodles, cooked and drained	9
6 tablespoons	butter	90 mL
2 tablespoons	oil	30 mL
1 cup	chopped onion	250 mL
1	garlic clove, minced	1
1	medium eggplant, peeled and diced	1
4	medium zucchini, sliced	4
2	green peppers, cut in thin strips	2
4	ripe medium tomatoes, chopped	4
1½ teaspoons	salt	7 mL
1 teaspoon	basil	5 mL
½ teaspoon	ground pepper	2 mL
6 tablespoons	flour	90 mL
3 cups	milk	750 mL
1 cup	grated Canadian Cheddar cheese	250 mL

Heat 2 tablespoons (30 mL) each butter and oil in a large frying pan. Add onion and garlic and cook until soft. Toss in eggplant, and brown 5 minutes. Add zucchini, peppers, tomatoes, salt, basil and pepper. Simmer covered 15 minutes then remove lid and cook over medium heat until nicely thickened (about 10 minutes). Set aside. Melt remaining butter in saucepan. Add flour and cook 2 minutes. Gradually whisk in milk. Bring to a boil stirring constantly until thickened. Stir in cheese. Season with additional salt and pepper. Spoon ¼ of vegetables in greased 9"x 13"x 2" (3 L) baking dish. Cover with a layer of sauce and noodles. Repeat layers, twice, ending with vegetables and sauce. Bake in 350°F (180°C) oven for 45 minutes.

Makes 6-8 servings.

Herbed mushroom quiche

1	deep unbaked 9" (23 cm) pie shell	1
1 cup	Swiss cheese, grated	250 mL
½ (10 ounce)	can mushroom slices, drained	½ (284 mL)
1 cup	cottage cheese	250 mL
1 cup	milk	250 mL
4	eggs	4
1 teaspoon	prepared mustard	5 mL
½ teaspoon	basil	2 mL
¼ teaspoon	pepper	1 mL

Line a deep 9" (23 cm) pie plate with pastry. Sprinkle Swiss cheese and mushroom slices in pie shell. In blender or food processor, blend cottage cheese, milk, eggs, mustard, basil and pepper until smooth. Pour mixture into pie shell. Bake 15 minutes at 425°F (220°C). Reduce heat and continue baking at 325°F (160°C) for 45 minutes longer, or until knife inserted in centre comes out clean. Let stand 5 minutes before serving.

Makes 6 servings.

HOT DEVILLED EGGS

3 tablespoons	butter	45 mL
3 tablespoons	flour	45 mL
½ teaspoon	dry mustard	2 mL
¾ teaspoon	salt	3 mL
1½ cups	milk	375 mL
1 teaspoon	prepared horseradish	5 mL
2 teaspoons	Worcestershire sauce	10 mL
1 tablespoon	chili sauce	15 mL
dash	Tabasco sauce	dash
¾ cup	cubed Canadian Cheddar cheese	175 mL
6	hard boiled eggs, sliced	6
½ cup	chopped ripe or green olives	125 mL

Melt butter in saucepan, blend in flour, mustard, and salt. Gradually stir in milk. Cook over medium heat, stirring constantly until smooth and thick. Stir in horseradish, Worcestershire sauce, chili sauce, and Tabasco sauce. Add cheese, stir until melted. Fold in eggs and olives; heat through. Serve over toast.
Makes 4-6 servings.

ITALIAN EGG NOODLE CASSEROLE

1 (375 g)	one package broad egg noodles	1 (375 g)
4	eggs	4
2½ cups	milk	625 mL
2 teaspoons	salt	10 mL
½ teaspoon	pepper	2 mL
1 pound	medium ground beef	500 g
1 (28-ounce)	can spaghetti sauce	1 (796 mL)

Cook noodles in boiling salted water; drain well. In a large bowl, combine eggs, milk, salt and pepper. Add noodles and toss together. Turn into a greased 2-quart (2 L) shallow, rectangular baking dish. Bake in a preheated 350°F (180°C) oven for 25-30 minutes or until knife inserted near centre comes out clean. Meanwhile, brown ground beef in a saucepan, drain off fat and stir in spaghetti sauce. Bring to a boil, reduce heat and simmer uncovered for 5 minutes.

To serve, cut noodles in squares, and spoon meat sauce over each serving.

Makes 6 servings.

LASAGNE SPECIAL

1 (12 ounce)	package lasagne noodles, cooked according to package directions	about 350 g
3-4 cups	diced cooked turkey, chicken or ham	750-1000 mL
1 (12 ounce)	package frozen spinach, cooked, drained and chopped	1 (340 g)
1½ cups	creamed cottage cheese	375 mL
1 tablespoon	chopped fresh parsley	15 mL
2	eggs, beaten	2
1½-2 cups	grated Cheddar, Swiss or Mozzarella cheese	375-500 mL

SAUCE

5 tablespoons	butter	75 mL
1	small onion, minced	1
5 tablespoons	flour	75 mL
¼ teaspoon	celery salt	1 mL
¼ teaspoon	poultry seasoning	1 mL
1 teaspoon	salt	5 mL
	freshly ground black pepper	
4 cups	milk	1 L

Butter a 9" x 13" (3.5L) baking dish or lasagne pan. Prepare sauce as follows: sauté onion in butter until golden. Blend in flour and seasonings and stir until bubbly. Gradually add milk and cook over medium heat until thickened. Mix spinach, cottage cheese, parsley and eggs together. Assemble lasagne as follows: spoon a thin layer of sauce in bottom of baking dish. Cover with a layer of lasagne, side by side. Add half the turkey, half the spinach-cheese mixture and spoon about one-third of the sauce over all. Top with another layer of lasagne, then turkey, spinach-cheese and sauce mixtures. Finish with a layer of lasagne, spoon remaining sauce on top and sprinkle with grated cheese. Refrigerate until ready to bake. Bake at 350°F (180°C) for about 1 hour. Let stand for 10 minutes before cutting in squares to serve.
   Makes 8 to 10 servings.

LINGUINE WITH HAM AND MUSHROOMS

3 tablespoons	butter	45 mL
$\frac{1}{2}$ cup	chopped onion	125 mL
$\frac{1}{4}$ cup	flour	50 mL
1 tablespoon	chicken bouillon mix	15 mL
$\frac{1}{4}$ teaspoon	dry mustard	1 mL
$\frac{1}{4}$ teaspoon	garlic powder	1 mL
2 $\frac{1}{2}$ cups	milk	675 mL
2 cups	chopped ham	500 mL
1 (10 ounce)	can sliced mushrooms, drained	1(284 mL)
450g	(half a 900g package) linguine	450g
	grated Canadian Parmesan cheese	

Melt butter in large saucepan. Sauté onion until tender. Blend in flour, bouillon mix, dry mustard and garlic powder. Gradually stir in milk. Cook over medium heat, stirring constantly, until mixture just comes to a boil and thickens. Stir in ham and mushrooms. Heat to serving temperature. Keep warm. Cook linguine according to package directions. Drain well. To serve, spoon sauce over individual servings of linguine; sprinkle with Parmesan cheese.

Makes 4 to 5 servings.

LINGUINE WITH CLAM SAUCE

1		garlic clove, minced	1
4		green onions, chopped	4
1	tablespoon	butter	15 mL
½	teaspoon	tarragon	2 mL
1	(5-ounce)	can baby clams	1 (150 g)
1	cup	clam juice plus white wine or water (or wine only)	250 mL
1	cup	peas	250 mL
2	tablespoons	flour	30 mL
1	cup	milk	250 mL
		salt to taste	
½	pound	linguine, cooked and drained	500 g

In a large heavy frying pan sauté garlic and onion in butter until softened. Stir in tarragon and drained clams. Toss well. Add clam juice, wine and peas and simmer 5 minutes. Blend flour with milk until smooth. Add to sauce. Stir and cook until nicely thickened. Spoon over linguine. Serve with grated Parmesan cheese.

Makes 4 servings.

Variations:

1. Add different vegetables for variety, such as: mushrooms, lightly tossed in butter, blanched carrots, broccoli, or snow peas.

2. Replace clams and clam juice with shrimp and ½ cup (125 mL) wine.

MACARONI AND CHEESE PARMIGIANO

1¾ cups	elbow macaroni	425 mL
3 tablespoons	butter	45 mL
3 tablespoons	finely-chopped onion	45 mL
3 tablespoons	flour	45 mL
1 teaspoon	salt	5 mL
½ teaspoon	basil leaves	2 mL
2½ cups	milk	625 mL
1½ cups	grated Canadian Parmesan cheese	375 mL
2	medium tomatoes, thinly sliced	2
2	slices bread	2
3 tablespoons	butter, melted	45 mL

Cook macaroni according to package directions; drain well. Melt 3 tablespoons (45 mL) butter in a large saucepan. Add onion; sauté until tender. Blend in flour, salt and basil. Gradually stir in milk. Cook over medium heat, stirring constantly, until mixture just comes to a boil and thickens. Remove from heat; add cheese and stir until well combined. Mix sauce with macaroni. Arrange half the tomato slices in bottom of 2-quart (2 L) casserole. Top with half the macaroni. Repeat layering. Tear bread into small pieces. Toss with 3 tablespoons (45 mL) melted butter; arrange over macaroni. Bake, uncovered in preheated 350°F (180°C) oven 30 minutes or until browned and bubbly.

Makes 6 servings.

MACARONI PIZZA

1¾ cups	elbow macaroni	425 mL
2	eggs	2
1¼ cups	milk	300 mL
pinch	pepper	pinch
1 teaspoon	oregano	5 mL
1 (7½-ounce)	can tomato sauce	1 (213 mL)
1½ cups	shredded Mozzarella or Provolone cheese	375 mL

Cook macaroni in boiling salted water according to package directions. Drain. In a large bowl beat together eggs, milk and pepper. Add macaroni. Pour into greased 9-inch square (2.5 L) pan. Bake in preheated 400°F (200°C) oven 10 minutes or until set. Remove from oven. Meanwhile mix oregano into tomato sauce. Spread over baked macaroni. Top with cheese. Return to oven for an additional 10 minutes. Cut into serving portions.

Makes 6 servings.

Variation:
Sausage Pizza: After adding tomato sauce, top with ½ pound (250 g) sausage meat, cooked and drained or ¼ pound (125 g) summer sausage, cut up. Cover with cheese and return to oven.

MACARONI AND CHEESE SOUFFLE

1½ cups	dry macaroni	375 mL
2 tablespoons	butter	30 mL
3 tablespoons	flour	45 mL
½ teaspoon	salt	2 mL
½ teaspoon	paprika	2 mL
⅛ teaspoon	cayenne	0.5 mL
1¼ cups	milk	300 mL
2 cups	sharp Cheddar cheese, grated	500 mL
3	eggs, separated	3

Grease a 6-cup (1.5 L) baking dish. Cook macaroni in boiling salted water according to directions. Drain well. Melt butter in large saucepan. Stir in flour, salt, paprika and cayenne. Gradually add milk, and cook over moderate heat, stirring constantly, until smooth and thickened. Add cheese and stir until melted. Remove from heat. Beat egg yolks until light. Add slowly to cheese sauce. Add cooked macaroni. Beat egg whites until stiff, but not dry. Fold gently into macaroni mixture. Pour into greased baking dish. Bake at 350°F (180°C) for 35 to 40 minutes until centre is firm when lightly touched.

Makes 6 servings.

MACARONI CASSEROLE WITH DOUBLE CHEESE

6 ounces	macaroni	175 g
¼ cup	butter	50 mL
¼ cup	all-purpose flour	50 mL
4 cups	milk	1 L
¼ teaspoon	Tabasco sauce	1 mL
½ teaspoon	Worcestershire sauce	2 mL
1 teaspoon	salt (or more to taste)	5 mL
¼ teaspoon	pepper	1 mL
¼ teaspoon	nutmeg	1 mL
2 teaspoons	dry mustard	10 mL
1½ cups	grated Swiss cheese (or Cheddar), packed	375 mL
¾ cup	grated Parmesan cheese	175 mL
1 cup	frozen peas defrosted – optional	250 mL
4 ounces	ham, diced – optional	125 g

TOPPING

2 cups	fresh breadcrumbs	500 mL
⅓ cup	butter, melted	75 mL
½ cup	grated Swiss cheese	125 mL

Bring at least 6 quarts (6 L) water to a boil. Add 1 tablespoon (15 mL) salt. Add macaroni. Cook until tender (about 10 minutes). Drain, rinse with cold water, shake out any excess moisture well. Reserve. For the sauce melt butter in a large saucepan. Whisk in flour. Cook, without browning. 5 minutes. Cool slightly. Whisk in milk. Bring to a boil. Reduce heat. Cook, stirring occasionally, 5 minutes. Stir in Tabasco, Worcestershire sauce, salt, pepper, nutmeg, mustard, Swiss cheese and Parmesan. Pat peas and ham dry with paper towels. Combine with well-drained macaroni. Stir in sauce. Taste mixture and add seasoning if necessary. Transfer to a buttered 3 quart (3 L) casserole. Combine ingredients for topping. Sprinkle over macaroni. Bake in a preheated 375°F (190°C) oven, uncovered, for 30 minutes or until hot and bubbling.

Makes 6 servings.

MACARONI RING

1 cup	elbow macaroni	250 mL
2	slices bread, cubed	2
1 cup	grated Canadian Cheddar cheese	250 mL
1½ cups	milk	375 mL
2 tablespoons	finely chopped parsley	30 mL
1 tablespoon	grated onion	15 mL
1 tablespoon	mayonnaise	15 mL
1 teaspoon	salt	5 mL

Cook macaroni noodles in boiling water until tender. Drain well and combine with remaining ingredients. Pour into lightly greased 1 quart ring mold (1 L). Place in a pan of hot water and bake in 350°F (180°C) oven for 40 minutes until set. Allow to stand 10 minutes before unmolding. Fill centre with Buffet Chicken or Shrimp à la King.*

Makes 4-6 servings.

*Macaroni may also be baked in a 1 quart (1 L) casserole dish in a pan of hot water. Sprinkle top with breadcrumbs and paprika and dot with butter before baking.

MANICOTTI WITH MUSHROOM SAUCE

8	manicotti shells	8
1 cup	cottage cheese	250 mL
1½ cups	shredded Canadian Mozzarella cheese	375 mL
½ cup	grated Canadian Parmesan cheese	125 mL
2	eggs, slightly beaten	2
1 teaspoon	Italian seasoning	5 mL
¾ teaspoon	salt	3 mL
2 tablespoons	butter	30 mL
½ cup	finely-chopped celery	125 mL
¼ cup	finely-chopped onion	50 mL
2 tablespoons	flour	30 mL
1 teaspoon	chicken bouillon mix	5 mL
1½ cups	milk	375 mL
1 (10-ounce)	can sliced mushrooms, drained	1 (284 mL)

Cook manicotti shells according to package directions; drain well. Beat cottage cheese until smooth. Add Mozzarella and Parmesan cheeses, eggs, Italian seasoning and ½ teaspoon (2 mL) of the salt. Stuff manicotti shells with equal amounts of cheese filling. Place in greased 2-quart (2 L) rectangular baking dish. Melt butter in a saucepan. Sauté celery and onion until tender. Blend in flour, chicken bouillon mix and remaining salt. Gradually stir in milk. Cook over medium heat, stirring constantly, until mixture just comes to a boil and thickens. Add mushrooms. Pour over manicotti. Cover and bake in preheated 350°F (180°C) oven 25 to 30 minutes or until hot and bubbly.

Makes 4 servings.

MONTE CRISTO SANDWICHES

4	slices buttered bread	4
2	slices cooked turkey	2
2	slices ham	2
2	slices Swiss cheese	2
1	egg	1
⅔ cup	milk	150 mL

Make 2 sandwiches using turkey, ham, and cheese. Beat egg and milk slightly. Dip sandwiches in egg mixture, turning to coat evenly. Fry in melted butter until evenly browned on both sides. Currant jelly, cranberry sauce or chili sauce add the finishing touch.

Makes 2 servings.

MUSHROOM AND RED PEPPER FRITTATA

3 tablespoons	butter	45 mL
1	onion, finely chopped	1
1	clove garlic, minced	1
1 pound	mushrooms, thinly sliced	500 g
2	red peppers, halved, seeded and diced	2
3 tablespoons	chopped fresh parsley	45 mL
8	eggs	8
1½ cups	milk	375 mL
2 cups	grated Cheddar cheese approx. 8 ounces (250 g)	500 mL
1 teaspoon	salt	5 mL
¼ teaspoon	black pepper	1 mL
½ cup	fresh breadcrumbs	125 mL
3 tablespoons	grated Parmesan cheese	45 mL

Melt butter in a large skillet; add onion and garlic. Cook without browning until tender and fragrant. Add mushrooms and red peppers; cook until all liquid from the mushrooms and peppers has evaporated. Mixture should be quite dry. Add parsley. Cool. Beat eggs with milk in a large bowl; add vegetables and remaining ingredients except Parmesan. Pour mixture into a buttered 9"x 13" (3 L) baking dish and smooth top. Sprinkle with Parmesan. Bake 35 to 40 minutes at 350°F (180°C) or until set. Rest 10 minutes before serving.

Note: If fresh red peppers are not available substitute green peppers or ¾ cup (175 mL) diced pimento. Recipe can be halved and baked in an 8" (2 L) square or round baking dish.

Makes 6-8 servings.

MUSHROOM ONION QUICHE

1	9" (23 cm) deep pie shell, unbaked	1
1	onion, thinly sliced	1
¼ cup	butter, melted	50 mL
½ pound	mushrooms, sliced	250 g
4	slices bacon, crisply fried and crumbled	4
1 cup	Cheddar cheese, grated	250 mL
3	eggs, beaten	3
¾ cup	milk	175 mL
¼ teaspoon	nutmeg	1 mL
½ teaspoon	salt	2 mL
¼ teaspoon	pepper	1 mL

Sauté onion in butter until soft, about 5 minutes. Add mushrooms and stir constantly over high heat for 3 minutes. Remove from heat, and set aside to cool. Place crumbled bacon in pie shell. Top with onion-mushroom mixture. Cover with grated cheese. Combine eggs with milk, nutmeg, salt and pepper, and pour over cheese. Bake at 425°F (220°C) for 15 minutes, then reduce heat to 350°F (180°C) and continue baking for 15 minutes longer, or until custard is set. Can be served hot or cold.

Makes 6 servings.

NIGHT-BEFORE SCRAMBLED EGGS

5 tablespoons	butter	75 mL
3 tablespoons	flour	45 mL
¾ teaspoon	salt	3 mL
2 cups	milk	500 mL
1½ cups	shredded old Canadian Cheddar cheese	375 mL
1 cup	sliced fresh mushrooms	250 mL
¼ cup	finely-chopped onion	50 mL
12	eggs, beaten	12
1 cup	fresh bread crumbs	250 mL

Melt 2 tablespoons (30 mL) of the butter in a saucepan. Blend in flour and salt. Gradually stir in milk. Cook over medium heat, stirring constantly, until mixture just comes to a boil and thickens. Remove from heat. Add cheese and stir until melted. Cover and set aside. Melt 2 more tablespoons (30 mL) of the butter in a large frypan. Sauté mushrooms and onion until tender and liquid has evaporated. Add eggs and continue cooking and stirring until mixture is just set. Stir cheese sauce into eggs. Spoon all into 1½-quart (1.5 L) shallow rectangular baking dish. Melt remaining 1 tablespoon (15 mL) butter. Toss with bread crumbs. Sprinkle over egg and sauce mixture. Cover and refrigerate overnight. Bake, uncovered, in preheated 350°F (180°C) oven 20 to 25 minutes or until heated through.

Makes 6 to 8 servings.

Noodles au Gratin

2 tablespoons	butter	30 mL
2 tablespoons	flour	30 mL
½ teaspoon	salt	2 mL
pinch	pepper	pinch
1½ cups	milk	375 mL
1½ cups	shredded Canadian Cheddar cheese	375 mL
1 (10-ounce)	can sliced mushrooms, drained	1 (284 mL)
187 g	(half a 375 g package) medium noodles	187 g

Melt butter in a medium saucepan. Blend in flour, salt and pepper. Gradually stir in milk. Cook over medium heat, stirring constantly, until mixture just comes to a boil and thickens. Remove from heat. Add cheese; stir until melted. Add mushrooms. Keep warm. Cook noodles according to package directions. Drain well; return noodles to saucepan. Add cheese sauce and toss lightly to coat.

Makes 6 side-dish servings.

NOODLE SQUARES MILANO

1 (375 g)	package wide egg noodles	1 (375 g)
4	eggs, slightly beaten	4
2½ cups	milk	625 mL
2 teaspoons	salt	10 mL
¼ teaspoon	pepper	1 mL
1 pound	bulk pork sausage*	500 g
1 (28-ounce)	can spaghetti sauce	1 (796 mL)

Cook noodles according to package directions; drain well. In a large bowl combine eggs, milk, salt and pepper. Add noddles; toss. Turn into a greased 2-quart (2 L) shallow rectangular baking dish. Bake in preheated 350°F (180°C) oven 30 to 35 minutes or until a knife inserted near centre comes out clean. Brown sausage meat in a saucepan. Drain off fat. Stir in spaghetti sauce. Bring to a boil. Reduce heat and simmer, uncovered, 5 minutes. To serve, cut noodles into squares and spoon meat sauce over each serving.

Makes 6 servings.

* 1 pound (500 g) ground beef may be substituted for sausage meat.

NOODLES ROMANOFF

1 cup	sour cream	250 mL
½ cup	grated Canadian Parmesan cheese	125 mL
1 tablespoon	finely-chopped green onion	15 mL
1 teaspoon	salt	5 mL
pinch	pepper	pinch
1	small clove garlic, minced	1
¾ cup	milk	175 mL
5 cups	broad egg noodles	1.25 L
2 tablespoons	butter	30 mL

In a small bowl blend together sour cream, cheese, onion, salt, pepper and garlic. Gradually stir in milk. Set mixture aside. Cook noodles according to package directions. Drain well; return to pan. Add butter and stir until melted. Pour milk mixture over noodles in pan and toss gently to coat. Serve immediately. Makes 6 servings.

Pasta Primavera au Gratin

3	tablespoons	butter	45 mL
3	tablespoons	all-purpose flour	45 mL
1	teaspoon	chicken bouillon mix	5 mL
½	teaspoon	salt	2 mL
2	cups	milk	500 mL
3	cups	shredded Canadian Cheddar cheese	750 mL
1	pound	linguine or spaghetti	500 g
2	tablespoons	butter	30 mL
1	teaspoon	dried basil	5 mL
4	cups	hot cooked vegetables	1 L

Melt butter in saucepan. Blend in flour, bouillon mix and salt. Gradually stir in milk. Cook and stir over medium heat until mixture comes to a boil. Remove from heat. Add cheese; stir until melted. Keep warm. Cook linguine according to package directions; drain well. Add 2 tablespoons (30 mL) butter, basil and hot cooked vegetables; toss lightly to combine. Pour hot cheese sauce over each serving.

Makes 4 servings.

Pasta and Sausage Fry Up

4	Italian style hot or mild sausages	4
3 tablespoons	butter	45 mL
1 tablespoon	oil	15 mL
1	clove garlic, minced	1
1	medium onion, chopped	1
1	head broccoli	1
1	red pepper, sliced in thin slivers	1
¼ pound	mushrooms, sliced	125 g
3 tablespoons	flour	45 mL
2 ½ cups	milk	625 mL
½ cup	grated Parmesan cheese	125 mL
6 ounces	cooked hot pasta	100 g

Prick sausages with fork. Simmer in boiling water until cooked through. Cool and cut in thin slices. Cook garlic and onion in 1 tablespoon (15 mL) each butter and oil until soft. Cut flowerettes of broccoli in bite size pieces, and tender part of upper stem. Add to onion with red pepper and mushrooms. Toss and cook until just tender. Season with salt. Remove and set aside. Cook sausages in same pan until nicely browned. Add to vegetables. Heat remaining butter in saucepan. Add flour. Cook and stir 1 minute. Whisk in milk. Cook and stir until well thickened. Stir in Parmesan cheese, and mix to blend. Toss sauce with vegetables and cooked pasta. Adjust seasoning.

Makes 6 servings.

PASTA WITH LOBSTER AND SHRIMP SAUCE

1 (5 ounce)	can lobster meat	1 (142 g)
1 (5 ounce)	can shrimp	1 (142 g)
⅓ cup	butter	75 mL
½ teaspoon	curry powder	2 mL
2 tablespoons	cognac (optional)	30 mL
4 cups	large pasta shells, uncooked	1 L
⅓ cup	flour	75 mL
¼ teaspoon	salt	1 mL
¼ teaspoon	pepper	1 mL
2½ cups	milk	625 mL
½ cup	Parmesan cheese, grated	125 mL

Drain lobster and shrimp, reserving the liquid. Cut into small pieces. Sauté seafood in butter with curry powder for 3 minutes. Add cognac, and simmer for 3 minutes. Blend in flour, salt and pepper. Gradually blend in milk. Cook, stirring constantly, until sauce is slightly thickened. Add reserved seafood liquid. Prepare pasta shells by boiling in large quantity of salted water. Drain well. In a greased 10"x 10" (25 cm x 25 cm) baking dish, arrange pasta shells. Pour prepared sauce over pasta. Sprinkle with Parmesan cheese. Bake at 350°F (180°C) for 30 minutes.

Makes 6 servings.

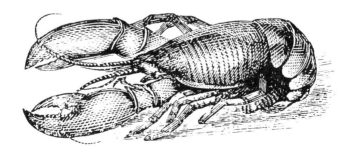

Pizza oven omelette

1 tablespoon	butter	15 mL
6	eggs	6
¾ cup	milk	175 mL
½ teaspoon	salt	2 mL
1 (7½-ounce)	can pizza sauce	1 (213 mL)
4 ounces	Canadian Mozzarella cheese, thinly sliced	125 g

Put butter in a 9-inch (1L) pie plate. Place in a preheated 425°F (220°C) oven until melted. Beat eggs well; stir in milk and salt. Pour into pie plate and bake 5 minutes. Reduce oven temperature to 350°F (180°C) and bake 10 to 12 minutes longer. Heat pizza sauce in a small saucepan. Arrange cheese on top of omelette. Broil until cheese is melted and bubbly. Cut omelette into wedges and spoon hot pizza sauce over individual servings.

Makes 4 to 6 servings.

POACHED EGGS IN SPINACH YOGURT SAUCE

1	onion, sliced	1
1	garlic clove, chopped	1
2 tablespoons	butter	30 mL
2 cups	milk or light cream	500 mL
6	eggs	6
2 tablespoons	flour	30 mL
1 (10 ounce)	package spinach, cooked, drained and chopped	1 (284 g)
1 cup	plain yogurt or sour cream	250 mL
6	hot baked potatoes, cut open	6

Sauté onion and garlic in butter. Add milk and bring to a boil. Reduce heat to a simmer and gently slip in eggs. Poach 3 to 5 minutes. Remove eggs with a slotted spoon and keep warm. Take ¼ cup (50 mL) sauce and whisk flour into it until smooth. Add to remaining sauce and stir constantly until thickened. Add spinach, yogurt and eggs. Heat through but do not boil. Serve 1 poached egg in each potato and top with sauce.

Makes 6 servings.

Note: This could be served over rice, omitting the potatoes.

QUICHE LORRAINE

1	unbaked 9″ (23 cm) pie shell	1
6	slices bacon, diced	6
3	eggs, separated	3
1 cup	milk	250 mL
½ teaspoon	salt	2 mL
¼ teaspoon	pepper	1 mL
1 cup	Gruyere cheese, grated	250 mL

Fry bacon until cooked, but not crisp. Drain and place in pastry shell. Beat egg yolks, milk, salt and pepper. In a separate bowl, beat egg whites until stiff but not dry. Fold into milk mixture. Add cheese. Pour mixture over the bacon and bake at 350°F (180°C) for 20 minutes, or until set. Let stand 5 minutes before serving.

Makes 6 servings.

SCRAMBLED EASTER EGG BRUNCH

5 tablespoons	butter	75 mL
1 cup	sliced fresh mushrooms	250 mL
1 tablespoon	flour	15 mL
1 teaspoon	salt	6 mL
1¾ cups	milk	425 mL
1 cup	shredded Canadian	250 mL
	Cheddar cheese	
6	eggs	6
6	puff pastry shells, warmed	6

Melt 2 tablespoons (30 mL) of the butter in a small saucepan. Sauté mushrooms until tender and liquid has evaporated. Blend in flour and ½ teaspoon (3 mL) of the salt. Gradually stir in 1 cup (250 mL) of the milk. Cook over medium heat, stirring constantly, until mixture just comes to a boil and thickens. Remove from heat. Add cheese and stir until melted. Cover and keep warm. Beat eggs well. Stir in remaining ¾ cup (175 mL) milk and remaining ½ teaspoon (3 mL) salt. Melt remaining 3 tablespoons (45 mL) butter in a frypan. Pour in egg mixture. Cook and stir until mixture is just set. Fill puff pastry shells with scrambled eggs and spoon mushroom sauce over top.

Makes 6 servings.

SEA SHELLS AND CHEESE

2 cups	small shell macaroni	500 mL
¼ cup	butter	50 mL
¼ cup	flour	50 mL
1 teaspoon	chicken bouillon mix	5 mL
¾ teaspoon	salt	3 mL
2½ cups	milk	625 mL
3 cups	shredded Canadian	750 mL
	Cheddar cheese	

Cook shell macaroni according to package directions; drain. Melt butter in a medium saucepan. Blend in flour, bouillon mix and salt. Gradually stir in milk. Cook over medium heat, stirring constantly, until mixture just comes to a boil and thickens. Remove from heat. Add 2½ cups (625 mL) of the cheese and stir until melted. Add macaroni and reheat to serving temperature. Sprinkle remaining cheese over individual servings.

Makes 5 or 6 servings.

Seafood surprise lasagne

2 (10 ounce)	packages fresh spinach	2 (284 g)
3 tablespoons	butter	45 mL
	salt, pepper to taste	
	dash nutmeg	
2	green onions, finely chopped	2
3 tablespoons	flour	45 mL
1½ cups	milk	375 mL
1 (6 ounce)	can crabmeat	1 (170 g)
2 tablespoons	dry sherry	30 mL
1 tablespoon	chopped dill	15 mL
8	lasagne noodles, cooked	8
1 cup	grated Swiss cheese	250 mL

Trim and wash spinach. Cook in a little water until tender. Drain well, chop and toss with 1 tablespoon (15 mL) butter. Season to taste with salt, pepper and nutmeg. Set aside. Melt remaining butter in medium saucepan. Cook green onions to soften. Add flour and cook, stirring over medium heat for one minute. Gradually whisk in milk. Cook over medium heat, stirring until smooth and thick. Drain crab and remove bits of shell. Add crabmeat to sauce with sherry, dill and ½ cup (125 mL) cheese. Season to taste with salt, and pepper. Lightly butter an 8″ (2 L) square shallow baking dish. Layer dish with noodles, spinach, crab. Repeat layers ending with noodles. Reserve a little of crab mixture for top. Sprinkle top with remaining cheese and crab. Cover with foil and bake in 400°F (200°C) oven for 30 minutes. Remove foil for last 10 minutes to allow browning. Let stand 5 minutes before serving.

Makes 6 servings.

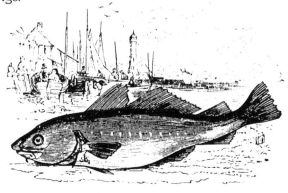

SPAGHETTINI WITH HERBED CHEESE SAUCE

3 tablespoons	butter	45 mL
⅓ cup	chopped green onion	75 mL
1	clove garlic, minced	1
1 tablespoon	flour	15 mL
1 teaspoon	salt	5 mL
½ teaspoon	basil leaves	2 mL
¼ teaspoon	oregano leaves	1 mL
¼ teaspoon	pepper	1 mL
1¾ cups	milk	425 mL
1 cup	shredded Canadian Mozzarella cheese	250 mL
1 cup	Canadian Ricotta cheese	250 mL
3 tablespoons	chopped fresh parsley	45 mL
250 g	(one quarter of 1 kg package) spaghettini	250 g

Melt butter in a large saucepan. Sauté onion and garlic until tender. Blend in flour, salt, basil, oregano and pepper. Gradually stir in milk. Cook over medium heat, stirring constantly, until mixture just comes to a boil and thickens. Remove from heat. Add cheeses and stir until melted and thoroughly combined. Stir in parsley. Keep warm. Cook spaghettini according to package directions; drain well. Add spaghettini to cheese sauce; toss well to coat. Serve immediately.
Makes 4 servings.

SPAGHETTI WITH CLAM SAUCE

1 (5-ounce)	can baby clams	1 (142 g)
6 tablespoons	butter	90 mL
¾ cup	chopped onion	175 mL
2	medium tomatoes, peeled and chopped	2
¾ teaspoon	basil leaves	3 mL
¾ teaspoon	salt	3 mL
¼ teaspoon	pepper	1 mL
2 tablespoons	chopped parsley	30 mL
2 tablespoons	flour	30 mL
2 teaspoons	chicken bouillon mix	10 mL
1¼ cups	milk	300 mL
¾ cup	grated Canadian Parmesan cheese	175 mL
250 g	(one quarter of 1 kg package) spaghetti	250 g

Drain clams, reserving liquid; set both aside. Melt 3 tablespoons (45 mL) of the butter in a saucepan. Sauté onion until tender. Add tomatoes, basil, salt and pepper. Bring to a boil. Cover and simmer 15 minutes, stirring occasionally. Stir in clams and parsley. Heat through and keep warm. Melt remaining 3 tablespoons (45 mL) butter in a saucepan. Blend in flour and bouillon mix. Gradually stir in milk. Cook over medium heat, stirring constantly, until mixture just comes to a boil and thickens. Remove from heat; add ½ cup (125 mL) of the cheese and stir until well combined. Keep warm. Cook spaghetti according to package directions, using liquid from clams to season boiling water. Drain well. Divide evenly among 4 plates; top with equal amounts of cheese sauce, then clam sauce. Sprinkle remaining cheese over top of each serving. Serve immediately.
   Makes 4 servings.

SPINACH FLAN

PASTRY

1	cup	all-purpose flour	250 mL
1	teaspoon	salt	5 mL
¼	cup	butter	50 mL
1		egg yolk	1
2	tablespoons	cold water	30 mL
1	teaspoon	lemon juice	5 mL

FILLING

3		eggs	3
1	cup	milk	250 mL
½	cup	grated Swiss cheese	125 mL
1	(10-ounce)	package fresh spinach, cooked, squeezed dry, finely chopped	1 (284 g)
1		clove garlic, crushed	1
3	tablespoons	fresh breadcrumbs	45 mL
		salt and pepper	

To prepare pastry, combine flour and salt in a mixing bowl. Cut in butter until mixture resembles coarse meal. Combine remaining ingredients, add to flour and combine until mixture forms a ball. Wrap and refrigerate 15 minutes.

Roll out pastry to fit an 8" (1 L) pie plate. Line pastry with a circle of waxed paper and half fill with dried beans. Bake in a preheated 375°F (190°C) oven 15 minutes. Remove paper and beans. Combine filling ingredients and pour into pie shell. Return to oven and bake an additional 30 minutes or until filling is set.

Makes 4 servings.

SPINACH QUICHE

2	9" (23 cm) pie shells, unbaked	2
¾ pound	fresh spinach	375 g
2 tablespoons	butter, melted	30 mL
4	green onions, chopped, white and green parts	4
1	clove garlic, minced	1
3	eggs, beaten	3
1¼ cups	milk	300 mL
1 teaspoon	salt	5 mL
1 teaspoon	basil	5 mL
½ teaspoon	celery salt	2 mL
1½ cups	brick cheese, grated	375 mL
2	tomatoes, sliced	2
1 tablespoon	dry breadcrumbs	15 mL
1 tablespoon	Parmesan cheese, grated	15 mL

Lightly cook spinach; drain well and chop. Sauté onions and garlic in butter until soft. Set aside. Combine eggs, milk, salt, basil and celery salt. Stir in brick cheese, spinach, and sautéed onions and garlic. Pour into pie shells. Bake at 425°F (220°C) for 15 minutes. Then reduce heat to 350°F (180°C) and bake for 15 minutes longer. Arrange tomato slices around the edge, and top with crumbs combined with the Parmesan cheese. Continue to bake for 10 minutes more, or until filling is set.

Makes 8 servings.

Springtime Crab and Asparagus Tart

1		9" (23 cm) pie crust,	1
		partially baked	
2	tablespoons	butter	30 mL
4		green onions, chopped finely	4
1	(6 ounce)	can crab or salmon	1 (200 g)
½	pound	asparagus, blanched	250 g
½	teaspoon	savoury	2 mL
1	tablespoon	sherry (optional)	15 mL
½	cup	grated Swiss cheese	125 mL
3		eggs	3
1	cup	milk	250 mL
		dash nutmeg and cayenne	

Set oven to 350°F (180°C).

Melt butter in large frying pan. Add onions and cook until softened. Stir in crab, asparagus, savoury and sherry. Heat through. Spoon into prepared pie crust. Top with cheese. Beat together eggs, milk, nutmeg and cayenne. Pour over crab mixture. Bake in pre-heated oven until set – about 35 minutes.

Makes 4 servings.

Tip: A partially-baked pie shell results in a crisper crust. Line pie plate with pastry, prick bottom of crust with fork. Line with waxed paper and fill with dried beans or rice. Bake in 400°F (200°C) oven for 8 minutes. Remove beans and paper and cool before filling.

STUFFED EGGS IN CHEESE SAUCE

¼ cup	butter	50 mL
¼ cup	flour	50 mL
½ teaspoon	salt	2 mL
2 cups	milk	500 mL
1½ cups	shredded Canadian Cheddar cheese	375 mL
6	hard boiled eggs	6
½ cup	chopped green onions	125 mL
¼ cup	finely chopped, drained mushroom slices	50 mL
1 teaspoon	Worcestershire sauce	5 mL
	chopped parsley	

Melt butter in saucepan. Add flour and salt, stir until smooth. Cook over low heat about 1 minute, stirring constantly. Remove from heat, add 1 cup (250 mL) milk. Blend well. Return to heat and stir until mixture begins to thicken. Stir in remaining milk. Heat to just boiling; cook about 2 minutes, stirring constantly until thick. Add 1 cup (250 mL) cheese and stir until cheese is melted. Cut hard boiled eggs in half lengthwise. Remove yolks, mash with fork and mix with ¼ cup (50 mL) of the cheese sauce. Add chopped mushrooms, green onions and Worcestershire sauce. Fill egg white halves.

Place in a shallow 1½ quart (1.5 L) baking dish, cover with cheese sauce. Top with ½ cup (125 mL) cheese.

Bake at 350°F (180°C) oven for 20 minutes or until bubbly. Garnish with chopped parsley, if desired.

Makes 6 servings.

SWISS CHEESE AND BACON SOUFFLE

3 tablespoons	butter	45 mL
3 tablespoons	flour	45 mL
½ teaspoon	dry mustard	2 mL
¼ teaspoon	salt	1 mL
pinch	pepper	pinch
1 cup	milk	250 mL
1½ cups	shredded Canadian Swiss cheese	375 mL
2 tablespoons	grated Canadian Parmesan cheese	30 mL
5	eggs, separated	5
6	slices cooked crisp bacon, finely crumbled	6

Melt butter in a medium saucepan. Blend in flour, mustard, salt and pepper. Gradually stir in milk. Cook over medium heat, stirring constantly, until mixture just comes to a boil and thickens. Remove from heat. Add cheeses and stir until melted. Beat egg yolks well. Stir a small amount of hot mixture into yolks; return to saucepan and blend thoroughly. Stir in bacon. Beat egg whites until stiff but not dry. Stir a large spoonful of beaten egg white into cheese sauce. Fold in remaining egg whites. Turn into a greased 8-cup (2 L) souffle dish. Bake in preheated 325°F (160°C) oven 45 to 55 minutes. Serve immediately.
Makes 4 servings.

TOMATO AND BROCCOLI QUICHE

1	9" (25 cm) quiche or deep pie plate lined with unbaked pastry	1
1	large onion, chopped	1
½ cup	sliced mushrooms	125 mL
½ teaspoon	oregano	2 mL
¾ teaspoon	salt	4 mL
¼ teaspoon	pepper	1 mL
2 tablespoons	butter	30 mL
8	bacon slices, diced	8
4 ounces	cream cheese	125 g
½ cup	grated Parmesan cheese	125 mL
1 cup	milk	250 mL
3	eggs, beaten	3
1 cup	chopped, cooked broccoli	250 mL
½ cup	soft bread crumbs (1 slice)	125 mL
4	tomatoes, sliced	4

Sauté onions, mushrooms and seasonings in butter. Fry bacon until crisp; drain and crumble. Cream together cheeses; blend in milk. Add beaten eggs. Fold in broccoli, bacon, bread crumbs, onions and mushrooms. Pour into pastry-lined shell. Arrange tomato slices around outer edge. Bake at 425°F (220°C) for 10 minutes. Lower temperature to 350°F (180°C) and bake 25 minutes longer. Let stand 10 minutes before serving.

Makes 6 servings.

TOTE ALONG CASSEROLE

1	tablespoon	all-purpose flour	15 mL
2	teaspoons	onion salt	10 mL
2	teaspoons	chili powder	10 mL
3½ cups	milk	875 mL
1¾ cups	macaroni	425 mL
1	cup	chopped green pepper	250 mL
2	cups	shredded Cheddar cheese	500 mL
1	(14-ounce)	can kidney beans	1 (398 mL)
4		cheese slices, cut in triangles	4

Combine flour, onion salt, and chili powder in a large skillet; gradually stir in milk. Add macaroni and green pepper. Cover and bring to simmering; reduce heat and simmer 15 minutes, stirring occasionally. Stir in cheese until melted, then add 1 cup (250 mL) beans. Place remaining beans in centre on top. Top with cheese triangles. Cover and heat 5-8 minutes longer.

Makes 8 servings.

VEGETABLE CREPES

CREPES

2	eggs	2
1 cup	flour	250 mL
2 tablespoons	butter, melted	30 mL
½ cup	milk	125 mL
½ cup	water	125 mL

FILLING

¼ cup	butter, melted	50 mL
1	medium onion, chopped	1
½ cup	green pepper, finely chopped	125 mL
½ cup	red pepper, finely chopped	125 mL
2	medium zucchini, coarsely grated	2
1½ cups	mushrooms, finely chopped	375 mL
½ teaspoon	basil	2 mL
½ teaspoon	oregano	2 mL
½ teaspoon	salt	2 mL
¼ teaspoon	pepper	1 mL
1 cup	Swiss cheese, grated	250 mL
¼ cup	Parmesan cheese, grated	50 mL

To make crepes, blend all ingredients thoroughly until batter is very smooth. Let stand for at least 2 hours. To cook, pour a scant ¼ cup (50 mL) batter into a 5″ (12 cm) greased or non-stick frying pan. Cook over medium heat until bottom is golden and top is dry. Turn out onto plate and repeat. Makes 16 crepes.

For filling, sauté onions and peppers in butter for 2 minutes. Add zucchini and mushrooms; cook 2 minutes longer. Stir in basil, oregano, salt, pepper and cheeses; stir until the cheese is melted.

continued on next page

VEGETABLE CREPES (continued)

SAUCE

3 tablespoons	butter, melted	45 mL
3 tablespoons	flour	45 mL
½ teaspoon	curry powder	2 mL
1½ cups	milk	375 mL
¼ teaspoon	salt	1 mL
dash	pepper	dash

For sauce, combine butter, flour and curry. Add milk gradually, and cook over medium heat, stirring constantly, until smooth and thickened. Add salt and pepper. Cover and set aside.

To assemble crepes, place the uncooked side of the crepes face up, and distribute filling equally. Roll up and place in greased 10" (25 cm) baking dish. Pour sauce over all. Bake at 350°F (180°C) for 15 minutes, or until heated through. Garnish with green pepper rings.

Makes 4 servings.

ASPARAGUS CHICKEN AND PASTA SUPPER

½ cup	uncooked pasta shells	125 mL
2	green onions, sliced	2
1 cup	fresh mushrooms, sliced	250 mL
5 tablespoons	butter, melted	75 mL
2 tablespoons	oil	30 mL
2	whole chicken breasts, boneless and skinless	2
2 tablespoons	flour	30 mL
¼ cup	white wine	50 mL
¾ cup	milk	175 mL
2 (12 ounce)	tins asparagus	2 (341 mL)
2 tablespoons	chopped parsley	30 mL

Cook pasta shells in boiling water. Drain well. Sauté green onions and mushrooms in butter and oil until soft. Remove vegetables. Add chicken and sauté for 10 minutes. Remove, slice thinly and keep warm. Add flour to pan, and blend until smooth. Add wine and milk, and continue to heat, stirring constantly, until smooth and thickened. Add pasta shells, green onions, mushrooms and chicken. Reheat to serving temperature. Heat asparagus in its own juice and arrange on serving platter. Surround with chicken and pasta mixture. Top with chopped parsley.
   Makes 4 servings.

Breaded Chicken Breasts with Spinach and Dill Sauce

3	chicken breasts, split, skinned and boned	3
1/3 cup	prepared mustard (preferably Dijon)	75 mL
1/3 cup	sour cream or unflavoured yogurt	75 mL
2 tablespoons	chopped fresh dill (or 1/2 teaspoon/2 mL dried)	30 mL
1/4 teaspoon	black pepper	1 mL
3 cups	fresh breadcrumbs	750 mL

SAUCE

2 tablespoons	butter	30 mL
3 tablespoons	all purpose flour	45 mL
3 cups	milk, hot	750 mL
1 teaspoon	salt	5 mL
1/4 teaspoon	black pepper	1 mL
10 ounce	spinach, cooked, squeezed dry and chopped	284 mL
1/3 cup	chopped fresh dill (or 2 teaspoons/10 mL dried)	75 mL
1/4 teaspoon	nutmeg	1 mL
1 cup	grated Swiss cheese (approx. 4 ounce/125 g)	250 mL

Pat chicken pieces dry. Combine mustard, sour cream, dill and pepper in a flat dish. Place breadcrumbs in another dish. Dip chicken into mustard mixture to coat; roll in breadcrumbs. Pat crumbs in well. Brush a cookie sheet lightly with oil and place in oven while preheating to 375°F (190°C). Place chicken on hot cookie sheet and bake 20 minutes. Turn and bake 15 to 25 minutes longer depending on thickness of chicken. While chicken is baking prepare sauce. Melt butter in large saucepan; whisk in flour. Cook without browning 5 minutes. Whisk in hot milk; bring to a boil. Add salt and pepper; cook over low heat, stirring occasionally, 10 minutes. Add spinach, dill, nutmeg; cook 5 minutes longer. Add cheese; cook until melted. Season to taste. Serve chicken breasts with sauce spooned over the top.

Makes 6 servings.

BREAST OF CHICKEN PRONTO

2 tablespoons	butter	30 mL
8	single chicken breasts, boned and slightly flattened	8
½ cup	water	125 mL
1 teaspoon	chicken bouillon mix	5 mL
1 teaspoon	salt	5 mL
¾ teaspoon	poultry seasoning	3 mL
2 tablespoons	flour	30 mL
1¼ cups	milk	300 mL
	chopped parsley	

Melt butter in a large frypan. Sauté chicken until lightly browned on both sides. Combine water, bouillon mix, salt and poultry seasoning. Pour around chicken. Bring to a boil. Reduce heat; cover and simmer 20 to 25 minutes or until chicken is cooked. Remove chicken to a heated platter; keep warm. Smoothly combine flour and milk. Stir into pan juices. Cook over medium heat, stirring constantly, until mixture just comes to a boil and thickens. Pour sauce over chicken and garnish with chopped parsley to serve.

Makes 6 to 8 servings.

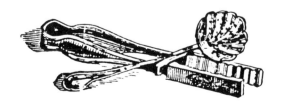

BUFFET CHICKEN AND SHRIMP A LA KING

⅓ cup	butter	75 mL
½ cup	finely-chopped onion	125 mL
½ cup	chopped celery	125 mL
½ cup	chopped green pepper	125 mL
¼ cup	flour	50 mL
1½ teaspoons	seasoned salt	7 mL
1 (10-ounce)	can cream of mushroom soup	1 (284 mL)
3 cups	milk	750 mL
3 cups	cubed, cooked, chicken	750 mL
1 (10-ounce)	can whole mushrooms, drained	1 (284 mL)
1 (4-ounce)	can medium deveined shrimp, drained	1 (113 g)
¼ cup	chopped pimiento	50 mL
	hot cooked rice, noodles or puff pastry shells	

Melt butter in a large saucepan. Sauté onion, celery and green pepper until tender. Blend in flour and salt. Combine soup and milk; gradually stir into flour mixture. Cook over medium heat, stirring constantly, until mixture just comes to a boil and thickens. Stir in chicken, mushrooms, shrimp and pimiento. Reheat to serving temperature. Transfer to a chafing dish. Serve over rice, noodles or puff pastry shells.

Makes 9 cups/2.25 L.

CARIBBEAN CHICKEN

3-4 pounds	chicken, cut into pieces	1.5-2 kg
2 tablespoons	butter	30 mL
2½ cups	milk	625 mL
1	large onion, sliced	1
3-4 tablespoons	flour	45-60 mL
3-4 tablespoons	butter	45-60 mL
1	large ripe mango, peeled and sliced, OR	1
1 cup	sliced fresh or canned peaches	250 mL
1 teaspoon	salt	5 mL
¼ teaspoon	pepper	1 mL
½ teaspoon	ginger, or to taste	2 mL
3 tablespoons	dark rum (optional)	45 mL

In a large casserole, brown chicken in 2 tablespoons (30 mL) butter. Cover chicken pieces in milk; add onion slices. Cover and bake in a 350°F (180°C) oven for 45 minutes or until chicken is tender; remove chicken and keep warm; reserve milk. Melt butter in a large saucepan; blend in flour and cook for 2 minutes. Gradually add milk and whisk until smooth. Cook over low heat, stirring constantly until thickened. Add fruit, seasonings, rum and chicken; heat an additional 10 minutes and serve.

Makes 4 to 6 servings.

CHEDDAR CHICKEN DIVAN

2 (10 ounce)'	boxes, frozen broccoli spears, parboiled, and drained	2 (284 g)
6	single chicken breasts, cooked, boned, and sliced	6
2 tablespoons	butter	30 mL
3 tablespoons	flour	45 mL
2	packets instant chicken bouillon powder	2
½ teaspoon	dry mustard	2 mL
2 cups	milk	500 mL
2 cups	Cheddar cheese, grated paprika	500 mL

Arrange cooked broccoli on bottom of a greased 9" x 13" (23 cm x 33 cm) baking dish. Top with cooked chicken slices. In a saucepan, melt butter, blend in flour, bouillon powder and mustard. Add milk, stirring constantly and cook over medium heat until smooth and thickened. Remove from heat. Add cheese and stir until melted. Pour sauce over chicken and broccoli. Sprinkle with paprika. Bake at 350°F (180°C) for 25 minutes, or until heated through.

Makes 6 servings.

CHEESY CHICKEN CASSEROLE

4	slices bread, crusts removed	4
2 cups	cooked chicken, diced	500 mL
3	eggs, beaten	3
2 cups	milk	500 mL
½ teaspoon	basil	2 mL
1 tablespoon	fresh parsley, chopped	15 mL
1 cup	Cheddar cheese, grated	250 mL

Cut bread into quarters. Place half on bottom of greased 6-cup (1.5 L) casserole. Arrange chicken over bread. Combine eggs, milk, basil and parsley. Pour half over the chicken. Sprinkle on half the cheese. Put on remaining bread. Pour on remaining milk mixture and rest of cheese. Press lightly with a fork to make bread absorb the milk. Let stand for at least 1 hour. Bake at 325°F (160°C) for 60 minutes until firm and golden.

Makes 4 servings.

Chicken Puff

1 (10 ounce)	can condensed cream of shrimp soup	1 (284 mL)
1 cup	milk	250 mL
2 cups	cooked chicken	500 mL
½ cup	frozen or canned peas	125 mL
3 tablespoons	butter	45 mL
¼ cup	flour	50 mL
¾ cup	milk	175 mL
4	eggs, separated	4
1 tablespoon	fresh parsley, chopped	15 mL
½ teaspoon	salt	2 mL
¼ teaspoon	pepper	1 mL
⅛ teaspoon	sage	0.5 mL

Mix the shrimp soup, milk, chicken and peas, and pour into a greased 8-cup (2 L) casserole. Melt the butter. Blend in flour. Add milk gradually, stirring constantly, and continue to cook until smooth and thickened. Beat egg yolks and add to sauce. Add parsley, salt, pepper and sage. Beat egg whites until stiff, but not dry. Fold gently into sauce. Spoon sauce over the chicken, and bake at 375°F (190°C) for 40 minutes.

Makes 6 servings.

CHICKEN PAPRIKA

3½ to 4 pound	cut-up stewing chicken*	1.75 to 2 kg
½ cup	all-purpose flour	125 mL
2 tablespoons	butter	30 mL
1 cup	finely-chopped onion	250 mL
1	clove garlic, minced	1
1 tablespoon	paprika**	15 mL
1 teaspoon	salt	5 mL
½ cup	water	125 mL
2 teaspoons	chicken bouillon mix	10 mL
1 cup	milk	250 mL
1 cup	sour cream	250 mL
	hot cooked noodles	

Coat chicken pieces lightly with flour; reserve 3 tablespoons (50 mL) of the flour. Melt butter in a large frypan. Brown chicken. Remove chicken from pan; sauté onion and garlic until tender. Stir in paprika, salt, water and bouillon mix. Return chicken to pan. Bring to a boil. Reduce heat; cover and simmer 1½ to 2 hours or until chicken is tender. Remove chicken. Smoothly combine reserved flour and milk. Stir into pan juices. Cook over medium heat, stirring constantly, until mixture just comes to a boil and thickens. Blend in sour cream. Add chicken and heat through. Do not boil. Serve with noodles.
Makes 4 to 5 servings.

* Cut-up broiler-fryer can be substituted. Reduce simmering time to 45 minutes - 1 hour.
** Increase paprika if desired.

CHICKEN VOL-AU-VENT

6	patty shells (vol-au-vent)	6
¼ cup	butter	50 mL
1 cup	sliced mushrooms	250 mL
½ cup	finely chopped celery	125 mL
1	apple, peeled, cored and finely chopped	1
¼ cup	chopped green pepper	50 mL
¼ cup	flour	50 mL
2 cups	milk	500 mL
1 teaspoon	salt	5 mL
	pepper, freshly ground	
½ teaspoon	curry powder, or to taste	2 mL
1 tablespoon	lemon juice	15 mL
2 tablespoons	mayonnaise	30 mL
½ cup	peas, frozen	125 mL
3 cups	cooked and diced chicken	750 mL

Melt butter in medium saucepan; add mushrooms, celery, apple and green pepper; cook until tender, about 5 minutes. Stir in flour until blended. Gradually add milk, stirring constantly until mixture thickens and comes to a boil. Add seasonings and lemon juice; simmer for 2 or 3 minutes. Add mayonnaise, peas and chicken; heat through. Spoon into and over patty shells.
Makes 6 servings.

CHICKEN IN WINE SAUCE

3 pound	fryer chicken	1.5 kg
6	cloves garlic	6
¼ cup	butter	50 mL
2 tablespoons	oil	30 mL
½ cup	dry white wine	125 mL
½ cup	water	125 mL
1 teaspoon	salt	5 mL
dash	pepper	dash
⅛ teaspoon	cayenne	0.5 mL
2 tablespoons	flour	30 mL
1 cup	18% cream	250 mL
	parsley for garnish	

Cut chicken into serving-sized pieces. Peel garlic cloves, but leave whole. In Dutch oven, melt butter with oil over medium heat. Brown chicken pieces well, turning occasionally. Remove when done. Add wine, water, garlic, salt, pepper and cayenne. Stir to loosen browned bits. Bring to boil. Return chicken to pan; cover and cook over low heat 45 minutes, or until tender. Remove chicken to a warm serving plate; cover with foil and keep warm. Remove garlic cloves. (If desired, garlic cloves may remain in the sauce, since they will have become tender and mild after this period of cooking.) Stir some sauce into flour, blending until smooth. Return flour to remaining sauce, stirring until evenly blended. Add cream, stirring constantly, and cook until thickened. Spoon some sauce on chicken. Serve remaining sauce separately. Garnish with parsley.
Makes 6 servings.

CHICKEN NOODLE DOODLE CASSEROLE

2 cups	medium egg noodles	500 mL
2 (10-ounce)	cans cream of mushroom soup	2 (284 mL)
1¼ cups	milk	300 mL
1½ cups	shredded Canadian Cheddar cheese	375 mL
2 (6½-ounce)	cans flaked chicken*, drained and broken up	2 (184 g)
2 cups	cooked peas	500 mL
1 cup	fresh bread crumbs	250 mL
2 tablespoons	butter, melted	30 mL

Cook noodles according to package directions; drain well. Place mushroom soup in a saucepan. Gradually stir in milk. Add cheese. Cook over medium heat, stirring constantly, until cheese is melted. Remove from heat. Stir in chicken, peas and cooked noodles. Divide mixture evenly into six 10-ounce (300 mL) custard cups. Combine bread crumbs and butter. Sprinkle over top of each casserole. Bake in preheated 325°F (160°C) oven 30 to 35 minutes or until hot and bubbly.
Makes 6 servings.

* 2 cups (500 mL) cooked, cut-up chicken may be substituted.

CHICKEN CREPES

CREPES

1 cup	water	250 mL
1 cup	milk	250 mL
4	eggs	4
½ teaspoon	salt	2 mL
2 cups	flour	500 mL
4 tablespoons	butter, melted	60 mL

CREAMED CHICKEN

5 tablespoons	butter, melted	75 mL
¼ cup	green pepper, chopped	50 mL
5 tablespoons	flour	75 mL
dash	lemon pepper	dash
2	packets instant chicken bouillon powder	2
3 cups	milk	750 mL
3 cups	cooked chicken, finely diced	750 mL

To prepare crepes: Blend all ingredients thoroughly in a blender until smooth. Let mixture stand for at least 2 hours. Grease a 6″ (15 cm) frypan lightly and pour a scant ¼ cup (50 mL) into pan. Swirl pan to coat evenly. Cook over medium heat until bottom is golden and top is dry (about 1 minute). Remove to plate. Repeat until all crepes are done (approximately 32 crepes).

For creamed chicken, sauté green pepper in butter until soft. Blend in flour, pepper and bouillon powder. Add milk gradually, stirring constantly, until smooth and thickened. Reserve one cup (250 mL) of the sauce for later use. Add chicken to the remainder of sauce. Let cool.

To assemble crepes, place crepe uncooked side up, and put a heaping spoonful of filling near edge. Fold in sides, and roll up. Place in greased baking dish seam side downward. Continue until all crepes are filled. Cover completely with reserved sauce. Bake at 350°F (180°C) for 20 minutes, or until bubbly and heated through. Both crepes and creamed chicken can be made and refrigerated one day ahead of assembling. Let crepes return to room temperature for ease of handling.

Makes 8 servings.

CHICKEN A LA KING

3 tablespoons	butter, melted	45 mL
1 cup	fresh mushrooms, sliced	250 mL
1 teaspoon	onion, finely chopped	5 mL
2 tablespoons	green pepper, finely chopped	30 mL
3 tablespoons	flour	45 mL
½ teaspoon	salt	2 mL
½ teaspoon	pepper	2 mL
2 cups	milk	500 mL
2 cups	cooked chicken, diced	500 mL
2 tablespoons	parsley, minced	30 mL
1	egg, beaten	1
2 teaspoons	lemon juice	10 mL
6	large ripe olives, sliced	6
¼ cup	pimento, sliced	50 mL

Sauté mushrooms, onion and green pepper in butter for 5 minutes until soft. Sprinkle in flour, salt and pepper. Blend well. Add milk gradually, and cook, stirring constantly, until smooth and thickened. Add chicken and parsley. Cover and cook until heated through. Add a little of the hot mixture gradually to the beaten egg, stirring constantly. Pour egg mixture back into the pan, blending well. Add lemon juice, olives and pimento. Stir and heat to serving temperature. Serve in patty shells or over toast points.
Makes 6 servings.

CHICKEN DRUMSTICKS ITALIANO

2 pounds	chicken drumsticks	1 kg
1 (14-ounce)	can tomatoes, undrained	1 (398 mL)
½ cup	water	125 mL
1 (1½-ounce)	package spaghetti sauce mix	1 (43 g)
½ teaspoon	salt	2 mL
½ cup	thinly-sliced green pepper	125 mL
3 tablespoons	flour	45 mL
1 cup	milk	250 mL
1 (10-ounce)	can sliced mushrooms, drained	1 (284 mL)
	hot cooked noodles	

Place chicken in a large frypan. Combine tomatoes, water, spaghetti sauce mix and salt. Pour over chicken. Add green pepper. Bring to a boil. Reduce heat, cover and simmer 30 to 35 minutes or until chicken is tender. Remove chicken from pan. Smoothly combine flour and milk. Stir into pan juices; add mushrooms. Cook over medium heat, stirring constantly, until mixture just comes to a boil and thickens. Return chicken to pan; heat through. Serve over noodles.

Makes 4 or 5 servings.

CHICKEN ESPAGNOL

3 tablespoons	butter	45 mL
2½ to 3 pound	cut-up broiler-fryer	1.25 to 1.5 kg
3 cups	sliced fresh mushrooms	750 mL
1 cup	chopped onion	250 mL
1	clove garlic, minced	1
2 teaspoons	chicken bouillon mix	10 mL
1 teaspoon	salt	5 mL
1 teaspoon	paprika	5 mL
¾ teaspoon	ground ginger	3 mL
½ teaspoon	chili powder	2 mL
1 (19-ounce)	can tomatoes	1 (540 mL
½ cup	water	125 mL
⅓ cup	flour	75 mL
1¼ cups	milk	300 mL
	hot cooked noodles	

Melt butter in a large frypan. Brown chicken. Remove chicken from pan; sauté mushrooms, onion and garlic until tender. Blend in bouillon mix, salt, paprika, ginger and chili powder. Add and break up undrained tomatoes. Stir in water. Return chicken to pan. Bring to a boil. Reduce heat; cover and simmer 40 minutes or until chicken is tender. Remove chicken to a serving dish; keep warm. Smoothly combine flour and milk. Gradually stir into frypan. Cook over medium heat, stirring constantly, until mixture just comes to a boil and thickens. Pour sauce over chicken and serve with noodles.

Makes 4 or 5 servings.

CHICKEN FRICASSEE

3½ to 4 pound	cut-up stewing chicken*	1.75 to 2 kg
½ cup	chopped onion	125 mL
1 cup	water	250 mL
1 teaspoon	salt	5 mL
1 tablespoon	chicken bouillon mix	15 mL
¼ teaspoon	rosemary, crushed	1 mL
¼ teaspoon	ground marjoram	1 mL
2	whole cloves	2
¼ cup	flour	50 mL
2 cups	milk	500 mL
2 teaspoons	lemon juice	10 mL
	hot mashed potatoes	

Place chicken pieces in a large saucepan. Add onion, water, salt, bouillon mix, rosemary, marjoram and cloves. Bring to boil. Reduce heat; cover and simmer 1½ to 2 hours or until chicken is tender. Remove chicken pieces; set aside. Discard cloves. Smoothly combine flour and milk. Stir into pan juices. Cook over medium heat, stirring constantly, until mixture just comes to a boil and thickens. Add lemon juice. Return chicken to pan. Heat through. Serve with mashed potatoes.

Makes 5 or 6 servings.

* Cut-up broiler-fryer can be substituted. Reduce simmering time to 45 minutes - 1 hour.

CHICKEN RICE BAKE

1 (10-ounce)	can cream of chicken soup	1 (284 mL)
1 (42 g)	envelope onion soup mix	1 (42 g)
1 (10-ounce)	can mushrooms, drained and chopped	1 (284 mL)
1⅓ cups	milk	325 mL
¾ cup	uncooked long grain rice	175 mL
1 (11-ounce)	package frozen peas and carrots, thawed	1 (312 g)
2½ to 3 pound	cut-up broiler-fryer	1.25 to 1.5 kg
	paprika	

Place cream of chicken soup in a bowl. Blend in onion soup mix and mushrooms. Gradually stir in milk. Reserve ½ cup (125 mL) of the soup mixture; set aside. Stir uncooked rice and vegetables into remaining soup mixture. Turn into a 2-quart (2 L) shallow rectangular baking dish. Arrange chicken pieces on top. Pour reserved soup mixture over chicken. Sprinkle with paprika. Cover with foil. Bake in preheated 375°F (190°C) oven 1 to 1¼ hours or until rice is tender.

Makes 4 to 6 servings.

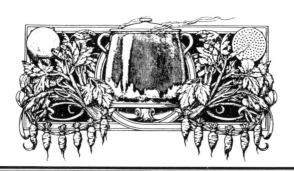

CHICKEN'N'SPOON BAKE

¾ cup	chopped celery	175 mL
¼ cup	chopped onion	50 mL
¼ cup	chopped green pepper	50 mL
2 tablespoons	butter	30 mL
2 tablespoons	all-purpose flour	30 mL
1½ teaspoons	Worcestershire sauce	7 mL
1	chicken bouillon cube	1
1⅔ cups	milk	400 mL
2 cups	chopped cooked chicken	500 mL
½ cup	cornmeal	125 mL
¾ teaspoon	salt	3 mL
¾ teaspoon	baking powder	3 mL
¾ cup	milk	175 mL
3	eggs, separated	3

Sauté vegetables in butter 5 minutes; stir in flour. Add Worcestershire sauce, bouillon cube and milk. Bring to a boil; cook 1 minute, stirring constantly. Add chicken. Spoon mixture into 4 buttered 2-cup (500 mL) individual casseroles or a 2-quart (2 L) casserole dish.

Meanwhile combine cornmeal, salt and baking powder in saucepan; stir in milk. Cook until thickened, stirring constantly. Remove from heat, stir in egg yolks. Beat egg whites until stiff peaks form; fold cornmeal mixture into egg whites. Spoon over chicken mixture. Bake in preheated 375°F (160°C) oven 25-30 minutes or until cornmeal topping is golden brown.

Makes 4 servings.

CHICKEN BREASTS SUPREME

6	large chicken breasts, skinless and boneless	6
2 (10 ounce)	cans condensed chicken broth	2 (284 mL)
½ cup	celery, chopped	125 mL
½ cup	onion, chopped	125 mL
1 cup	carrots, chopped	250 mL

SAUCE

3 tablespoons	butter, melted	45 mL
½	green pepper, diced	½
3 tablespoons	flour	45 mL
1 cup	milk	250 mL
2 tablespoons	pimento, diced	30 mL

Place chicken breasts in dutch oven. Add broth, celery, onion and carrots. Cover and simmer gently until tender (about 1 hour). Remove chicken from broth. Cut breasts in half, and cover to keep warm. Boil broth, uncovered, until reduced to half the volume. Purée broth and vegetables in blender or food mill. Measure 1½ cups (375 mL). (Remainder can be kept for other uses.) In dutch oven, sauté green pepper in butter for 5 minutes. Blend in flour. Gradually add broth purée and milk, stirring constantly, until smooth and thickened. Stir in pimento and simmer 2 minutes. Arrange chicken breasts in a shallow baking dish. Cover with sauce, and heat at 350°F (180°C) for 20 minutes, or until bubbly. Serve with fluffy, cooked rice.

Makes 8 servings.

CHICKEN WILD RICE CASSEROLE

1 (170 g)	package long grain and wild rice	1 (170 g)
3 tablespoons	butter	45 mL
2 cups	sliced fresh mushrooms	500 mL
1 (10-ounce)	can cream of mushroom soup	1 (284 mL)
1 cup	milk	250 mL
¼ cup	chopped parsley	50 mL
1 teaspoon	poultry seasoning	5 mL
1 teaspoon	onion salt	5 mL
2 cups	cut-up cooked chicken*	500 mL
	paprika	

Prepare rice according to package directions. Melt butter in frypan. Add mushrooms and sauté until tender and liquid has evaporated; stir into cooked rice. Combine soup, milk, parsley, poultry seasoning and onion salt. Layer half of the rice mixture, chicken and soup mixture in a 2-quart (2 L) casserole. Repeat layers. Sprinkle top with paprika. Bake in preheated 350°F (180°C) oven 30 to 35 minutes.

Makes 6 servings.

* Leftover turkey may be substituted.

CHICKEN CASSEROLE WITH SAVOURY BISCUITS

1		large chicken	1
6		slices bacon, cut in 1" (3 cm) slices	6
¼	cup	butter	50 mL
6		small onions, peeled	6
4		carrots, cut in chunks	4
½	pound	mushrooms, sliced	250 g
3		celery stalks, sliced	3
¼	cup	flour	50 mL
1	teaspoon	salt	5 mL
2	cups	milk	500 mL
1		chicken bouillon cube	1
2	tablespoons	dry sherry (optional)	30 mL

Cut chicken into serving pieces. Flour lightly. Sauté bacon until crisp; drain on paper towelling. Brown chicken in bacon fat; remove to deep pie-plate or casserole dish.

Add butter to frypan; stir-fry onion, carrots, mushrooms and celery lightly. Add to casserole along with bacon. Blend flour and salt into liquid remaining in frypan; gradually stir in milk. Add bouillon cube; stir until thickened and bubbly. Add sherry; pour over casserole. Bake at 350°F (180°C) for 35 minutes.

Remove from oven. Top with biscuits.

SAVOURY BISCUITS

2	cups	flour	500 mL
4	teaspoons	baking powder	20 mL
1	teaspoon	salt	5 mL
½	teaspoon	paprika	2 mL
½	teaspoon	savoury	2 mL
¼	cup	shortening	50 mL
¼	cup	chopped parsley	50 mL
1	cup	milk	250 mL

Mix dry ingredients. Cut in shortening until mixture resembles coarse bread crumbs. Add parsley. Add milk and mix quickly to form a soft dough. Drop by spoonfuls on top of casserole. Bake at 425°F (220°C) for 25 minutes or until lightly browned.

CHICKEN VEGETABLE AND RICE CASSEROLE

4 cups	cooked long grain white rice	1 L
3 cups	grated carrots	750 mL
2 cups	grated Canadian Cheddar cheese	500 mL
½ cup	grated Mozzarella cheese	125 mL
¾ cup	milk	175 mL
2 tablespoons	butter	30 mL
2 tablespoons	finely chopped onion	30 mL
1 teaspoon	salt	5 mL
¼ teaspoon	pepper	1 mL
2 cups	cubed cooked chicken	500 mL
1 cup	frozen green peas	250 mL
1 (14 ounce)	can tomatoes, drained and chopped	1 (388 mL)

Combine all the ingredients. Mix well. Pour into a lightly greased 9"x 13"x 2" (3.5 L) baking pan. Cover and bake in 350°F (180°C) oven for one hour.

Makes 6 servings.

CHICKEN AND BROCCOLI CASSEROLE

6	tablespoons	butter	90 mL
1	tablespoon	finely chopped green onion	15 mL
6		chicken breasts, skinned & boned	6
1	teaspoon	lemon juice	5 mL
3	tablespoons	flour	45 mL
2	cups	milk	500 mL
½	teaspoon	basil	2 mL
½	teaspoon	salt	2 mL
¼	teaspoon	pepper	1 mL
1	tablespoon	chopped parsley	15 mL
1	cup	grated Swiss or Canadian Cheddar cheese	250 mL
½	pound	wide noodles, cooked & drained	250 g
2		tomatoes, sliced (optional)	2
2	cups	trimmed & blanched broccoli	500 mL

Heat butter in a large frypan; add green onions and chicken breasts; sprinkle with lemon juice. Cook chicken over medium heat 6 minutes, turning once; remove chicken. Add flour to pan juices; cook and stir two minutes. Remove from heat; whisk in milk.

Bring sauce to a boil, stirring constantly; cook until thickened. Stir in basil, salt, pepper, parsley and half the cheese.

Place cooked noodles in a buttered baking dish; add half the sauce; arrange tomatoes, broccoli and chicken on top. Cover with remaining sauce; sprinkle with remaining cheese. Bake in 350°F (180°C) oven for 35 minutes.

Makes 6 servings.

CHICKEN LIVERS ON TOAST POINTS

⅓ cup	flour	75 mL
1 pound	chicken livers	500 g
¼ cup	butter, melted	50 mL

SAUCE:		
¼ cup	butter, melted	50 mL
½	onion, chopped	½
½	green pepper, chopped	½
½ cup	fresh mushrooms, sliced	125 mL
¼ cup	flour	50 mL
1 teaspoon	salt	5 mL
½ teaspoon	thyme	2 mL
⅛ teaspoon	pepper	0.5 mL
1 cup	milk	250 mL
1 cup	chicken broth	250 mL
6	slices toast	6

Dredge chicken livers in ⅓ cup (75 mL) flour. Sauté in butter for about 15 minutes until fully cooked and golden. Keep warm. For sauce, sauté onion, green pepper and mushrooms in butter for 5 minutes until soft. Stir in flour, salt, thyme and pepper. Add milk and broth gradually, stirring constantly over medium heat, until smooth and thickened. Cut toast diagonally. Divide chicken livers evenly over the toast. Spoon sauce over livers. Serve immediately.

Makes 6 servings.

CHICKEN RING WITH CRANBERRY SAUCE

RING:

4	packets instant chicken bouillon powder	4
2 cups	milk	500 mL
1 tablespoon	butter	15 mL
½ teaspoon	poultry seasoning	2 mL
⅛ teaspoon	pepper	0.5 mL
3 cups	cooked chicken, diced	750 mL
1 cup	dry bread crumbs	250 mL
1 cup	cooked rice	250 mL
¾ cup	celery, diced	175 mL
½ cup	onion, diced	125 mL
1 tablespoon	fresh parsley, chopped	15 mL
3	eggs, beaten	3

CRANBERRY SAUCE:

2 (6-½ ounce)	cans whole berry cranberry sauce	2 (184 mL)
1 teaspoon	orange extract	5 mL

For ring, heat bouillon powder and milk in a small saucepan until dissolved. Add butter, poultry seasoning and pepper. Cool slightly. In a large bowl, combine chicken, bread crumbs, rice, celery, onion and parsley. Stir eggs into milk mixture, and add to chicken. Mix lightly until blended. Pour into a very well-greased 6-cup (1.5 L) ring mold, and place in a pan of hot water. Bake at 350°F (180°C) for 1 hour and 15 minutes, or until knife inserted at centre comes out clean. Remove from water and let stand 10 minutes. Meanwhile, heat cranberries over medium heat, stirring constantly, until hot. Stir in orange extract. Unmold ring onto serving platter. Slice and serve with hot sauce.

Makes 6 servings.

CHICKEN-APPLE CURRY

1 tablespoon	butter, melted	15 mL
1 cup	apple, peeled and finely chopped	250 mL
1 cup	celery, sliced	250 mL
½ cup	onion, chopped	125 mL
1	clove garlic, minced	1
2 tablespoons	cornstarch	30 mL
2 teaspoons	curry powder	10 mL
1½ cups	chicken broth and mushroom juice	375 mL
1 cup	milk	250 mL
2 cups	cooked chicken, sliced	500 mL
½ (10 ounce)	can sliced mushrooms (reserve juice)	½ (284 mL)

In a saucepan, sauté apple, celery, onion and garlic in butter. Cook about 5 minutes until soft. Combine cornstarch, curry powder, chicken broth, mushroom juice and milk. Add to vegetable mixture, and cook, stirring constantly, until mixture thickens and bubbles. Stir in chicken and mushrooms and heat thoroughly. Serve over hot, fluffy rice.

Makes 6 servings.

COCONUT CURRIED CHICKEN

2 cups	coconut, shredded	500 mL
2 cups	milk	500 mL
½ cup	onion, chopped	125 mL
1 teaspoon	ginger, chopped	5 mL
1 tablespoon	butter, melted	15 mL
1 teaspoon	garlic powder	5 mL
2-4 teaspoons	curry powder	10-20 mL
¼ teaspoon	salt	1 mL
dash	pepper	dash
2 tablespoons	flour	30 mL
2 cups	cooked chicken, diced	500 mL
1 tablespoon	lemon juice	15 mL
¼ cup	raisins	50 mL

Soak coconut in milk for 1 hour. Sauté onions and ginger in butter for 5 minutes until soft. Add garlic powder, curry powder, salt and pepper. Sprinkle with flour. Add coconut mixture gradually, stirring constantly, until smooth. Simmer 15 minutes, stirring frequently. Add chicken and simmer 15 minutes longer. Add lemon juice and raisins. Serve with fluffy rice.

Makes 4 servings.

COLD DAY CHICKEN CURRY

⅓ cup	butter	75 mL
2 (3 pound)	cut-up chickens	2 (1.5 kg)
1 cup	finely chopped onion	250 mL
¼ cup	thinly sliced crystallized ginger	50 mL
2 tablespoons	curry	30 mL
1 tablespoon	salt	15 mL
1 teaspoon	sugar	5 mL
pinch	pepper	pinch
pinch	ground cloves	pinch
3 cups	milk	750 mL
3 tablespoons	butter	45 mL
3 tablespoons	flour	45 mL
½ cup	lime juice	125 mL
6 cups	cooked white rice	1.5 L
	chopped parsley	

In a 6 quart (6 L) dutch oven, brown the chicken pieces a few at a time, on both sides, in melted butter.

Add onion, ginger, curry, salt, sugar, pepper, cloves and milk; bring to a boil. Reduce heat, cover and simmer for 1 hour until meat is tender. With a slotted spoon, remove the chicken and arrange on a large serving platter. Cover and keep warm.

In a large saucepan melt 3 tablespoons (45 mL) butter, add flour. Cook over medium heat, stirring constantly. Add reserved sauce. Cook until mixture thickens slightly.

Add lime juice and reheat gently. Pour sauce over chicken. Mound rice in centre of platter and sprinkle with parsley.

Makes 6-8 servings.

CORDON BLEU CHICKEN CASSEROLE

6	single chicken breasts, cooked	6
6	slices ham, cooked	6
6	slices Swiss cheese	6
4 tablespoons	butter	60 mL
4 tablespoons	flour	60 mL
¼ teaspoon	thyme	1 mL
1	egg, beaten	1
1¾ cups	milk	425 mL
¼ cup	dry breadcrumbs	50 mL
2 teaspoons	butter, melted	10 mL

In a greased 8"x 8" (20 cm x 20 cm) baking pan, arrange a layer of chicken, top with a layer of ham, then a layer of cheese. To prepare sauce, melt 4 tablespoons (60 mL) butter, and add flour and thyme. Combine egg and milk, and add slowly to flour mixture. Continue cooking, stirring constantly, until smooth and thickened. Pour over chicken. Combine 2 teaspoons (10 mL) melted butter with dry breadcrumbs. Spread over sauce. Bake at 350°F (180°C) for 30 minutes, or until heated through.
Makes 6 servings.

CREAMED CHICKEN IN NOODLE RING

NOODLE RING

1 pound	broad egg noodles	500 g
3	eggs, beaten	3
1 cup	milk	250 mL
½	green pepper, chopped	½
½ teaspoon	salt	2 mL
¼ teaspoon	pepper	1 mL
2 tablespoons	ketchup	30 mL
1 cup	Cheddar cheese, grated	250 mL
1½ teaspoons	Worcestershire sauce	7 mL

CREAMED CHICKEN

4 tablespoons	butter	60 mL
4 tablespoons	flour	60 mL
1½ cups	milk	375 mL
½ teaspoon	salt	2 mL
¼ teaspoon	pepper	1 mL
¼ teaspoon	garlic powder	1 mL
1 tablespoon	parsley, chopped	15 mL
¼ cup	Cheddar cheese, grated	50 mL
4 cups	cooked chicken, diced	1 L

Cook noodles in boiling water and drain well. Combine eggs, milk, green pepper, salt, pepper, ketchup, cheese and Worcestershire sauce. Add noodles and pour into a well-greased ring mold. Set in pan of hot water and bake at 350°F (180°C) for 45 minutes. Let stand for 5 minutes, and unmold onto a large platter.

For creamed chicken, melt butter in a large saucepan. Add flour and blend well. Add milk, stirring constantly, and continue cooking over medium heat, until smooth and thickened. Stir in salt, pepper, garlic powder, parsley and cheese. Add chicken and heat thoroughly. Fill centre of noodle ring with creamed chicken. Garnish with parsley sprigs and cherry tomatoes.

Makes 8 servings.

CREAMY CHICKEN CHOW MEIN

3 tablespoons	butter, melted	45 mL
2	stalks celery, thinly sliced	2
½	onion, chopped	½
3 cups	cooked chicken, chopped	750 mL
1 (10 ounce)	box frozen Chinese vegetables, thawed and drained	1 (284 g)
½ (10 ounce)	can mushroom pieces, drained	½ (284 mL)
1 (10 ounce)	can condensed cream of chicken soup	1 (284 mL)
1 cup	milk	250 mL
1 tablespoon	soy sauce	15 mL
⅛ teaspoon	Tabasco sauce	0.5 mL
	crisp chow mein noodles	
	chopped almonds	

Sauté celery and onion in butter until tender. Stir in chicken, vegetables, mushroom pieces, condensed soup, milk, soy sauce and Tabasco sauce. Heat to boiling over medium heat, stirring constantly. Reduce heat to low, and simmer, uncovered, stirring frequently, for 5 minutes. Serve with chow mein noodles. Sprinkle with chopped almonds.

Makes 6 servings.

CRUNCHY CHICKEN

1 cup	corn flake crumbs	250 mL
½ cup	Parmesan cheese, finely grated	125 mL
½ cup	walnuts, finely chopped	125 mL
¾ teaspoon	salt	3 mL
2	eggs, beaten	2
1 cup	buttermilk	250 mL
3 pound	broiler-fryer chicken, cut up	1.5 kg
¼ cup	butter, melted	50 mL

Combine crumbs, cheese, walnuts and salt. In a separate bowl, combine eggs and buttermilk. Dip chicken pieces in crumb mixture, then egg mixture, then in crumbs again. Place in a 9"x 13" (23 cm x 33 cm) baking dish. Drizzle with butter. Bake at 375°F (190°C) for 1 hour and 10 minutes, or until chicken is tender.

Makes 6 servings.

CRUNCHY CHICKEN SALAD MOLD

2	envelopes unflavoured gelatin	2
1½ cups	milk	375 mL
¾ teaspoon	salt	3 mL
2 cups	coarsely chopped cooked chicken	500 mL
1½ cups	small curd cottage cheese	375 mL
1 cup	mayonnaise or salad dressing	250 mL
½ cup	bottled Italian salad dressing	125 mL
½ cup	finely chopped celery	125 mL
½ cup	finely chopped green pepper	125 mL
⅓ cup	finely chopped green onion	75 mL

In saucepan, sprinkle gelatin over milk; let stand for 10 minutes to soften. Cook over low heat, stirring constantly, until gelatin has dissolved; cool. Stir in salt. Chill until mixture mounds on a spoon.

Fold chicken, cottage cheese, mayonnaise, Italian dressing, celery, green pepper and onion into gelatin mixture. Spoon into 6 cup (1.5 L) mold. Chill until firm.

Makes 6 servings.

EAST INDIAN CHICKEN

¼ cup	poppyseeds	50 mL
½ cup	unsweetened coconut	125 mL
4	cloves garlic	4
1 cup	buttermilk	250 mL
4 pounds	chicken pieces	2 kg
¼ cup	oil	50 mL
4	onions, coarsely chopped	4
1 teaspoon	cumin seed	5 mL
4	bay leaves	4
1 teaspoon	coriander seed	5 mL
½	cinnamon stick	½
3	whole cloves	3
1 cup	yogurt	250 mL
1 tablespoon	salt	15 mL

In blender, crush poppyseeds and coconut. Add garlic cloves. Purée until crushed. Slowly add buttermilk while machine is running. Wash chicken pieces and place in a shallow bowl. Pour buttermilk mixture over chicken to marinate. Cover and refrigerate at least 4 hours, but overnight if possible. Combine cumin, bay leaves, coriander, cinnamon and cloves in a cheesecloth bag. When ready to cook chicken, heat ¼ cup (50 mL) of oil in a Dutch oven. Sauté onions with spice bag for 5-10 minutes until onions are soft. Scrape marinade from chicken and reserve. Cook chicken with onions for 45 minutes until chicken is tender. Remove spice bag and discard. Remove chicken pieces, and keep warm. Stir marinade into onions and cook for 5 minutes, stirring constantly. Add yogurt and heat until hot, but do not boil. Replace chicken pieces and serve immediately.

Makes 8 servings.

Easy Chicken Divan

3 tablespoons	butter	45 mL
3 tablespoons	flour	45 mL
2 teaspoons	chicken bouillon mix	10 mL
½ teaspoon	dry mustard	2 mL
2 cups	milk	500 mL
¾ cup	shredded Canadian Swiss cheese	175 mL
2 cups	cut-up cooked chicken	500 mL
2 (10-ounce)	packages frozen broccoli spears	2 (283 g)
	paprika	

Melt butter in a medium saucepan. Blend in flour, bouillon mix and mustard. Gradually stir in milk. Cook over medium heat, stirring constantly, until mixture just comes to a boil and thickens. Remove from heat. Add ½ cup (125 mL) of the cheese and stir until melted. Add chicken to sauce and keep warm. Cook broccoli according to package directions; drain. Place in bottom of 2-quart (2 L) shallow rectangular baking dish. Pour hot chicken and sauce over broccoli. Top with remaining cheese and sprinkle lightly with paprika. Place dish under broiler 2 or 3 minutes or until cheese is melted and bubbly.

Makes 4 or 5 servings.

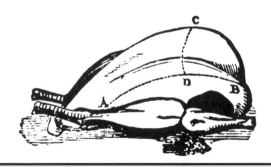

HASTY TASTY CHICKEN CHOP SUEY

3 tablespoons	butter	45 mL
¾ cup	celery, sliced ¼-inch/ 6 mm thick	175 mL
½ cup	chopped onion	125 mL
3 cups	cut-up, cooked, chicken	750 mL
1 (19-ounce)	can chop suey vegetables, drained	1 (540 mL)
1 (10-ounce)	can cream of chicken soup	1 (284 mL)
1 cup	milk	250 mL
2 tablespoons	soy sauce	30 mL
½ teaspoon	salt	2 mL
	hot cooked instant rice	

Melt butter in a saucepan. Sauté celery and onion until just tender. Stir in chicken, vegetables, soup, milk, soy sauce and salt. Cook and stir over medium heat until mixture just comes to a boil. Reduce heat and simmer, uncovered, 5 minutes. Serve over rice.

Makes 6 servings.

HAWAIIAN CHICKEN

⅓ cup	grated coconut	75 mL
1 cup	milk	250 mL
6	halves of chicken breasts	6
6	canned pineapple spears (retain juice)	6
1 cup	flour	250 mL
1 teaspoon	salt	5 mL
¼ teaspoon	pepper	1 mL
2 tablespoons	butter	30 mL
2 tablespoons	shortening	30 mL
½ cup	grated coconut	125 mL
1 tablespoon	flour	15 mL

Blend ⅓ cup (75 mL) coconut with milk and set aside. Bone chicken breasts, leaving skin on. Roll each half around a pineapple spear and fasten with a toothpick. Then roll in flour seasoned with salt and pepper. Dip into coconut milk and roll in flour again. Chill for 1 hour or longer. Measure remaining coconut milk, and add enough pineapple juice to make 1 cup (250 mL). Melt butter and shortening together in a skillet. Brown chicken pieces on both sides in fat. Transfer to a greased shallow baking dish. Sprinkle with ½ cup (125 mL) grated coconut. Meanwhile add 1 tablespoon (15 mL) flour to pan drippings and stir in the remaining coconut milk and pineapple juice. Cook, stirring constantly, until smooth and thickened. Pour sauce over chicken. Cover and bake at 400°F (200°C) for 30 minutes. Remove toothpicks. Serve over hot rice.
　　Makes 6 servings.

HOME-STYLE CHICKEN STEW

3½ to 4 pound	cut-up stewing chicken*	1.75 to 2 kg
2 cups	water	500 mL
2 teaspoons	chicken bouillon mix	10 mL
1 teaspoon	salt	5 mL
¾ teaspoon	poultry seasoning	3 mL
pinch	ground thyme	pinch
2 cups	thinly-sliced carrots	500 mL
¾ cup	sliced celery	175 mL
5	small onions, halved	5
½ cup	flour	125 mL
1½ cups	milk	375 mL
1½ cups	frozen peas	375 mL

Place chicken pieces in a large saucepan. Add water, bouillon mix, salt, poultry seasoning and thyme. Bring to a boil. Reduce heat; cover and simmer 1½ to 2 hours or until chicken is tender. Remove chicken. Add carrots, celery and onion to broth; simmer, covered, 20 minutes longer or until vegetables are just tender. Remove skin and bones from chicken. Cut chicken into bite-size pieces. Smoothly combine flour and milk. Gradually stir into saucepan. Add peas. Cook over medium heat, stirring constantly, until mixture just comes to a boil and thickens. Return cut-up chicken to pan and heat through.

Makes 6 to 8 servings.

* Cut-up broiler-fryer can be substituted. Reduce simmering time to 45 minutes - 1 hour.

JELLIED CHICKEN LOAF

2	envelopes unflavoured gelatin	2
2 cups	milk	500 mL
2	packets instant chicken bouillon powder	2
1 teaspoon	salt	5 mL
2 cups	creamy yogurt, unflavoured	500 mL
1 cup	mayonnaise or salad dressing	250 mL
4 cups	cooked chicken, diced	1 L
1	stalk celery, chopped	1
½	green pepper, chopped	½
4	green onions, chopped	4
2 tablespoons	parsley, chopped	30 mL
1 teaspoon	dill weed	5 mL

Sprinkle gelatin over milk. Let stand for 10 minutes to soften. Add instant chicken bouillon powder and salt. Cook over low heat, stirring constantly, until gelatin and bouillon powder are thoroughly dissolved. Remove from heat and cool. Combine yogurt and mayonnaise. Gradually stir in the gelatin mixture. Chill until softly set. Fold in chicken, celery, green pepper, green onion, parsley and dill. Pour into 9"x 5" (23 cm x 13 cm) loaf pan that has been rinsed with cold water. Chill until set. Unmold at serving time, and garnish with green pepper rings and parsley.

Makes 8 servings.

LEMON CHICKEN PILAF

4		chicken pieces	4
1	teaspoon	thyme	5 mL
1	teaspoon	grated lemon rind	5 mL
1	teaspoon	paprika	5 mL
1	tablespoon	each oil and butter	15 mL
1		clove garlic, minced	1
1		onion, chopped	1
1		carrot, sliced	1
1	cup	raw long grain converted rice	250 mL
2	cups	milk	500 mL
		salt and pepper to taste	
		finely chopped parsley	

Rub the chicken pieces with thyme, lemon rind and paprika. Heat the oil and butter in a large heavy frying pan. Brown chicken pieces well. Remove and set aside. Add the garlic, onion, carrot and rice to the same pan. Toss well over medium heat for two minutes. Add the milk. Stir well. Return the chicken pieces to the pan. Cover and simmer until tender, about 25 minutes. Season to taste; sprinkle parsley on top.

Makes 4 servings.

PICNIC OVEN FRIED CHICKEN

2½ to 3 pound	broiler-fryer chicken (cut into serving pieces)	1.25 to 1.5 kg
¾ cup	bread crumbs	175 mL
½ cup	grated Parmesan cheese	125 mL
½ cup	finely chopped nuts	125 mL
¾ teaspoon	salt	3 mL
2	eggs, beaten	2
1 cup	buttermilk	250 mL
¼ cup	butter, melted	50 mL

Coat chicken with combined crumbs, cheese, nuts and salt. Dip in combined eggs and buttermilk mixture. Coat with crumbs again. Place in 9" x 13" x 2" (3.5 L) rectangular pan. Drizzle with butter. Bake at 375°F (190°C) 45 minutes to 1 hour, or until chicken is tender.

Makes 4 servings.

Poached Chicken with Mushrooms

4	chicken breasts, boned and skinned	4
1 cup	chicken stock	250 mL
1	bouquet garni (celery top, bay leaf, parsley sprig)	1
2 teaspoons	butter	10 mL
4	green onions, finely chopped	4
8 ounces	mushrooms, sliced	250 g
1 teaspoon	cornstarch	5 mL
1 cup	milk	250 mL
½ teaspoon	dried tarragon	2 mL
	freshly ground pepper	
	fresh chopped parsley	

Place chicken breasts, stock and bouquet garni in a saucepan. Heat slowly, cover pot and simmer until chicken is cooked and just tender, about 10 minutes. Remove chicken and keep warm. Discard bouquet garni. Set pan over high heat and reduce stock by two-thirds; set aside. Meanwhile melt butter in another pan, add green onions and mushrooms and toss over moderate heat for several minutes. Dissolve cornstarch in milk and add to mushroom mixture; stir constantly while mixture comes to a boil. Flavour sauce with tarragon, pepper and the reduced stock. Arrange chicken breasts on a warm platter and cover with mushroom sauce and chopped parsley.

Makes 4 servings.

Savoury Coated Chicken and Sauce

2 cups	packaged seasoned stuffing mix	500 mL
¼ teaspoon	poultry seasoning	1 mL
1¾ cups	milk	425 mL
1 (10-ounce)	can cream of chicken soup	1 (284 mL)
3 to 3½ pounds	chicken pieces	1.5 to 1.75 kg
1 tablespoon	butter	15 mL
1 cup	sliced, fresh mushrooms	250 mL
2 tablespoons	flour	30 mL
¼ teaspoon	salt	1 mL

Place stuffing mix and poultry seasoning in blender container. Cover and blend until fine crumbs are formed. Pour into pie plate. Mix ¼ cup (50 mL) of the milk and ½ cup (125 mL) of the soup in a bowl. Dip chicken pieces in milk mixture, then in crumb mixture. Arrange in greased 9x13x2-inch (3.5 L) baking dish. Bake in preheated 400°F (200°C) oven 50 to 60 minutes or until tender. Melt butter in a saucepan. Sauté mushrooms until tender and liquid has evaporated. Blend in flour and salt. Stir in remaining soup and milk. Cook over medium heat, stirring constantly, until mixture just comes to a boil and thickens. Serve sauce with chicken.

Makes 6 servings.

SHERRIED CHICKEN AND MUSHROOMS IN TOAST CUPS

4	slices thinly sliced sandwich bread	4
½ pound	fresh mushrooms, sliced	250 g
⅓ cup	onion, finely chopped	75 mL
4 tablespoons	butter, melted	60 mL
4 tablespoons	flour	60 mL
⅛ teaspoon	pepper	0.5 mL
2 cups	milk	500 mL
2	packets instant chicken bouillon powder	2
¼ cup	sour cream	50 mL
1 tablespoon	dry sherry	15 mL
2 cups	cooked chicken, diced	500 mL
2 tablespoons	fresh parsley, chopped	30 mL
	paprika	

To make toast cups, remove crusts from thinly sliced sandwich bread. Lightly butter both sides of slice. Press slices into muffin tins, allowing the corners to protrude. Bake at 275°F (140°C) for 35 minutes, or until crisp and golden. When removed from tin, bread will retain cup shape. Sauté mushrooms and onion in butter for 5 minutes, until soft. Blend in flour and pepper. Gradually add milk, stirring constantly, over medium heat, until smooth and thickened. Add instant bouillon powder, stirring until dissolved. Stir in sour cream and sherry. Add chicken and parsley. Reheat to serving temperature. Spoon into bread cups, allowing filling to spill over cup onto plate. Sprinkle with paprika.

Makes 4 servings.

NOTES

BAKED SALMON A LA RUSSE

6	medium potatoes	6
¼ cup	butter, melted	50 mL
2	large onions, chopped	2
6 ounces	smoked salmon, sliced	170 g
¼ teaspoon	pepper	1 mL
1 teaspoon	dillweed	5 mL
3	eggs	3
2 cups	milk	500 mL

Peel potatoes and slice as thinly as possible. Rinse and dry. Sauté the onions in butter for 5 minutes until soft but not brown. Arrange ⅓ of the potatoes over the bottom of a greased 8-cup (2 L) baking dish. Arrange half of the onions over the potatoes, and top with half of the salmon slices. Sprinkle salmon with pepper and dillweed. Repeat the layering, ending with potatoes. Beat eggs and milk together. Pour over the layers in the baking dish. Bake at 325°F (160°C) for 1¼ hours, or until the custard is set.

Makes 6 servings.

Bubbly Baked Scallops

12 ounces	fresh or frozen scallops, thawed	340 g
2 tablespoons	butter, melted	30 mL
1	onion, finely chopped	1
2	stalks celery, chopped	2
4 ounces	fresh mushrooms, sliced	115 g
1	green pepper, chopped	1
3 tablespoons	flour	45 mL
1½ cups	milk	375 mL
¼ cup	dry bread crumbs	50 mL
¼ cup	Cheddar cheese, grated	50 mL

Rinse scallops, and pat dry. Sauté onion, celery, mushrooms and green pepper in butter until soft, about 10 minutes. Sprinkle with flour. Add milk gradually, stirring constantly, and continue to cook over medium heat until smooth and thickened. Combine scallops with sauce and pour into a 6 cup (1.5 L) casserole. Top with breadcrumbs and grated cheese. Bake at 350°F (180°C) for 30 minutes until hot and bubbly. Serve in a patty shell or over fluffy rice.

Makes 6 servings.

CHEESY SEAFOOD CASSEROLE

2 cups	cooked rice	500 mL
½ (10 ounce)	can condensed cheese soup	½ (284 mL)
1 pound	firm-fleshed fish fillets, cut into chunks	500 g
½ teaspoon	salt	2 mL
¼ teaspoon	lemon pepper	1 mL
1 cup	zucchini, thinly sliced	250 mL
½ teaspoon	basil	2 mL
2	eggs, beaten	2
1 cup	milk	250 mL

Mix rice with condensed cheese soup. Put ½ of mixture on bottom of greased 6-cup (1.5 L) casserole. Combine fish chunks, salt, lemon pepper, zucchini and basil. Pour over rice. Place remaining rice on fish. Beat eggs and milk together. Pour over casserole. Let stand 10 minutes for liquid to spread through evenly. Bake, uncovered, at 350°F (180°C) for 60 minutes.

Makes 6 servings.

CHOWDER CASSEROLE

8 ounces	smoked cod or haddock (finnan haddie)	250 g
3 tablespoons	butter	45 mL
2	onions, coarsely chopped	2
2	carrots, coarsely chopped	2
2	ribs celery, coarsely chopped	2
3 tablespoons	all-purpose flour	45 mL
3 cups	milk, hot	750 mL
½ teaspoon	dried thyme (or 1 teaspoon (5 mL) fresh)	2 mL
1	bay leaf	1
	salt, pepper and Tabasco sauce to taste	
1 cup	frozen peas	250 mL
1 pound	white fleshed fish fillets (haddock, cod, turbot etc.) cut in 1″ (2.5 cm) pieces	500 g

TOPPING

6	medium potatoes, peeled & quartered	6
2 tablespoons	butter	30 mL
1	egg	1
2 tablespoons	milk	30 mL

Place smoked cod in a bowl and cover with water; soak for 30 minutes. Drain well. Remove skin and bones; cut flesh into 1″ (2.5 cm) pieces. Reserve. Melt butter in a large saucepan. Add onions, carrots, celery; cook 5 minutes. Sprinkle with flour; cook gently without browning 5 minutes. Stir in hot milk; bring to a boil. Add seasonings. Reduce heat; cook, stirring occasionally, 15 minutes. Add peas, smoked fish and fresh fish. Cook 8 minutes. Remove bay leaf.

continued on next page

CHOWDER CASSEROLE (continued)

TOPPING

To prepare topping, boil potatoes in water until tender. Drain well and mash with butter. Season to taste. Beat egg with milk; reserve 2 tablespoons (30 mL) mixture and add remaining egg to potatoes. Place fish mixture in a buttered 2 quart (2 L) casserole. Pipe or spoon potatoes over the top. Brush with reserved egg mixture. Bake at 375°F (190°C) for 30 minutes until bubbly and brown. Let rest 10 minutes before serving. Note: An equal amount of fresh fish can be used in place of the smoked fish.

Makes 6 servings.

COD AU GRATIN

2 tablespoons	butter	30 mL
2 tablespoons	flour	30 mL
¼ teaspoon	salt	1 mL
¾ teaspoon	dry mustard	3 mL
1 cup	milk	250 mL
1½ cups	shredded Canadian Cheddar cheese	375 mL
1 (16-ounce)	package frozen cod fillets, thawed chopped parsley	1 (454 g)

Melt butter in a saucepan. Blend in flour, salt and dry mustard. Gradually stir in milk. Cook over medium heat, stirring constantly, until mixture just comes to a boil and thickens. Remove from heat. Add cheese and stir until melted. Spread half the sauce in the bottom of 1½-quart (1.5 L) rectangular baking dish. Wipe any excess water from fish fillets and cut in serving size pieces. Arrange on top of sauce. Spread remaining cheese sauce over fish. Bake in preheated 400°F (200°C) oven 20 minutes or until fish flakes easily with a fork. Garnish with chopped parsley if desired.

Makes 4 servings.

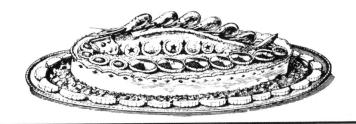

COTTAGE CURRIED SHRIMP

3 tablespoons	butter	45 mL
1 tablespoon	curry powder	15 mL
2 tablespoons	onion, finely chopped	30 mL
2 ½ tablespoons	flour	37 mL
½ teaspoon	salt	2 mL
¼ teaspoon	savory	1 mL
dash	cayenne	dash
1 cup	milk	250 mL
2 cups	cottage cheese	500 mL
1 (10 ounce)	box frozen peas, cooked and drained	1 (280 mL)
1 pound	cooked shrimp	500 g
¼ cup	peeled apple, finely chopped	50 mL

Melt butter in a large saucepan. Add curry powder and onion, and sauté until soft, about 5 minutes. Blend in flour, salt, savory and cayenne. Remove from heat, and gradually add milk and cottage cheese. Heat to boiling, stirring constantly, until thickened. Boil and stir 1 minute. Stir in peas, shrimp and apple, and reheat to serving temperature. Serve over fluffy rice.

Makes 6 servings.

CRAB IMPERIAL

12 ounces	crabmeat, canned, or frozen and thawed	340 g
2 tablespoons	butter, melted	30 mL
2 tablespoons	flour	30 mL
½ teaspoon	salt	2 mL
⅛ teaspoon	pepper	0.5 mL
1 teaspoon	dry mustard	5 mL
dash	cayenne	dash
1 cup	milk	250 mL
1 cup	18% cream	250 mL
1	egg, well beaten	1
2 teaspoons	lemon juice	10 mL
⅓ cup	soft breadcrumbs	75 mL
1 tablespoon	butter, melted	15 mL

Drain crabmeat. Remove any bits of shell or cartilage. Cut into bite-sized chunks. Blend flour into 2 tablespoons (30 mL) melted butter; add salt, pepper, mustard and cayenne. Beat egg and lemon juice into cream. Add to milk; then add milk mixture gradually to flour mixture. Cook until thickened, stirring constantly. Remove from heat. Add crabmeat. Divide into 6 well greased individual baking shells. Combine crumbs with 1 tablespoon (15 mL) melted butter and sprinkle over crab mixture. Place under broiler about 6" (15 cm) from heat for about 5 minutes, or until bubbly and crumbs are lightly browned.

Makes 6 appetizers.

CRAB CUSTARD BALTIMORE

1 pound	crabmeat, fresh or frozen and thawed	500 g
2 tablespoons	butter, melted	30 mL
4	slices white bread, crusts removed	4
1 cup	Swiss cheese, grated	250 mL
2 tablespoons	fresh parsley, minced	30 mL
4 teaspoons	onion, diced	20 mL
2 cups	milk	500 mL
3	eggs, well-beaten	3
1 teaspoon	salt	5 mL
½ teaspoon	Worcestershire sauce	2 mL
¼ teaspoon	pepper	1 mL
	paprika	

Drain crabmeat. Remove any bits of shell or cartilage, and cut into ½" (1 cm) pieces. Pour melted butter into an 8"x 8" (20 cm x 20 cm) baking dish. Arrange bread slices in a single layer. Place crabmeat on top. Cover with cheese, parsley and onion. Combine milk, eggs, salt, Worcestershire sauce and pepper. Pour over crabmeat. Sprinkle with paprika, and set in a larger pan containing 1" (2.5 cm) of hot water. Bake at 350°F (180°C) for 50 to 60 minutes, or until a knife inserted in the centre comes out clean.

Makes 6 servings.

CRABMEAT LUNCHEON DISH

8	slices white bread, crusts removed	8
2 (5 ounce)	cans crabmeat, drained	2 (142 g)
½ cup	mayonnaise	125 mL
1 cup	onion, finely chopped	250 mL
1 cup	celery, finely chopped	250 mL
2 cups	milk	500 mL
4	eggs, beaten	4
1 cup	Swiss cheese, grated	250 mL

Break bread into small pieces and place half in a greased 9"x 9" (23 cm x 23 cm) baking dish. Combine crabmeat, mayonnaise, onion and celery, and pour over bread. Add remaining bread pieces. Combine milk and eggs well, and pour over all. Refrigerate overnight. When ready to bake, sprinkle cheese on top. Bake at 350°F (180°C) for 55 minutes, or until puffy and golden. Serve immediately.

Makes 6 servings.

CRISPY-TOPPED FISH

2 pounds	frozen cod fillets, thawed	1 kg
2 tablespoons	butter	30 mL
2 tablespoons	all-purpose flour	30 mL
½ teaspoon	salt	2 mL
pinch	nutmeg	pinch
½ teaspoon	dry mustard	2 mL
1 cup	milk	250 mL
¾ cup	shredded Canadian Cheddar cheese	175 mL
1 teaspoon	lemon juice	5 mL
1½ cups	coarsely crushed potato chips	375 mL

Separate fillets; place in a broad shallow baking dish. Bake in preheated 400°F (200°C) oven 8 to 10 minutes, or until fish will flake easily with a fork. Melt butter in saucepan, blend in flour, salt, nutmeg and mustard. Gradually stir in milk. Cook over medium heat, stirring constantly until thickened. Stir in cheese and lemon juice. Drain liquid from baked fish. Pour sauce over fish, sprinkle with potato chips.

Bake 10 minutes longer.

Makes 6 servings.

CURRIED TUNA

4 tablespoons	butter	60 mL
½ teaspoon	curry powder	2 mL
4 tablespoons	flour	60 mL
¼ teaspoon	salt	1 mL
dash	pepper	dash
2 cups	milk	500 mL
2 (6 ½ ounce)	cans flaked tuna	2 (184 g)
1 (14 ounce)	can pineapple chunks, drained	1 (398 mL)
1 ½ cups	celery, chopped	375 mL
½ cup	toasted, slivered almonds	125 mL

In a large saucepan, melt butter. Add curry powder, flour, salt and pepper, blending well. Add milk gradually, stirring constantly, and continue to cook over medium heat until smooth and thickened. Add tuna, pineapple, celery and almonds. Reheat to serving temperature. Serve over fluffy rice.

Makes 6 servings.

FISH IN A POUCH WITH MUSHROOM SAUCE

3	tablespoons	butter	45 mL
½	pound	mushrooms, thinly sliced	250 g
2		green onions, finely chopped	2
¼	teaspoon	thyme	1 mL
2	tablespoons	lemon juice	30 mL
2	tablespoons	dry sherry (optional)	30 mL
3	tablespoons	flour	45 mL
1½	cups	milk	375 mL
1	tablespoon	chopped parsley	15 mL
		salt	
		freshly ground pepper	
1½	pounds	white fish fillets, fresh or thawed	750 g

Pre-heat oven to 450°F (230°C).

Melt butter in a small saucepan; add mushrooms, green onions, thyme, lemon juice and sherry. Simmer, covered, five minutes; remove lid; cook a little longer for liquid to reduce. Stir in flour; cook two minutes. Remove from heat; whisk in milk. Return to heat and bring to a boil, stirring constantly; cook until very thick. Stir in parsley; season to taste.

Lightly butter six pieces of foil or parchment paper. Place fish fillets on foil or parchment, season with a little salt and pepper and cover with sauce. Wrap securely and bake in oven about 10-15 minutes or until fish flakes when tested.

Pouches may be individually served or the fish may be removed from packages and arranged on a serving platter.

Makes 4 servings.

Fresh Salmon with Dilly Cucumber Sauce

2 tablespoons	corn starch	30 mL
2 cups	milk	500 mL
¼ cup	butter	50 mL
1 teaspoon	salt	5 mL
½ teaspoon	dill weed	2 mL
2 tablespoons	lemon juice	30 mL
1 cup	peeled, seeded and chopped cucumber poached, grilled or baked fresh salmon	250 mL

Smoothly combine corn starch and milk in a saucepan. Add butter, salt and dill weed. Cook over medium heat, stirring constantly, until mixture comes to a boil and thickens. Reduce heat and cook 2 minutes longer, stirring occasionally. Stir in lemon juice and cucumber. Heat through. Serve spooned over poached, grilled or baked fresh salmon.

Makes 3 cups/750 mL sauce.

INDIVIDUAL CLAM BAKES

1½ cups	pared diced potatoes	375 mL
⅓ cup	chopped onion	75 mL
1 (10-ounce)	can baby clams, undrained	1 (284 mL)
2 tablespoons	butter	30 mL
2 tablespoons	flour	30 mL
¼ teaspoon	salt	1 mL
¼ teaspoon	celery salt	1 mL
1 cup	milk	250 mL
1 (7-ounce)	can whole kernel corn, drained	1 (199 mL)
4	slices buttered toast paprika	4

Cook potatoes and onion in boiling salted water until tender; drain well. Drain clams, reserving ½ cup (125 mL) liquid. Melt butter in a saucepan. Blend in flour, salt and celery salt. Gradually stir in milk and reserved clam liquid. Cook over medium heat, stirring constantly, until mixture just comes to a boil and thickens. Stir in potato-onion mixture, clams and corn. Heat through. Cut each slice of toast into 4 triangles. To serve, line 4 shallow baking dishes with toast triangles; spoon clam mixture into centre. Sprinkle with paprika.

Makes 4 servings.

Macaroni Tuna Casserole

1 ½ cups	dry elbow macaroni, cooked and drained	375 mL
3 tablespoons	butter, melted	45 mL
1 tablespoon	onion, chopped	15 mL
½	green pepper, chopped	½
2	stalks celery, chopped	2
3 tablespoons	flour	45 mL
½ teaspoon	celery salt	2 mL
⅛ teaspoon	pepper	0.5 mL
¼ teaspoon	basil	1 mL
2 cups	milk	500 mL
1 (6 ½ ounce)	can flaked tuna, drained	1 (184 g)
¼ cup	Parmesan cheese, grated	50 mL

Sauté onion, green pepper and celery in butter for 5 minutes until soft. Blend in flour, celery salt, basil and pepper. Add milk gradually, and continue to cook, stirring constantly, until smooth and thickened. Stir in tuna and cooked macaroni. Pour into a greased 6-cup (1.5 L) casserole and sprinkle Parmesan cheese over top. Bake, uncovered, at 350°F (180°C) for 30 minutes, until bubbly and heated through.

Makes 6 servings.

NOODLES NEPTUNE

3 cups	dry broad egg noodles	750 mL
3 tablespoons	butter, melted	45 mL
$\frac{1}{2}$	onion, chopped	$\frac{1}{2}$
$\frac{1}{2}$	green pepper, chopped	$\frac{1}{2}$
4	green onions, chopped	4
$\frac{1}{2}$ teaspoon	salt	2 mL
$\frac{1}{4}$ teaspoon	pepper	1 mL
dash	sage	dash
dash	thyme	dash
1 tablespoon	fresh parsley, chopped	15 mL
1 teaspoon	lemon juice	5 mL
1 cup	tomato sauce	250 mL
1 tablespoon	mayonnaise	15 mL
1 cup	milk	250 mL
1 pound	fish fillets, sliced 1" (2.5 cm) thick	500 g
2 tablespoons	flour	30 mL
$\frac{1}{2}$ cup	Cheddar cheese, grated	125 mL

Cook noodles in boiling, salted water. Drain, rinse and keep warm. Sauté onion, green pepper and green onions in butter until soft, about 5 minutes. In a large pot, blend together salt, pepper, sage, thyme, parsley, lemon juice, tomato sauce, mayonnaise and milk. Bring to a boil, add fish slices, and simmer gently for 10-15 minutes, or until fish is opaque and flakes easily with a fork. Remove $\frac{1}{2}$ cup (125 mL) of the liquid, and add flour to it, mixing well to break up lumps. Return to the pot, and boil gently, stirring constantly, until thickened. Remove from heat. Add cheese, and stir until melted. Serve over hot noodles.

Makes 4 servings.

Poached fish with vegetable sauce

4 tablespoons	butter, melted	60 mL
1	stalk celery	1
1	large onion, chopped	1
1	clove garlic, minced	1
2 cups	milk	500 mL
2 pounds	frozen fish fillets, thawed	1 kg
1 (10 ounce)	can sliced mushrooms	1 (284 mL)
4 tablespoons	flour	60 mL
½ teaspoon	salt	2 mL
¼ teaspoon	pepper	1 mL
1 teaspoon	paprika	5 mL
1 teaspoon	dill weed	5 mL

In a large pot, sauté celery, onion and garlic in butter until soft, about 5 minutes. Add milk and bring to a boil. Thaw fillets enough to slice into serving-sized pieces and add to the milk mixture. Simmer gently until fish is opaque, about 10 minutes. (Do not boil, as this may toughen fish.) When done, remove fish and keep warm. Drain mushrooms, and combine liquid with the flour. Add this to the cooking liquid in the pot, stirring constantly until well blended. Add mushrooms, salt, pepper, paprika and dill weed. Stir constantly over medium heat until thickened. Pour over warm fillets. Serve with hot, fluffy rice and a crisp salad.

Makes 8 servings.

POACHED FISH PARMESAN

1¼ pounds	fillets of sole, turbot or haddock	625 g
2 cups	milk	500 mL
3 tablespoons	butter	45 mL
3 tablespoons	flour	45 mL
½ cup	dry white wine	125 mL
½ cup	grated Parmesan cheese	125 mL
2 tablespoons	chopped fresh chives or green onions	30 mL
	salt and pepper to taste	
1	red or green pepper, thinly sliced	1
1 pound	fettuccine or broad egg noodles	500 g

Cut fillets into serving size portions. In a large covered frypan, simmer fillets gently in milk for 5 to 10 minutes, or until fish flakes easily with a fork. Carefully transfer fish to a platter and keep warm; reserve milk. Melt butter in a medium saucepan; add flour and cook for 2 minutes. Gradually add reserved hot milk; cook over low heat, stirring constantly, until thickened. Add wine, Parmesan cheese, chives, salt, pepper and red pepper; heat an additional 10 minutes. Meanwhile, cook fettuccine in boiling, salted water until tender; drain. Arrange fish and sauce over pasta.

Makes 4 servings.

PRONTO SHRIMP CURRY

2 tablespoons	butter	30 mL
2 tablespoons	finely-chopped onion	30 mL
2 tablespoons	finely-chopped celery	30 mL
3 tablespoons	flour	45 mL
1 teaspoon	curry powder*	5 mL
1 teaspoon	chicken bouillon mix	5 mL
½ teaspoon	salt	2 mL
2 cups	milk	500 mL
2½ cups	cooked shrimp**	750 mL
	hot cooked rice	
	assorted condiments	

Melt butter in a saucepan. Sauté onion and celery until tender. Blend in flour, curry powder, bouillon mix and salt. Gradually stir in milk. Cook over medium heat, stirring constantly, until mixture just comes to a boil and thickens. Stir in shrimp and heat through. Serve with rice and assorted condiments such as yogurt, coconut, chutney, raisins, etc.

Makes 4 to 6 servings.

* Increase curry powder if desired.
** Leftover cooked meat such as chicken or lamb may be substituted.

QUICK FISH BAKE

1 (16-ounce)	package frozen cod fillets	1 (454 g)
1 tablespoon	butter	15 mL
1 tablespoon	lemon juice	15 mL
¼ teaspoon	salt	1 mL
	hot cooked rice or noodles	

With a serrated knife cut frozen fish crosswise into 4 equal pieces. Place in greased 1½-quart (1.5 L) shallow rectangular baking dish. Dot with butter; sprinkle with lemon juice and salt. Bake in preheated 450°F (230°C) oven 20 to 25 minutes or until fish flakes easily with a fork. Place each serving on a bed of rice or noodles and top with Hot Tartar Sauce.

Makes 4 servings.

HOT TARTAR SAUCE

2 tablespoons	butter	30 mL
2 tablespoons	flour	30 mL
¼ teaspoon	salt	1 mL
1⅓ cups	milk	325 mL
¼ cup	mayonnaise or salad dressing	50 mL
3 tablespoons	sweet pickle relish	45 mL
2 tablespoons	minced onion	30 mL
1 tablespoon	lemon juice	15 mL

Melt butter in a small saucepan. Blend in flour and salt. Gradually stir in milk. Cook over medium heat, stirring constantly, until mixture just comes to a boil and thickens. Stir in mayonnaise, relish, onion, and lemon juice.

Makes about 1½ cups/375 mL.

SALMON SOUFFLE

1 (15 ½ ounce)	can salmon	1 (439 g)
6	slices white bread, crusts removed	6
2 cups	Cheddar cheese, grated	500 mL
3	eggs	3
2 cups	milk	500 mL
1 teaspoon	Worcestershire sauce	5 mL
½ teaspoon	dry mustard	2 mL
¼ teaspoon	pepper	1 mL
	chopped parsley for garnish	

Drain excess liquid from the salmon. Grease a 6-cup (1.5 L) casserole. Break the bread into bite-sized pieces. On bottom of casserole, place half of the bread, then half of the salmon and half of the cheese. Repeat layers. In a large bowl, combine eggs, milk, Worcestershire sauce, mustard and pepper. Pour over layers. Cover and refrigerate at least 8 hours or overnight. Bake at 350°F (180°C) for 55 minutes, or until puffy and golden. Sprinkle with parsley and serve immediately.

Makes 6 servings.

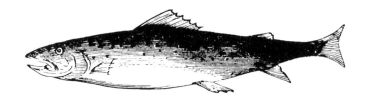

SALMON LOAF WITH EGG SAUCE

1 pound	can of salmon	500g
2	eggs	2
$^2/_3$ cups	milk	150 mL
1 tablespoon	minced onion	15 mL
1 teaspoon	lemon juice	5 mL
$^1/_4$ teaspoon	salt	1 mL
pinch	pepper	pinch
3 cups	cornflakes	750 mL
$^1/_4$ cup	butter	50 mL
$^1/_4$ cup	flour	50 mL
1 cup	milk	250 mL
$^1/_4$ cup	salmon liquid	50 mL
3	hard cooked eggs, peeled and finely chopped	3
2 tablespoons	lemon juice	30 mL
$^1/_4$ cup	finely chopped parsley	50 mL

Drain salmon, and reserve juice. Remove skin and bones, and flake finely. In medium bowl, beat eggs well and add to salmon with $^2/_3$ cup (150mL) milk, onion, 1 tsp (5mL) lemon juice, salt and pepper. Mix thoroughly. Add cereal, and mix again. Pour into well buttered 9" x 5" (23 cm x 12 cm) loaf pan, and bake in preheated 325°F (160°C) oven for 1 hour.

While salmon is baking prepare sauce. Melt butter, stir in flour and cook over low heat until mixture begins to bubble. Add 1 cup (250mL) milk and salmon liquid, stirring constantly until mixture has thickened. Remove from heat, stir in chopped eggs and lemon juice.

Loosen the salmon loaf with a sharp knife and invert onto a heated platter.

Cut into generous slices, and top with egg sauce and parsley.

Makes 6 servings

SALMON SUPREME

4 tablespoons	butter, melted	60 mL
½ cup	onion, chopped	125 mL
2 cups	fresh mushrooms, sliced	500 mL
4 tablespoons	flour	60 mL
1 teaspoon	salt	5 mL
1 cup	milk	250 mL
1 cup	10% cream	250 mL
1	egg, beaten	1
1 tablespoon	dry sherry (optional)	15 mL
2 (7½ ounce)	cans salmon, mashed with juice	2 (213 g)
½ cup	crushed corn flakes	125 mL
1 tablespoon	butter, melted	15 mL

In a large saucepan, sauté onion and mushrooms in 4 tablespoons (60 mL) butter until soft, about 5 minutes. Blend in flour and salt. Add milk and cream gradually, stirring constantly. Continue to cook and stir over medium heat until smooth and thickened. Combine egg and sherry with salmon. Stir into sauce. Pour into a greased 6-cup (1.5 L) casserole. Combine crushed corn flakes with 1 tablespoon (15 mL) melted butter. Sprinkle flakes over top of casserole. Bake, uncovered, at 350°F (180°C) for 40 minutes until firm.

Makes 6 servings.

SALMON FLORENTINE

1 (12-ounce)	package frozen chopped spinach	1 (340 g)
1 (15½-ounce)	can salmon	1 (439 g)
	water	
3 tablespoons	butter	45 mL
3 tablespoons	flour	45 mL
1 teaspoon	chicken bouillon mix	5 mL
1 teaspoon	dry parsley flakes	5 mL
¾ cup	milk	175 mL
2 tablespoons	lemon juice	30 mL

Cook spinach according to package directions; drain well. Line 4 individual shallow baking dishes or shells with spinach. Drain salmon, reserving liquid; add water to make ¾ cup (175 mL). Remove skin and bones from salmon, if desired, and break up into large pieces. Melt butter in a saucepan. Blend in flour, bouillon mix and parsley. Gradually stir in milk and reserved salmon liquid. Cook over medium heat, stirring constantly, until mixture just comes to a boil and thickens. Stir in lemon juice. Fold 1 cup (250 mL) hot sauce into salmon. Divide evenly among prepared dishes. Top with remaining hot sauce. Broil about 4 inches (10 cm) from source of heat 8 minutes or until browned and bubbly.

Makes 4 servings.

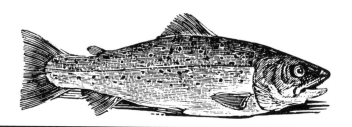

SALMON CHEDDAR MACARONI

1¼ cups	elbow macaroni	300 mL
3 tablespoons	butter	45 mL
½ cup	finely-chopped onion	125 mL
3 tablespoons	flour	45 mL
1 teaspoon	garlic salt	5 mL
½ teaspoon	Worcestershire sauce	2 mL
1½ cups	milk	375 mL
1½ cups	shredded old Canadian Cheddar cheese	375 mL
¾ cup	sour cream	175 mL
2 (7¾-ounce)	cans salmon, drained and flaked	2 (220 g)
½ cup	fresh bread crumbs	125 mL
1 tablespoon	butter, melted	15 mL

Cook macaroni according to package directions. Drain. Melt 3 tablespoons (45 mL) butter in a large saucepan. Sauté onion until tender. Blend in flour, garlic salt and Worcestershire sauce. Gradually stir in milk. Cook over medium heat, stirring constantly, until mixture just comes to a boil and thickens. Remove from heat. Add cheese and stir until melted. Stir in sour cream. Add and mix in salmon and macaroni. Turn into a 2-quart (2 L) casserole. Combine bread crumbs and 1 table-spoon (15 mL) melted butter. Spoon crumbs around outside edge of casserole. Bake in preheated 350°F (180°C) oven 25 to 30 minutes.

Makes 6 to 8 servings.

SAUCY SALMON IN TOAST CUPS

1	(350 g)	package frozen peas	1	(350 g)
¼	cup	butter		50 mL
1	cup	chopped onion		250 mL
⅓	cup	all-purpose flour		75 mL
½	teaspoon	dry mustard		2 mL
1	cup	tomato juice		250 mL
¾	cup	milk		175 mL
1½	cups	shredded Canadian Cheddar cheese		375 mL
1	(1 pound)	can of salmon, drained and flaked	1	(350 g)

Cook peas according to package directions reserving ¼ cup (50 mL) liquid. In 3 quart (3 L) saucepan melt butter; sauté onion. Stir in flour and mustard. Cook over low heat until mixture is smooth. Remove from heat; stir in tomato juice, then gradually add milk and reserved cooking liquid. Heat to boiling; stirring constantly. Boil and stir 1 minute. Remove from heat; stir in cheese until melted. If necessary, return to low heat to finish melting cheese (do not boil). Add salmon and peas. Heat to serving temperature. Serve in toast cups.

Makes 6 to 8 servings.

For toast cups see page 415.

SCALLOPED SCALLOPS

⅓ cup	butter	75 mL
½ cup	all-purpose flour	125 mL
1 tablespoon	garlic salad dressing mix	15 mL
2 cups	milk	500 mL
1 (10-ounce)	can sliced mushrooms	1 (284 mL)
2 (12-ounce)	packages scallops, thawed and well drained	2 (340 g)
1 tablespoon	grated lemon peel	15 mL
1½ cups	shredded Canadian Cheddar cheese	375 mL

Melt butter in a large skillet; stir in flour and dressing mix until thoroughly blended. Remove from heat; stir in milk to blend. Add mushrooms, scallops and lemon peel. Heat to simmer. Cook and stir until mixture is thick and scallops are opaque. Sprinkle on cheese, remove from heat and let stand until cheese is melted. To serve: Place a thick slice of tomato on top of two small toasted English muffin halves and top with about ¾ cup (175 mL) scallop mixture.

Note: If using lemon and garlic salad dressing mix, use 1 teaspoon (5 mL) grated lemon peel.

Makes approximately 4 cups/1 L.

Scalloped Fish Casserole

4	large potatoes, cooked and sliced	4
1	small onion, finely sliced	1
1 pound	fish fillets	450 g
1 teaspoon	dried thyme	5 mL
1 teaspoon	salt	5 mL
½ teaspoon	pepper	2 mL
¼ cup	butter	50 mL
¼ cup	flour	50 mL
2 cups	milk	500 mL
1 cup	grated Canadian Cheddar cheese	250 mL
½ teaspoon	dry mustard	2 mL

Lay sliced potatoes in shallow buttered baking dish. Top with onion slices and fish. Sprinkle with thyme, and half of the salt and pepper. Melt butter in small saucepan. Add flour and cook 2 minutes. Off heat, whisk in milk. Return to heat. Bring to boil, stirring constantly 5 minutes until sauce thickens. Stir in cheese, dry mustard and remaining salt and pepper. Pour sauce over fish. Bake in preheated 400°F (200°C) oven approximately 30 minutes.

Makes 4 servings.

SCALLOPS ON TOAST

1 ½ cups	fresh mushrooms, sliced	375 mL
6 tablespoons	butter, melted	90 mL
½ cup	flour	125 mL
1 tablespoon	garlic salad dressing mix	15 mL
2 cups	milk	500 mL
1 ½ pounds	scallops, fresh or frozen and thawed	750 g
1 teaspoon	lemon juice	5 mL
½ cup	Mozzarella cheese, grated	125 mL
2	tomatoes	2
8	English muffins, sliced and toasted	8

Sauté mushrooms in butter until soft, about 5 minutes. Stir in flour and salad dressing mix. Gradually add milk, stirring constantly, until evenly blended. Add scallops and lemon juice. Cook over low heat, stirring constantly, until mixture is thickened and scallops are opaque. Remove from heat and add cheese, stirring until cheese is melted. To serve, place a thick slice of tomato on top of 2 toasted English muffin halves. Top with about ¾ cup (175 mL) scallop mixture.

Makes 8 servings.

Sea captain's casserole

3 tablespoons	butter, melted	45 mL
1	onion, finely chopped	1
2 cups	fresh mushrooms, sliced	500 mL
1 (10 ounce)	can condensed cream of mushroom soup	1 (284 mL)
½ cup	creamy yogurt, unflavoured	125 mL
¼ cup	mayonnaise	50 mL
½ cup	milk	125 mL
½ teaspoon	Worcestershire sauce	2 mL
1 pound	shrimp or scallops	500 g
1 (5 ounce)	can crabmeat	1 (142 g)
1 cup	Mozzarella cheese, grated	250 mL
¼ cup	dry bread crumbs	50 mL

Sauté onions and mushrooms in butter until soft, about 5 minutes. In a bowl, blend soup, yogurt, mayonnaise, milk and Worcestershire sauce. Stir in seafood and sautéed vegetables. Place in a greased 6-cup (1.5 L) casserole. Combine cheese and bread crumbs and sprinkle over top. Bake at 350°F (180°C) for 45 minutes. Serve with hot, fluffy rice.

Makes 6 servings.

Seafood Mornay with Parsley Rice

3 tablespoons	butter	45 mL
3 tablespoons	flour	45 mL
1 tablespoon	chicken bouillon mix	15 mL
½ teaspoon	dry mustard	2 mL
½ teaspoon	salt	2 mL
2 cups	milk	500 mL
½ cup	shredded Canadian Swiss cheese	125 mL
3 cups	cooked mixed seafood	750 mL

Melt butter in a large saucepan. Blend in flour, bouillon mix, mustard and salt. Gradually stir in milk. Cook over medium heat, stirring constantly, until mixture just comes to a boil and thickens. Remove from heat. Add cheese and stir until melted. Add seafood; heat through. Serve over PARSLEY RICE.

Makes 6 servings.

PARSLEY RICE

2½ cups	boiling water	625 mL
1 teaspoon	chicken bouillon mix	5 mL
1 cup	regular long grain rice	250 mL
3 tablespoons	chopped parsley	45 mL
2 tablespoons	butter	30 mL

Combine water and bouillon mix in a saucepan; bring to a boil. Add rice. Cover tightly and cook over low heat until all water is absorbed, about 25 minutes. Stir in parsley and butter. Toss lightly to combine.

Makes 3½ cups/875 mL.

Seafood Crepes

CREPES

3	eggs	3
3 tablespoons	butter, melted	45 mL
1½ cups	flour	375 mL
¾ cup	milk	175 mL
¾ cup	water	175 mL

SAUCE

2 tablespoons	butter, melted	30 mL
2 tablespoons	flour	30 mL
1 cup	milk	250 mL
½ teaspoon	thyme	2 mL
½ teaspoon	Worcestershire sauce	2 mL
½ cup	white wine	125 mL
¼ cup	mayonnaise	50 mL

To make crepes, combine all ingredients thoroughly until batter is very smooth. Refrigerate for at least 2 hours. To cook, pour ¼ cup (50 mL) batter into a greased or non-stick 5″ (12 cm) frypan. Cook over medium heat until bottom is golden and top is dry. Turn out and repeat until 16 crepes are made. They can be made up to 2 days ahead and stored in the refrigerator.

To make sauce, combine flour and butter. Add milk gradually, and cook over medium heat, stirring constantly, until smooth and thickened. Add thyme, Worcestershire sauce and wine.

continued on next page

SEAFOOD CREPES (continued)

FILLING

1 tablespoon	butter, melted	15 mL
1 tablespoon	onion	15 mL
1 (4 ounce)	can baby shrimp	1 (112 g)
1 (6 ½ ounce)	can tuna	1 (184 g)
1 (5 ounce)	can crabmeat	1 (142 g)

To make filling, sauté onion in butter until soft, about 5 minutes. Blend in seafood. Stir in 1 cup (250 mL) of sauce. Let cool, as it is easier to work with filling when cool. Fill each crepe with a scant ¼ cup (50 mL) of filling. Fold in sides and roll up. Place in a shallow baking dish, seam side downward. Combine remaining sauce with mayonnaise. Spread mixture over crepes. Bake at 400°F (200°C) for 15-20 minutes or until bubbly and golden.

Makes 16 crepes.

Seafood Newburg

¼	cup	butter	50 mL
¼	cup	all-purpose flour	50 mL
¾	teaspoon	salt	3 mL
½	teaspoon	paprika	2 mL
2	cups	milk	500 mL
2	tablespoons	sherry	30 mL
2	tablespoons	fresh lemon juice	30 mL
4	cups	cooked seafood: lobster, shrimp, and crab meat	1 L

In large skillet melt butter; blend in flour, salt and paprika. Cook over low heat until mixture is smooth. Remove from heat; stir in milk. Heat to boiling, stirring constantly. Boil and stir 1 minute. Add sherry and lemon juice. Stir in seafood and heat to serving temperature. Serve over buttered rice.

Makes approximately 4 cups/1 L.

NOTE:

The following will yield 4 cups (1 L) seafood: 2 (8-ounce/ 250 g) packages frozen lobster tails, 1 (6-ounce/200 g) frozen crabmeat and 2 cups (500 mL) frozen deveined shrimp.

Variation: Substitute 1 tablespoon (15 mL) Worcestershire sauce for sherry; omit lemon juice. Use 4 cups (1 L) cooked cut-up chicken for seafood and add ½ cup (125 mL) sliced ripe olives. Garnish with grated Parmesan cheese.

SEAFOOD KEBABS WITH LEMON SAUCE

8	shrimp, peeled and deveined	8
8	scallops	8
4	small fish fillets	4
1	green pepper, cut into 1" (3 cm) squares	1
½	Spanish onion, sectioned	½
8	cherry tomatoes	8

MARINADE

½	cup	vegetable oil	125 mL
¼	cup	lemon juice	50 mL
1	tablespoon	chopped green onions	15 mL
½	teaspoon	tarragon	2 mL

LEMON SAUCE

2	tablespoons	butter	30 mL
1	tablespoon	flour	15 mL
1	cup	milk	250 mL
2	teaspoons	Dijon mustard	10 mL
½	teaspoon	salt	2 mL
		pepper, freshly ground	
2	tablespoons	lemon juice	30 mL
½	teaspoon	grated lemon rind	2 mL
		parsley, chopped	

Cut fish fillets into strips about 1" x 5" (3 cm x 10 cm); roll up and thread on skewers, along with other seafood and vegetables. (If fillets are too thick to roll, cut into chunks.) Marinate kebabs for at least 2 hours in mixture of oil, lemon juice, green onions and tarragon.

To make sauce: Melt butter in small saucepan over medium heat. Add flour to make a smooth paste; blend in milk, stirring constantly. Cook until thick and bubbly. Add mustard, salt, pepper, lemon juice and rind. Heat through. Sprinkle with parsley.

Broil or barbecue kebabs about 3 minutes on each side, basting frequently with marinade. Serve with Lemon Sauce.

Makes 1 ¼ cups (300 mL) sauce and 4 kebabs.

SEAFOOD SCALLOP

½ pound	firm-fleshed fish fillets	250 g
½ pound	scallops, fresh or frozen and thawed	250 g
2 ½ cups	milk	625 mL
1 (5 ounce)	can oysters, drained	1 (142 g)
6 tablespoons	butter	90 mL
½ cup	flour	125 mL
1 (10 ounce)	can condensed cream of mushroom soup	1 (284 mL)
1 ⅓ cups	cooked rice	325 mL
1 cup	dry breadcrumbs	250 mL
3 tablespoons	butter, melted	45 mL
2 tablespoons	chopped parsley	30 mL

Place fillets and scallops in a large saucepan. Cover with milk and heat slowly until milk is warm. Drain and reserve milk. Break fillets into bite-sized pieces, and combine with scallops and drained oysters. Melt butter in a saucepan and blend in flour. Stir in reserved milk gradually. Add mushroom soup and stir constantly while heating, until thickened and smooth. Pour a layer of sauce into a greased 6-cup (1.5 L) casserole. Add a layer of rice, then a layer of seafood. Repeat, ending with sauce. Combine breadcrumbs with melted butter and parsley, and sprinkle on top. Bake, uncovered, at 350°F (180°C) for 1 hour.

Makes 8 servings.

Seafood Casserole

6 cups	dry broad egg noodles	1.5 L
4 tablespoons	butter, melted	60 mL
½	onion, chopped	½
2	stalks celery, chopped	2
2 tablespoons	pimento	30 mL
1	green pepper, chopped	1
1 (10 ounce)	can condensed clam chowder	1 (284 mL)
2 (6½ ounce)	cans flaked tuna	2 (184 g)
2 (4 ounce)	cans shrimp, drained	2 (113 g)
2½ cups	milk	625 mL
½ cup	sherry (optional)	125 mL
2 tablespoons	lemon juice	30 mL
¼ teaspoon	pepper	1 mL
2 cups	frozen green peas, thawed	500 mL
2 (10 ounce)	cans condensed cream of mushroom soup	2 (284 mL)
1 cup	dry bread crumbs	250 mL
¼ cup	Cheddar cheese, grated	50 mL
2 tablespoons	butter, melted	30 mL

Boil noodles in salted water until just soft. Drain and rinse. Sauté onion, celery, pimento and green pepper in 4 tablespoons (60 mL) butter until soft, about 5 minutes. In a large bowl, combine clam chowder, tuna, shrimp, milk, sherry, lemon juice, pepper, peas and mushroom soup. Add noodles and sautéed vegetables. Stir well and pour into a greased 9"x 13" (23 cm x 33 cm) baking dish. Combine bread crumbs, cheese and 2 tablespoons (30 mL) butter and sprinkle over casserole. Bake at 350°F (180°C) for 30 minutes, or until golden brown and bubbly.

Makes 10 servings.

Shrimp Fromage Casserole

1-2 cups	cooked shrimp pieces	250-500 mL
1	clove garlic, minced	1
½ cup	fresh mushrooms, chopped	125 mL
½	onion, chopped	½
½ cup	butter, melted	125 mL
4 tablespoons	flour	60 mL
½ teaspoon	salt	2 mL
⅛ teaspoon	pepper	0.5 mL
2 cups	milk	500 mL
1 cup	sharp Cheddar cheese, grated	250 mL
½ cup	Mozzarella cheese, grated	125 mL
2 tablespoons	Parmesan cheese, grated	30 mL
½ cup	creamed cottage cheese	125 mL
½ cup	sour cream	125 mL
2½ cups	dry elbow macaroni, cooked and drained	625 mL
¼ cup	buttered bread crumbs	50 mL

Sauté garlic, mushrooms and onion in ¼ cup (50 mL) butter until soft, about 5 minutes. Set aside. Combine remaining butter with flour, salt and pepper. Add milk gradually, stirring constantly, and cook over medium heat until smooth and thickened. Stir in the Cheddar, mozzarella and Parmesan cheeses until melted. Remove from heat and blend in the cottage cheese and sour cream. Fold in the sautéed vegetables, shrimp and cooked macaroni. Turn into a greased 10" x 10" (25 cm x 25 cm) baking pan. Top with bread crumbs. Bake at 350°F (180°C) for 30 minutes, or until golden and bubbly.

Makes 10 servings.

Sole supreme

2 pounds	sole fillets	1 kg
1 teaspoon	salt	5 mL
2 tablespoons	finely-chopped green onion	30 mL
3 tablespoons	lemon juice	45 mL
3 tablespoons	butter	45 mL
3 tablespoons	flour	45 mL
1 teaspoon	chicken bouillon mix	5 mL
1½ cups	milk	375 mL
¼ cup	grated Canadian Parmesan cheese	50 mL
	chopped parsley	

Fold fillets in half. Place in a buttered large shallow baking dish. Sprinkle with ½ teaspoon (2 mL) of the salt, onion and lemon juice. Bake, covered, in preheated 400°F (200°C) oven 20 to 25 minutes or until fish flakes easily with a fork. Drain fish, reserving ½ cup (125 mL) of pan liquid. Melt butter in a saucepan. Blend in flour, bouillon mix and remaining ½ teaspoon (3 mL) salt. Gradually stir in milk and reserved pan liquid. Cook over medium heat, stirring constantly, until mixture just comes to a boil and thickens. Add 3 tablespoons (45 mL) of cheese; stir until combined. Pour sauce over fish in baking dish. Sprinkle remaining cheese over top. Return to oven and bake 10 to 15 minutes longer or until hot and bubbly. Serve garnished with chopped parsley.

Makes 6 to 8 servings.

SUMMER TUNA SALAD MOLD

2	envelopes unflavoured gelatin	2
1½ cups	milk	375 mL
¾ teaspoon	salt	3 mL
2 (6½-ounce)	cans drained, flaked tuna	2 (184 g)
1½ cups	fine curd cottage cheese	375 mL
1 cup	mayonnaise or salad dressing	250 mL
½ cup	bottled Italian salad dressing	125 mL
½ cup	finely-chopped celery	125 mL
½ cup	finely-chopped green pepper	125 mL
⅓ cup	finely-chopped green onion	75 mL

Sprinkle gelatin over milk in a saucepan. Let stand 10 minutes to soften. Cook over low heat, stirring constantly, until gelatin has dissolved. Cool. Stir in salt. Chill until mixture mounds from a spoon. Fold tuna, cottage cheese, mayonnaise, Italian salad dressing, celery, green pepper and green onion into gelatin mixture. Spoon into a 6-cup (1.5 L) mold. Chill until firm.

Makes 6 servings.

THERMIDOR EN CASSEROLE

3 tablespoons	butter	45 mL
3 tablespoons	flour	45 mL
¼ teaspoon	salt	1 mL
¼ teaspoon	paprika	1 mL
pinch	cayenne pepper	pinch
2 cups	milk	500 mL
1½ cups	shredded Canadian Cheddar cheese	375 mL
2 teaspoons	lemon juice	10 mL
¼ teaspoon	Worcestershire sauce	1 mL
3 cups	cooked fish or seafood*	750 mL
¼ cup	fine, dry, bread crumbs	50 mL
1 tablespoon	butter, melted	15 mL

Melt 3 tablespoons (45 mL) butter in a saucepan. Blend in flour, salt, paprika and cayenne pepper. Gradually stir in milk. Cook over medium heat, stirring constantly, until mixture just comes to a boil and thickens. Remove from heat. Add cheese and stir until melted. Stir in lemon juice, Worcestershire sauce and fish or seafood. Spoon into a 1½-quart (1.5 L) round casserole. Combine crumbs and melted butter. Sprinkle around edge of casserole. Bake in preheated 350°F (180°C) oven 30 to 35 minutes.

Makes 6 servings.

* Use a variety of canned or frozen fish and/or seafood, eg. tuna, salmon, lobster, scallops, shrimp, etc.

TOMATO, TUNA, EGG, MOZZARELLA CASSEROLE

2 tablespoons	butter, melted	30 mL
2 tablespoons	flour	30 mL
1 cup	milk	250 mL
1 (7 ounce)	can chunk tuna	1 (198 g)
¼ teaspoon	salt	1 mL
¼ teaspoon	pepper	1 mL
4	hard-cooked eggs	4
2	medium tomatoes, sliced	2
¼ pound	mozzarella cheese, sliced	125 g
	paprika	

Combine butter and flour well. Gradually add milk, stirring constantly, and continue to cook over medium heat until smooth and thickened. Add tuna, salt and pepper and mix well. Cut eggs in half lengthwise. Arrange eggs and tomato slices in a greased 9" (23 cm) pie plate. Pour tuna mixture over eggs and tomatoes. Top with cheese slices. Sprinkle with paprika. Broil 4" (10 cm) from heat for 5 minutes, until cheese is melted and lightly browned.

Makes 6 servings.

TUNA CELERY CRUNCH

¼ cup	butter	50 mL
3 cups	sliced celery	750 mL
¼ cup	flour	50 mL
½ teaspoon	salt	2 mL
1¾ cups	milk	425 mL
1 cup	shredded Canadian Cheddar cheese	250 mL
2 (6½-ounce)	cans tuna, drained and broken up	2 (184 g)
	Chow Mein Noodles*	

Melt butter in a saucepan. Sauté celery until crisp-tender, about 5 minutes. Blend in flour and salt. Gradually stir in milk. Cook over medium heat, stirring constantly, until mixture just comes to a boil and thickens. Remove from heat. Add cheese and stir until melted. Add tuna and heat through. To serve spoon tuna mixture over Chow Mein Noodles.

Makes 4 to 5 servings.

* Can also be served over hot cooked rice or noodles or in puff pastry shells.

TUNA BROCCOLI CASSEROLE

1	bunch broccoli, in spears	1
4 tablespoons	butter	60 mL
4 tablespoons	flour	60 mL
1 cup	milk	250 mL
1 cup	chicken broth	250 mL
½ teaspoon	salt	2 mL
dash	pepper	dash
¼ cup	grated Swiss cheese	50 mL
1 (7 ounce)	can flaked tuna, drained	1 (198 g)
½ cup	fresh mushrooms, sliced	125 mL
2	hard-cooked eggs, sliced	2

Cook broccoli spears until just tender. Arrange in bottom of an 8"x 8" (20 cm x 20 cm) baking dish. Melt butter; blend in flour. Stir in milk and chicken broth to form a smooth sauce. Add cheese, salt and pepper. Cook, stirring constantly, until sauce is thickened and cheese is melted. Arrange tuna, mushrooms and eggs evenly over the broccoli. Pour sauce over all. Bake at 300°F (150°C) for 20 minutes; then place under broiler for 2 minutes to brown. Serve over crisp toast points or a bed of rice.

Makes 4 servings.

TUNA À LA KING

3 tablespoons	butter, melted	45 mL
½	small onion, chopped	½
4	mushrooms, chopped	4
2 tablespoons	flour	30 mL
⅛ teaspoon	thyme	0.5 mL
½ teaspoon	salt	2 mL
⅛ teaspoon	pepper	0.5 mL
1 cup	milk	250 mL
1 (6½ ounce)	can tuna, flaked and drained	1 (184 g)
½ cup	green peas, cooked	125 mL
1 tablespoon	pimento	15 mL

Sauté onion and mushrooms in butter until soft, about 5 minutes. Stir in flour, thyme, salt and pepper. Add milk gradually, stirring constantly, and continue to cook over medium heat until smooth and thickened. Stir in tuna, peas and pimento and reheat to serving temperature. Serve in patty shells or over toast points.

Makes 4 servings.

TUNA CASSEROLE DELUXE

2 cups	wide egg noodles	500 mL
1 (10-ounce)	package frozen chopped broccoli	1 (283 g)
¼ cup	butter	50 mL
½ cup	finely-chopped onion	125 mL
¼ cup	flour	50 mL
1½ teaspoons	salt	7 mL
¾ teaspoon	basil leaves	3 mL
3 cups	milk	750 mL
¼ cup	grated Canadian Parmesan cheese	50 mL
2 (7-ounce)	cans solid tuna, drained and broken up	2 (198 g)
½ cup	fine dry bread crumbs	125 mL
2 tablespoons	butter, melted	30 mL

Cook noodles according to package directions; drain. Cook broccoli according to package directions until crisp-tender; drain well. Melt ¼ cup (50 mL) butter in a saucepan. Sauté onion until tender. Blend in flour, salt and basil. Gradually stir in milk. Cook over medium heat, stirring constantly, until mixture just comes to a boil and thickens. Remove from heat; add cheese and stir until well combined. Fold in noodles, broccoli and tuna. Turn into a 2-quart (2 L) shallow rectangular casserole. Combine bread crumbs and 2 tablespoons (30 mL) melted butter. Sprinkle around outside edge of casserole. Bake in preheated 350°F (180°C) oven 20 to 25 minutes.

Makes 4 or 5 servings.

TUNA IN A TUB

2 tablespoons	butter	30 mL
¼ cup	thinly-sliced celery	50 mL
2 tablespoons	flour	30 mL
¼ teaspoon	salt	1 mL
1 cup	milk	250 mL
¾ cup	shredded Canadian Cheddar cheese	175 mL
1 (6½-ounce)	can tuna, drained and broken up	1 (184 g)
1 cup	cooked peas and carrots	250 mL

Melt butter in a saucepan. Sauté celery until tender. Blend in flour and salt. Gradually stir in milk. Cook over medium heat, stirring constantly, until mixture just comes to a boil and thickens. Remove from heat. Add cheese and stir until melted. Add tuna, peas and carrots; heat through. Spoon about ¼ cup (50 mL) tuna mixture into each TOAST CUP.

Makes 4 servings.

TOAST CUPS

8	slices whole wheat bread	8
3 tablespoons	butter, melted	45 mL

Trim crusts from bread. Brush both sides with melted butter. Press 1 slice into each of eight 6-ounce (150 mL) custard cups. Place on cookie sheet. Bake in preheated 350°F (180°C) oven 15 to 20 minutes or until bread is toasted.

Makes 8 toast cups.

BOBOTIE

1	large onion, chopped	1
1 tablespoon	butter	15 mL
1 pound	ground lamb	500 g
1 tablespoon	curry powder	15 mL
1 teaspoon	turmeric	5 mL
1 teaspoon	salt	5 mL
4 teaspoons	vinegar	20 mL
1 tablespoon	sugar	15 mL
1 tablespoon	chutney	15 mL
¼ cup	seedless raisins	50 mL
2	slices soft bread, cubed	2
1 cup	milk	250 mL
1	egg, beaten	1
12	whole almonds, blanched	12

Sauté onions in butter over medium heat until soft. Add meat and cook until red colour disappears, breaking up the meat to avoid lumps. Add curry powder, turmeric and salt, and cook for 2 minutes. Add vinegar, sugar, chutney, raisins, and bread, stirring well. Place in a greased 6 cup (1.5 L) casserole, smoothing the top surface. Combine milk and egg well, and pour over the surface of the meat – do not stir. Arrange the almonds over the top. Bake uncovered at 350°F (180°C) for 1 hour. Serve with plain boiled rice and chutney.

Makes 6 servings.

LAMB LOAF

4	slices white bread, cubed	4
1 cup	milk	250 mL
1	egg, beaten	1
2 pounds	ground lamb	1 kg
½	green pepper, finely chopped	½
1 cup	carrot, finely grated	250 mL
1	small stalk celery, diced	1
2 teaspoons	Angostura bitters	10 mL
1 teaspoon	dried basil	5 mL
1½ teaspoons	salt	7 mL
¼ teaspoon	pepper	1 mL
1 tablespoon	butter	15 mL

Place bread cubes in a bowl and cover with milk. When soft, mash and blend well with fork. Add beaten egg, ground lamb, green pepper, carrot and celery. Mix thoroughly. Add bitters, basil, salt and pepper. Stir until well mixed. Place in an ungreased 9"x 5"x 3" (23 cm x 12 cm x 7 cm) loaf pan. Dot with butter. Bake at 350°F (180°C) for 1 hour. Serve hot or cold.
    Makes 8 servings.

Lamb Stew

1½ pounds	stewing lamb, cut in bite-size pieces	750 g
2	cloves garlic, minced	2
2 teaspoons	salt	10 mL
1 teaspoon	celery seeds	5 mL
1 teaspoon	rosemary, crushed	5 mL
3 cups	water	750 mL
6	medium carrots, peeled and quartered	6
4	potatoes, peeled and quartered	4
3	medium onions, peeled and quartered	3
5 tablespoons	flour	75 mL
1¼ cups	milk	300 mL
1 teaspoon	Worcestershire sauce	5 mL

Combine lamb, garlic, salt, celery seeds, rosemary and water in a large saucepan. Bring to a boil. Reduce heat; cover and simmer 1 to 1½ hours. Add carrots, potatoes and onions and simmer an additional 25 to 30 minutes or until vegetables are tender. Smoothly combine flour, milk and Worcestershire sauce. Stir into meat mixture. Cook over medium heat, stirring constantly, until mixture just comes to a boil and thickens.

Makes 6 to 8 servings.

MIDDLE EAST MEAT PIE

1	large eggplant, thickly sliced	1
½ cup	flour	125 mL
2	eggs, beaten	2
	oil for frying	
1½ pounds	ground lamb	750 g
1	onion, diced	1
2	stalks celery, diced	2
1	clove garlic, minced (optional)	1
1 teaspoon	salt	5 mL
¼ teaspoon	pepper	1 mL
1 teaspoon	dry mustard	5 mL
½ cup	catsup	125 mL
6	medium potatoes, thinly sliced	6
1½ cups	milk	375 mL
2	eggs, beaten	2

Cut eggplant into ⅜" (1 cm) slices. Salt well, and let stand for 30 minutes. Rinse and pat dry. Dip each slice first into flour, and then into eggs. Brown in oil, about 5 minutes for each side. Set aside. Brown the lamb with onion, celery, garlic, salt, pepper and mustard. Stir in the catsup and simmer 5 minutes. In a greased 9"x 13" (23 cm x 33 cm) baking pan, layer half the eggplant slices, then lamb, then potatoes, and end with remaining eggplant. Combine milk and eggs. Pour over layered ingredients. Bake, uncovered, at 350°F (180°C) for 1 hour.

Makes 8 servings.

MOUSSAKA

4	medium eggplants, sliced ¼" (0.5 cm) thick	4
	salt	
¾ cup	butter, divided	175 mL
2	large onions, chopped	2
2	garlic cloves, minced (optional)	2
1½ pounds	lamb	750 g
½ cup	tomato sauce	125 mL
	salt and pepper, to taste	
	dash of nutmeg	
½ cup	flour	125 mL
4½ cups	hot milk	1 L
3	egg yolks, well-beaten	3
1¼ cups	grated Parmesan cheese	300 mL

Sprinkle eggplant generously with salt and let stand while preparing meat and milk sauce. Sauté onion and garlic in 3 tablespoons (45 mL) butter; add lamb and brown; drain off some fat. Add tomato sauce, salt and pepper; simmer for 5 to 10 minutes. Set aside. In a saucepan, melt remaining butter, blend in flour. Gradually add milk, stirring constantly; cook over low heat until smooth and thickened. Add salt, pepper and nutmeg to taste. Add a small amount of the milk sauce to egg yolks and whisk into milk sauce. Cook an additional 2 minutes, stirring constantly; reserve. Rinse eggplant and pat dry. Brown eggplant slices on both sides under broiler. Place ⅓ of the eggplant slices in a greased 9" x 13" (3.5 L) baking pan; cover with half of the lamb mixture and sprinkle with ½ cup (125 mL) cheese. Repeat layering. Top with eggplant. Cover with milk sauce, sprinkle with remaining cheese and bake at 350°F (180°C) for 1 hour.

Makes 6 to 8 servings.

QUICK CASHEW CURRY

1	medium onion, chopped	1
1-2	garlic cloves, finely chopped	1-2
2 tablespoons	butter	30 mL
1½ pounds	boneless lamb, cut into small cubes	750 g
1½ teaspoons	curry powder, or to taste	7 mL
1 tablespoon	flour	15 mL
1½ cups	milk	375 mL
½ teaspoon	salt	2 mL
⅛ teaspoon	cayenne pepper	0.5 mL
½ cup	chopped cashews	125 mL
2	medium tomatoes, chopped	2

Sauté onion and garlic in butter; add meat and brown. Blend in curry powder and flour, stirring constantly. Add milk and whisk until smooth; add seasonings. Cook, stirring constantly, over medium heat until thickened. Add cashews and tomatoes. Heat an additional 10 minutes, or until meat is tender. Serve on a bed of rice or broad egg noodles.

Makes 4 servings.

BAKED SAUSAGE PIZZA

3	eggs, beaten	3
¾ cup	flour	175 mL
¾ cup	milk	175 mL
¼ teaspoon	salt	1 mL
1 teaspoon	chili powder	5 mL
1 tablespoon	butter	15 mL
1 pound	English breakfast sausage	500 g
1 cup	carrots, cut in ¼″ (0.5 cm) slices	250 mL
2	small zucchini, cut in ¼″ (0.5 cm) slices	2
½ cup	stuffed olives, sliced	125 mL
½ teaspoon	garlic salt	2 mL
½ teaspoon	dried savory	2 mL
1 cup	sharp Cheddar cheese, grated	250 mL

Heat a 10″ (25 cm) oven-proof flan dish on the lowest shelf in a 350°F (180°C) oven. Meanwhile, combine eggs, flour, milk, salt and chili powder. Remove dish from oven. Add butter and swirl to coat the dish evenly. Add batter immediately. Bake on lowest shelf for 20 minutes or until golden. Remove from oven and keep warm. Meanwhile, sauté sausage in a large skillet until browned and cooked through. Remove to a bowl. In remaining fat, sauté carrots and zucchini until just soft, about 5 minutes. Add sausage, olives, garlic salt, and savory. Combine well. Spoon vegetable and sausage mixture onto the pancake. Sprinkle grated cheese over all. Return to the oven and heat until cheese melts. Cut into wedges and serve immediately.

Makes 6 servings.

Braised Pork Loin

1	tablespoon	butter	15 mL
2½-3 pounds	pork loin	1.5 kg
2	cups	milk	500 mL
½	teaspoon	salt	2 mL
½	teaspoon	pepper	2 mL
1		bay leaf	1
1	teaspoon	dried sage	5 mL

Heat butter in large casserole and brown pork on all sides. Add milk, seasoning and herbs. Adjust heat to medium-low, partially cover pot and cook meat slowly for about 2 hours until tender and juices run clear – 170°F (75°C) on meat thermometer.

Remove pork to warm platter. Skim off any fat in pot. Add a few tablespoons of water and stir over high heat to loosen caramelized juices and form a tasty sauce.

Makes 6 servings.

COUNTRY HAM CASSEROLE

2 tablespoons	butter	30 mL
1	medium onion, chopped	1
2	green onions, finely chopped	2
½ pound	ham cubed	250 g
¼ teaspoon	dried thyme	1 mL
4 cups	cooked rice	1 L
1 tablespoon	finely chopped parsley	15 mL
	salt and pepper to taste	
1½ cups	grated Mozzarella cheese	375 mL
1 (10 ounce)	can cream of celery soup	1 (284 mL)
¾ cup	milk	175 mL
½ cup	sour cream	125 mL
½ cup	breadcrumbs	125 mL
1 tablespoon	melted butter	15 mL
1 tablespoon	grated Parmesan cheese	15 mL

Melt butter in a large frying pan. Add chopped onions. Cook until softened. Stir in ham and thyme. Heat through. Combine with cooked rice, parsley, salt and pepper to taste and grated Mozzarella. Mix soup, milk and sour cream together in a small saucepan. Heat until just blended. Mix with rice. Pour into greased 9″ square (2.5 L) baking dish. Sprinkle with combined breadcrumbs, butter and Parmesan. Bake in 350°F (180°C) oven until hot and bubbly. Let rest 10 minutes before serving.

Makes 4-6 servings.

COUNTRY STYLE PORK CHOPS

6	loin pork chops	6
1 tablespoon	butter	15 mL
1 (10-ounce)	can beef broth	1 (284 mL)
¼ teaspoon	salt	1 mL
¼ teaspoon	ground sage	1 mL
pinch	ground thyme	pinch
2	medium onions, sliced	2
2 cups	diagonally-sliced carrots	500 mL
1 cup	diagonally-sliced celery	250 mL
¼ cup	flour	50 mL
1¼ cups	milk	300 mL
	hot mashed potatoes	

Trim any excess fat from chops. Melt butter in a large frypan. Brown chops on both sides. Combine beef broth, salt, sage and thyme. Pour over chops in pan. Place onions, carrots and celery on top of meat. Bring to a boil. Reduce heat; cover and simmer 45 to 50 minutes or until meat and vegetables are tender. Remove chops from pan; keep warm. Smoothly combine flour and milk. Stir into vegetables and pan juices. Cook over medium heat, stirring constantly, until mixture just comes to a boil and thickens. Spoon vegetables and gravy over chops and hot mashed potatoes to serve.

Makes 6 servings.

CREAMED PORK TENDERLOIN

8	slices pork tenderloin	8
¼ cup	flour	50 mL
1 teaspoon	salt	5 mL
½ teaspoon	pepper	2 mL
¼ cup	butter	50 mL
1 (10 ounce)	package frozen broccoli, thawed	1 (280 g)
1	onion, chopped	1
1	carrot, chopped	1
1 cup	hot water	250 mL
2	packets instant chicken bouillon	2
½ teaspoon	thyme	2 mL
2 teaspoons	lemon rind	10 mL
3 tablespoons	lemon juice	45 mL
1	small bay leaf	1
¼ cup	flour	50 mL
1 cup	milk	250 mL

Pound pork tenderloin to ¼″ (0.5 cm) thick. Combine ¼ cup (50 mL) flour, 1 teaspoon (5 mL) salt, and ½ teaspoon (2 mL) pepper to make seasoned flour. Dip pork into seasoned flour, and brown in butter. Place in an 8″x 8″ (20 cm x 20 cm) oven dish and cover with thawed broccoli. Add onion and carrots to fat remaining in pan and cook until soft, about 5 minutes. Add hot water, bouillon powder, thyme, lemon rind and juice, and bay leaf. Heat until powder is dissolved. Pour over broccoli and pork tenderloin. Bake, covered, for 45 minutes at 350°F (180°C). When done, strain liquid left in dish into a small saucepan. Whisk ¼ cup (50 mL) flour into 1 cup (250 mL) milk, and add to the hot liquid. Heat and stir constantly until thick. Pour over pork and broccoli. Serve with hot, buttered noodles.

Makes 8 servings.

CREAMY HERBED PORK CHOPS

4		pork chops	4
2		small onions, sliced	2
2	tablespoons	flour	30 mL
1½	cups	milk	375 mL
½	teaspoon	salt	2 mL
		pepper, freshly ground	
½	teaspoon	dry sage or	2 mL
		1 tablespoon (15 mL) chopped	
		fresh sage	
½	teaspoon	dry thyme or 1 tablespoon	2 mL
		(15 mL) chopped fresh thyme	
		nutmeg to taste	

Place pork chops in large, cold frypan over medium high heat. Brown chops on each side; reduce heat. Add onion slices and 1 tablespoon (15 mL) water. Cover and cook over low heat until chops are completely cooked, about 25 minutes. Remove chops to warm platter.

To 2 tablespoons (30 mL) pan drippings, add flour and blend until smooth. Over medium heat, add milk, stirring constantly until gravy thickens and boils. Blend in salt, pepper, herbs and nutmeg. Simmer for 2 minutes, to blend flavours. Pour over pork chops.

Makes 4 servings.

FRAN'S FRUGAL CHEDDAR AND HAM FLAN

2 tablespoons	butter	30 mL
½ cup	chopped onion	125 mL
½ cup	sliced, fresh mushrooms	125 mL
1½ cups	shredded Canadian Cheddar cheese	375 mL
1 cup	cubed ham	250 mL
1	9-inch (1 L) unbaked pie shell	1
2	eggs, beaten	2
1¼ cups	milk	300 mL
¼ cup	flour	50 mL
½ teaspoon	salt	2 mL
¼ teaspoon	pepper	1 mL
	tomato slices	
	parsley	

Melt butter in a frypan; sauté onions and mushrooms until tender. Sprinkle cheese, ham, onions and mushrooms in bottom of pie shell. Combine eggs and milk. Gradually blend into a mixture of flour, salt and pepper. Pour into pie shell. Bake in preheated 375°F (190°C) oven 30 to 40 minutes or until a knife inserted near centre comes out clean. Let stand 5 minutes. Serve hot or cold, garnished with tomato slices and parsley.

Makes 6 servings.

DELMONICO POTATOES AND HAM

6 cups	1/8-inch/6 mm thick raw potato slices	1.5 L
1 cup	thinly-sliced onion	250 mL
2 cups	water	500 mL
1 teaspoon	salt	5 mL
1 (10-ounce)	can cream of mushroom soup	1 (284 mL)
1 cup	milk	250 mL
1 tablespoon	prepared mustard	15 mL
1 tablespoon	chopped parsley	15 mL
1/4 teaspoon	pepper	1 mL
2 cups	diced cooked ham	500 mL

Place potatoes, onion, water and salt in a large saucepan. Bring to a boil. Reduce heat; cover and simmer until potatoes are tender. Drain well. Combine soup, milk, mustard, parsley and pepper. Layer half of the potatoes, onions and ham in a 2-quart (2 L) shallow rectangular baking dish. Repeat layers. Bake in preheated 350°F (180°C) oven 35 to 40 minutes or until hot and bubbly.

Makes 6 servings.

HAM AND BROCCOLI ROYALE

1½ cups	cooked rice	375 mL
1 (10 ounce)	package frozen broccoli, thawed	1 (280 g)
3 tablespoons	butter, melted	45 mL
1 cup	dried bread crumbs	250 mL
1 cup	onion, chopped	250 mL
1½ tablespoons	flour	25 mL
⅛ teaspoon	pepper	0.5 mL
1½ cups	milk	375 mL
2 cups	cooked ham, diced	500 mL
1 cup	Cheddar cheese, grated	250 mL

Spoon cooked rice into greased 6-cup (1.5 L) casserole. Layer the broccoli over the rice. Melt the butter and pour 1 tablespoon (15 mL) over the bread crumbs in a bowl; Set aside. Sauté the onion in the remaining butter until soft. Blend in the flour and pepper. Slowly blend in the milk, stirring constantly over medium heat until smooth and thickened. Add ham; heat until bubbly. Pour onto greased casserole. Sprinkle the buttered crumbs over the top. Cover and bake at 350°F (180°C) for 35 minutes. Uncover and sprinkle grated cheese over the top. Bake 10 minutes longer.

Makes 6 servings.

Ham and Noodle Supper Supreme

2½ cups	dry broad noodles	625 mL
1 cup	celery, chopped	250 mL
½ cup	green pepper, chopped	125 mL
½ cup	onion, chopped	125 mL
4 tablespoons	butter	60 mL
¼ cup	flour	50 mL
2½ cups	milk	625 mL
¼ teaspoon	pepper	1 mL
4 cups	cooked ham, diced	1 L
3 cups	creamed cottage cheese	750 mL
½ cup	dry bread crumbs	125 mL
⅓ cup	butter, melted	75 mL
1 cup	Cheddar cheese, grated	250 mL

Cook noodles in boiling water. Drain. Sauté celery, green pepper and onion in butter for 5 minutes until soft. Remove from heat and blend in flour. Stir in milk gradually, and heat, stirring constantly, until mixture becomes smooth and thickened. Add pepper, ham, cottage cheese and noodles. Turn into a 9"x 13" (23 cm x 33 cm) greased casserole. Combine crumbs and melted butter and sprinkle over casserole. Top with cheese. Bake at 350°F (180°C) for 60 minutes. Let stand 10 minutes before serving. This can be prepared and sliced one day ahead, since it reheats well. It can also be served cold.

Makes 10-12 servings.

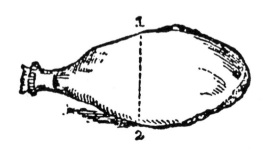

Ham on English Muffins

¼	cup	butter	50 mL
½	cup	chopped green pepper	125 mL
3	tablespoons	all-purpose flour	45 mL
¼	teaspoon	salt	1 mL
1	teaspoon	dry mustard	5 mL
2	cups	milk	500 mL
2	cups	cubed cooked ham	500 mL
½	cup	sour cream, at room temperature	125 mL
6		English Muffins, split and toasted	6

In 2-quart saucepan melt butter; sauté green pepper until just tender. Stir in flour, salt and mustard. Cook over low heat until mixture is smooth. Remove from heat; stir in milk. Add ham. Heat to boiling, stirring constantly. Boil and stir 1 minute. Remove from heat and stir in sour cream.

Serve over English Muffins.

Makes 6 servings.

Ham Tetrazzini

1¾ cups	broken spaghetti	375 mL
⅓ cup	butter	75 mL
3 cups	sliced, fresh mushrooms	750 mL
1 cup	thinly-sliced green pepper	250 mL
⅓ cup	flour	75 mL
2 teaspoons	salt	10 mL
2 teaspoons	chicken bouillon mix	10 mL
½ teaspoon	oregano leaves	2 mL
4 cups	milk	1 L
5 cups	ham, cut in thin strips (about 1½ lbs/750 g)	1.25 L
2 tablespoons	grated Canadian Parmesan cheese	30 mL

Cook spaghetti according to package directions; drain well. Melt butter in a saucepan. Sauté mushrooms and green pepper until tender and mushroom liquid has evaporated. Blend in flour, salt, bouillon mix and oregano. Gradually stir in milk. Cook over medium heat, stirring constantly, until mixture just comes to a boil and thickens. Add ham and cooked spaghetti. Place in a greased 3-quart (3 L) shallow rectangular baking dish. Sprinkle with Parmesan cheese. Bake in preheated 350°F (180°C) oven 25 to 30 minutes or until hot and bubbly.

Makes 8 to 10 servings.

Ham and Broccoli Spuds

2 tablespoons	butter	30 mL
2 tablespoons	flour	30 mL
1 teaspoon	chicken bouillon mix	5 mL
pinch	dry mustard	pinch
1½ cups	milk	375 mL
1½ cups	shredded Canadian Cheddar cheese	375 mL
1 cup	chopped, cooked broccoli	250 mL
1 cup	diced, cooked ham	250 mL
6	hot baked potatoes butter	6

Melt butter in a medium saucepan. Blend in flour, bouillon mix and dry mustard. Gradually stir in milk. Cook over medium heat, stirring constantly, until mixture just comes to a boil and thickens. Remove from heat. Add cheese and stir until melted. Stir in broccoli and ham. Heat through; do not boil. Split each baked potato in half. Spread with butter. Spoon meat and vegetable mixture over each half. Serve immediately.

Makes 6 servings.

HAM AND EGG CASSEROLE

8	hard-cooked eggs, peeled and finely chopped	8
1½ cups	cooked ham, chopped	375 mL
1 tablespoon	parsley, finely chopped	15 mL
1 cup	cracker crumbs	250 mL
½ cup	soft butter	125 mL
3 tablespoons	flour	45 mL
2 cups	milk	500 mL
	pinch of pepper	

Preheat oven to 350°F (180°C).

Combine the chopped eggs, ham and parsley, and set aside. Place half the butter in a medium frypan over moderate heat, and add the cracker crumbs. Sauté the crumbs until golden brown, remove from heat and set aside. Place the remaining butter in a saucepan over moderate heat. Stir in flour and continue stirring until smooth. Gradually add the milk and pepper, and continue stirring until the sauce has thickened. Butter a 8-cup (2 L) casserole dish. Layer one third each of the crumbs, ham and egg mixture, and the cream sauce. Repeat to make two more layers. Bake for 25 minutes or until top is browned. Serve at once.

Makes 6 servings.

ORIENTAL BARBECUED PORK WITH PINEAPPLE PEANUT SAUCE

1½ pounds	lean pork	750 g

SOY MARINADE

¼ cup	oil	50 mL
¼ cup	soy sauce	50 mL
2 tablespoons	lemon juice	30 mL
1 teaspoon	sugar	5 mL

PINEAPPLE PEANUT SAUCE

½ cup	shredded coconut	125 mL
1 cup	hot milk	250 mL
2 tablespoons	butter	30 mL
1	onion, chopped	1
1	garlic clove, minced	1
1 teaspoon	curry powder	5 mL
½ teaspoon	ginger	2 mL
½ cup	crushed, drained pineapple	125 mL
¼ cup	crunchy peanut butter	50 mL

Slice pork in thin strips or cubes; toss with soy marinade; let stand one hour or more. Pour milk over coconut; let stand 15 minutes. Heat butter in a saucepan, add onion and garlic; cook until softened. Stir in curry powder and ginger. Add milk and coconut, pineapple and peanut butter; cook, stirring constantly, until thickened. Thread pork on small skewers. Barbecue or broil until well-browned and cooked (approximately 5 minutes). Serve with warm sauce.

Makes 4 servings.

PEAMEAL BACON AND RICE CASSEROLE

¼ cup	butter	50 mL
½ cup	chopped onion	125 mL
1	clove garlic, minced	1
¼ cup	flour	50 mL
1 teaspoon	salt	5 mL
1 teaspoon	chicken bouillon mix	5 mL
2½ cups	milk	625 mL
2 cups	shredded Canadian Cheddar cheese	500 mL
1 (12-ounce)	package frozen chopped spinach, thawed and drained	1 (340 g)
3 cups	cooked rice	750 mL
8	slices peameal bacon, cooked	8

Melt butter in a saucepan. Sauté onion and garlic until tender. Blend in flour, salt and bouillon mix. Gradually stir in milk. Cook over medium heat, stirring constantly, until mixture just comes to a boil and thickens. Remove from heat. Add cheese and stir until melted. Stir in spinach. Spoon half of the rice into a greased 2-quart (2 L) round casserole. Top with half the cheese sauce. Repeat layers. Arrange bacon slices on top. Bake in preheated 350°F (180°C) oven 30 to 35 minutes or until heated through.

Makes 8 servings.

PIQUANT HAM CROQUETTES

4 tablespoons	butter	60 mL
6 tablespoons	flour	90 mL
1 cup	milk	250 mL
2	eggs	2
1	small onion, finely chopped	1
2	mushrooms, finely chopped	2
1 tablespoon	dijon-style mustard	15 mL
2 tablespoons	chopped parsley	30 mL
1 pound	smoked ham, finely chopped	500 g
	salt	
	freshly ground pepper	
	dry breadcrumbs	

Melt 3 tablespoons (45 mL) butter in saucepan. Stir in flour and cook for 2 minutes. Remove from heat gradually blend in milk. Return to heat and stir constantly until mixture comes to a boil and is smooth and thick. Remove from heat and add one lightly beaten egg yolk. Heat remaining butter in a frying pan and gently sauté onion and mushroom until tender. Add to sauce with mustard, parsley and finely chopped ham. Season to taste. Refrigerate until cold and firm. Shape into croquettes. Lightly beat remaining egg white and egg. Dust croquettes lightly with flour, dip in egg and coat in breadcrumbs. Fry in hot butter or oil until crisp and brown. Drain on paper towel and serve either hot or cold.

Makes 4 main dish servings or appetizers for 8.

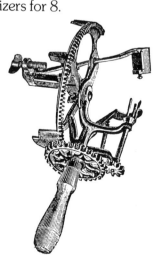

Pork 'N' Pineapple Stirfry

2	tablespoons	butter	30 mL
1		garlic clove, chopped finely	1
4		green onions	4
1	pound	pork tenderloin, thinly sliced	500 g
1		green or red pepper, cut in strips	1
1	cup	fresh or canned pineapple, cut in chunks	250 mL
1	teaspoon	ground ginger	5 mL
½	teaspoon	dried thyme	2 mL
¼	teaspoon	salt	1 mL
1	tablespoon	flour	15 mL
1	cup	milk	250 mL

Melt butter in large, heavy frying pan. Slice onions on bias into ½" (1.5 cm) pieces; toss with garlic in butter over medium heat until just softened. Add pork and brown lightly, 2-3 minutes per side. Set aside. Toss peppers, pineapple, ginger, thyme and salt into pan. Cook briefly.

Sprinkle with flour and add milk; stir well. Bring just to a boil, lower heat and cook, stirring, for a few minutes until sauce is smooth and lightly thickened. Return pork to pan; heat through. Serve at once with hot steamed rice. For a spectacular presentation, cut a fresh pineapple in half, use the flesh for the recipe and pile the Pork 'n' Pineapple into the pineapple shells to serve.

Makes 4 servings.

PORK CHOP CASSEROLE

6	pork chops	6
4 tablespoons	butter	60 mL
2 tablespoons	flour	30 mL
1 teaspoon	salt	5 mL
½ teaspoon	pepper	2 mL
1 teaspoon	caraway seed	5 mL
1 tablespoon	parsley, chopped	15 mL
2 cups	potatoes, sliced	500 mL
2 cups	milk	500 mL

Slash fat at edges of pork chops. Brown chops in butter in a hot frying pan. Set aside. Add flour to fat remaining in frypan and blend well. Add milk gradually, stirring constantly, and heat until smooth and thickened. Add salt, pepper, caraway seed and parsley. Place 3 chops in a greased casserole, cover with half the potatoes, and half of the sauce. Repeat with remaining chops, potatoes and sauce. Cover and bake at 350°F (180°C) for 40 minutes. Uncover and bake 20 minutes longer.

Makes 6 servings.

PORK CORDON BLEU

6	rectangular pieces pork schnitzel	6
6	slices Swiss cheese	6
6	slices cooked ham	6
3 tablespoons	flour	45 mL
1 teaspoon	salt	5 mL
⅛ teaspoon	pepper	0.5 mL
1	egg, beaten	1
2 tablespoons	water	30 mL
1½ cups	fine breadcrumbs	375 mL
1 tablespoon	oil	15 mL
1 tablespoon	butter	15 mL
¼ cup	chopped onion	50 mL
½ (10 ounce)	can condensed cream of mushroom soup	½ (284 mL)
1 cup	milk	250 mL

Pound slices of pork until thin rectangles about 4"x 8" (10 cm x 20 cm). Place a piece of cheese and ham on each slice, and fold over to cover filling. Dip in flour seasoned with salt and pepper. Then dip in egg diluted with the water. Coat with breadcrumbs. Let dry at least 1 hour, then brown in oil and butter combined. Place in a shallow casserole. Combine onion, mushroom soup and milk. Pour over the pork. Bake, uncovered, at 350°F (180°C) for 1 hour.

Makes 6 servings.

Pork Pastitsio

1 pound	ground pork	500 g
1 cup	onion, chopped	250 mL
1 (14 ounce)	can tomato sauce	1 (398 mL)
1 teaspoon	salt	5 mL
$\frac{1}{2}$ teaspoon	oregano	2 mL
$\frac{1}{4}$ teaspoon	cinnamon	1 mL
$\frac{1}{4}$ teaspoon	pepper	1 mL
1	clove garlic, minced	1
2 cups	macaroni	500 mL
1 cup	grated Cheddar cheese	250 mL
1	egg, beaten	1

CHEESE SAUCE

3 tablespoons	butter	45 mL
3 tablespoons	flour	45 mL
1 $\frac{1}{2}$ cups	milk	375 mL
$\frac{1}{2}$ teaspoon	salt	2 mL
1 cup	grated Cheddar cheese	250 mL
1	egg, beaten	1

In heavy pot, brown pork until crumbly. Add onions. Sauté, stirring occasionally, until onions are tender. Drain fat. Add tomato sauce, salt, oregano, cinnamon, pepper and garlic. Cover and simmer for 15 minutes. Meanwhile, cook macaroni in boiling water. Rinse under cold water and drain well. Mix macaroni with grated cheese and egg. Spread half the macaroni in a greased 9" x 13" (23 cm x 33 cm) baking pan. Cover with meat sauce, and top with remaining macaroni. Prepare cheese sauce by melting butter in a saucepan. Stir in flour until well combined. Add milk and continue to cook, stirring constantly, until smooth and thickened. Remove from heat and stir in salt and cheese. Whisk hot sauce into beaten egg. Pour cheese sauce over macaroni and bake at 350°F (180°C) for 30 minutes. Let stand for 15 minutes before serving. Cut into squares.
Makes 8 servings.

SAUCY FRUIT AND CHOPS

2 tablespoons	butter	30 mL
1½ pounds	pork shoulder butt chops	750 g
1	beef bouillon cube	1
¾ cup	hot water	175 mL
1 teaspoon	salt	5 mL
2	large apples	2
½ cup	raisins	125 mL
3 tablespoons	flour	45 mL
1⅓ cups	milk	325 mL

Melt butter in a large frypan. Brown chops on both sides. Dissolve bouillon cube in hot water. Add salt. Pour over chops in pan. Cut apples into ½-inch (12 mm) wedges. Arrange apple wedges on meat. Sprinkle raisins over all. Bring to a boil. Reduce heat; cover and simmer 25 to 30 minutes or until meat is tender. Remove apples and chops from pan. Smoothly combine flour and milk. Stir into pan juices. Cook over medium heat, stirring constantly, until mixture just comes to a boil and thickens. Return meat and fruit to pan and reheat to serving temperature.

Makes 4 servings.

SAVOURY STUFFED SAUSAGE LOAF

2 pounds	bulk pork sausage	1 kg
1 cup	finely-chopped carrot	250 mL
¾ cup	finely-chopped green pepper	175 mL
1 cup	milk	250 mL
1	egg	1
1 cup	fine dry bread crumbs	250 mL
1½ teaspoons	salt	7 mL
¼ teaspoon	pepper	1 mL
1 tablespoon	butter	15 mL
½ cup	finely-chopped onion	125 mL
2 cups	packaged seasoned stuffing mix	500 mL

Combine sausage meat, carrot, green pepper, milk, egg, bread crumbs, salt and pepper in a large bowl. Melt butter in a saucepan. Sauté onion until tender. Add stuffing mix and blend well. Press half of the sausage mixture into a lightly greased 2-quart (2 L) rectangular casserole. Spread stuffing evenly over sausage. Add remaining sausage mixture and press down firmly. Bake in preheated 350°F (180°C) oven 1¼ hours or until loaf shrinks from sides of casserole. Pour off any drippings. Unmold onto serving platter and slice to serve.

Makes 8 servings.

Savoury Franks and Cabbage

¼ cup	butter	50 mL
1 teaspoon	caraway seeds	5 mL
2 cups	sliced onion rings	500 mL
8 cups	coarsely-shredded cabbage	2 L
¼ cup	flour	50 mL
1½ teaspoons	salt	7 mL
2 cups	milk	500 mL
3 tablespoons	prepared mustard	45 mL
1 pound	weiners, cut in 1-inch/ 2.5 cm pieces	500 g

Melt butter in a large saucepan or Dutch oven. Add caraway seeds and onion. Sauté until onion is tender. Add cabbage; cover and cook over low heat, stirring occasionally, 5 to 10 minutes or until cabbage is tender. Blend in flour and salt. Stir milk into mustard; gradually add to cabbage. Cook over medium heat, stirring constantly, until mixture just comes to a boil and thickens. Add weiners to cabbage and cook, stirring occasionally until weiners are heated through.
Makes 6 to 8 servings.

SKILLET SAUSAGE AND MACARONI SUPPER

2 tablespoons	butter	30 mL
½ cup	finely-chopped onion	125 mL
½ cup	chopped green pepper	125 mL
1 (500 g)	package frozen sausage meat, thawed	1 (500 g)
¾ teaspoon	salt	3 mL
¼ teaspoon	ground oregano	1 mL
pinch	garlic powder	pinch
1¾ cups	water	425 mL
1 (10-ounce)	can beef broth	1 (284 mL)
1¾ cups	elbow macaroni	425 mL
3 tablespoons	flour	45 mL
2 cups	milk	500 mL
⅓ cup	grated Canadian Parmesan cheese	75 mL
2 tablespoons	chopped parsley	30 mL

Melt butter in a large frypan. Sauté onion and green pepper 5 minutes. Add meat to pan and continue cooking until meat is lightly browned. Blend in salt, oregano and garlic powder. Stir in water and beef broth. Add macaroni. Bring to a boil. Reduce heat; cover and simmer 15 to 20 minutes or until macaroni is tender. Blend in flour. Gradually stir in milk. Cook over medium heat, stirring constantly, until mixture just comes to a boil and thickens. Stir in cheese and parsley.

Makes 6 servings.

TENDER PORK AU LAIT

2 pounds	boneless pork stew	1 kg
1 teaspoon	salt	5 mL
½ teaspoon	pepper	2 mL
¾ cup	milk	175 mL
7	slices white bread, cubed	7
3 tablespoons	onion, chopped	45 mL
3 tablespoons	butter, melted	45 mL
½ teaspoon	salt	2 mL
¼ teaspoon	pepper	1 mL
¼ teaspoon	celery salt	1 mL
2 tablespoons	milk	30 mL

Pound pork pieces until thin. Season both sides with salt and pepper. Arrange on the bottom of an 8″x 8″ (20 cm x 20 cm) baking pan. Pour the ¾ cup (175 mL) milk over the pork. Combine bread cubes, onion, butter, salt, pepper and celery salt. Moisten with the 2 tablespoons (30 mL) of milk, and arrange evenly on top of pork. Cover and bake at 350°F (180°C) for 70 minutes. Uncover, and cook for 10 minutes longer.

Makes 6 servings.

Versatile Freezer Pork Dinners

BASIC PORK MIXTURE

2 pounds	lean boneless pork shoulder	1 kg
3 tablespoons	cooking oil	45 mL
1 cup	water	250 mL
1½ cups	chopped onion	375 mL
¾ cup	chopped celery	175 mL
2	cloves garlic, minced	2
½ teaspoon	salt	2 mL

Cut meat into ½-inch (12 mm) cubes. In a large saucepan sauté meat in oil, part at a time until browned. Return all to saucepan and add water, onion, celery, garlic and salt. Bring to a boil. Reduce heat; cover and simmer 1 hour, or until meat is tender. Cool. Divide into two 2 cup (500 mL) portions. Use one portion to make Curried Pork or Deep Dish Pork Pie. Freeze remaining portion; thaw and use as needed.

Makes 4 cups/1L.

CURRIED PORK

2 tablespoons	butter	30 mL
2 tablespoons	flour	30 mL
1 teaspoon	curry powder	5 mL
1 teaspoon	chicken bouillon mix	5 mL
¼ teaspoon	salt	1 mL
1½ cups	milk	375 mL
1 (2 cup)	container Basic Pork Mixture, thawed	1 (500 mL)

Melt butter in a saucepan. Blend in flour, curry powder, bouillon mix and salt. Gradually stir in milk. Cook over medium heat, stirring constantly, until mixture just comes to a boil and thickens. Add container of pork mixture and heat through. Serve over hot cooked rice.

Makes 4 servings.

continued on next page

VERSATILE FREEZER PORK (continued)

DEEP DISH PORK PIE

1½ cups	cubed potatoes	375 mL
1 cup	sliced carrots	250 mL
3 tablespoons	flour	45 mL
½ teaspoon	salt	2 mL
¼ teaspoon	ground thyme	1 mL
¼ teaspoon	ground marjoram	1 mL
2 cups	milk	500 mL
1 (2 cup)	container Basic Pork Mixture, thawed	1 (500 mL)
1 cup	frozen peas	250 mL
	sufficient pastry for single crust pie	

Cook potatoes and carrots in small amount of boiling salted water; drain. Smoothly combine flour, salt, thyme, marjoram and milk. Stir into Basic Pork Mixture. Cook over medium heat, stirring constantly, until mixture just comes to a boil and thickens. Add potatoes, carrots and peas. Spoon into 6-cup (1.5 L) shallow casserole. Roll out pastry to fit the surface of casserole. Fit pastry over meat and vegetable mixture; press edges with fork to seal. Make several slits in crust and bake in preheated 400°F (200°C) oven 30 to 40 minutes or until pastry is brown and pie heated through.

Makes 4 servings.

TURKEY BUFFET CASSEROLE

2	cups	medium egg noodles	125 g
1	(300 g)	frozen broccoli spears	1 (300 g)
3	tablespoons	butter	45 mL
3	tablespoons	all-purpose flour	45 mL
1	teaspoon	salt	5 mL
¼	teaspoon	prepared mustard	1 mL
¼	teaspoon	pepper	1 mL
2	cups	milk	500 mL
1	cup	shredded Cheddar cheese	250 mL
2	cups	cut-up cooked turkey	500 mL
½	cup	slivered, toasted almonds	125 mL

Cook noodles in boiling salted water until tender; drain. Cook broccoli as label directs until just tender. Drain, dice broccoli stems but leave florets whole. Meanwhile, in saucepan over low heat melt butter, stir in flour, salt, mustard, pepper and milk. Cook stirring constantly, until thick and smooth. Remove from heat and stir in cheese until melted. In greased shallow casserole or 8″ (2L) square baking dish, arrange noodles, broccoli stems and turkey, pour cheese sauce over all. Arrange broccoli florets on top, pressing them lightly into sauce; sprinkle with almonds. Bake uncovered, at 350°F (180°C) for 15 minutes, or until bubbling hot.

Makes 4 to 6 servings.

TURKEY CASSEROLE

6	slices bacon	6
¼ cup	flour	50 mL
1 teaspoon	salt	5 mL
¼ teaspoon	pepper	1 mL
3 cups	milk	750 mL
¼ pound	smoked Gruyere cheese, cubed	125 g
2 cups	cooked turkey, diced	500 mL
½	green pepper, chopped	½
3 cups	cooked rice	750 mL
	paprika	

Cook bacon until crisp. Remove slices and crumble. Discard all but ¼ cup (50 mL) of bacon fat from frypan and stir in the flour, salt and pepper. Gradually add the milk, stirring constantly over low heat, until smooth and thickened. Add the cheese and stir until melted. Stir in the turkey and green pepper, and heat thoroughly. Stir in the rice and bacon. Pour the mixture into a greased 8-cup (2 L) casserole. Sprinkle top with paprika. Bake at 325°F (160°C) for 40 minutes.

Makes 6 servings.

TURKEY DIVAN

2 (10 ounce)	boxes frozen broccoli spears, thawed	2 (284 g)
3 cups	cooked turkey, diced	750 mL
¾ cup	milk	175 mL
1 (10 ounce)	can condensed cheddar cheese soup	1 (284 mL)
1 tablespoon	butter, melted	15 mL
⅓ cup	dry bread crumbs	75 mL
½ teaspoon	paprika	2 mL

Place thawed broccoli on bottom of greased 9"x 13" (23 cm x 33 cm) baking dish. Top with cooked turkey. Blend milk and condensed cheese soup. Pour over the turkey. Combine butter, breadcrumbs and paprika. Sprinkle over sauce. Bake at 350°F (180°C) for 35 minutes, or until heated through.

TURKEY FRENCH TOAST SANDWICH

1 cup	cooked turkey, finely chopped	250 mL
2 tablespoons	onion, finely chopped	30 mL
¼ cup	celery, finely diced	50 mL
¼ cup	mayonnaise	50 mL
½ teaspoon	prepared mustard	2 mL
½ teaspoon	salt	2 mL
dash	pepper	dash
8	slices bread	8
3	eggs, beaten	3
6 tablespoons	milk	90 mL
2 tablespoons	butter	30 mL

For filling, combine turkey, onion, celery, mayonnaise, mustard, salt and pepper. Spread bread evenly with filling to make 4 sandwiches. Cut sandwiches in half. Combine eggs and milk. Dip each half into egg mixture and fry in butter until golden brown (2 to 3 minutes each side). Top with cheese sauce.

CHEESE SAUCE

2 tablespoons	butter, melted	30 mL
2 tablespoons	flour	30 mL
1 cup	milk	250 mL
1 cup	Cheddar cheese, grated	250 mL

Combine butter and flour. Add milk gradually, stirring constantly over low heat, until smooth and thickened. Add cheese and stir until melted. Pour over sandwiches.
Makes 4 servings.

TURKEY LEFTOVERS WITH ALMONDS

½ cup	butter, melted	125 mL
3	onions, finely chopped	3
3 cups	mushrooms, sliced	750 mL
⅔ cup	flour	150 mL
½ teaspoon	salt	2 mL
⅛ teaspoon	pepper	0.5 mL
¾ teaspoon	curry powder	4 mL
¾ teaspoon	dry mustard	4 mL
dash	cayenne	dash
3 cups	chicken or turkey broth	750 mL
1 cup	milk	250 mL
6 cups	cooked turkey, diced	1.5 L
1 cup	blanched almonds, sliced & toasted	250 mL
½ cup	sherry	125 mL
2 tablespoons	chopped parsley	30 mL

Sauté onions and mushrooms in butter until soft. Blend in flour, salt, pepper, curry powder, mustard and cayenne. Gradually add the broth and milk and cook over medium heat, stirring constantly, until smooth and thickened. Add turkey and almonds and blend well. Reheat to serving temperature. Add sherry and sprinkle with chopped parsley. Serve with Chinese noodles or fluffy rice.

Makes 8 servings.

Turkey noodle casserole

⅓ cup	onion, chopped	75 mL
⅓ cup	celery, diced	75 mL
¼ cup	green pepper, diced	50 mL
2 cups	mushrooms, sliced	500 mL
3 tablespoons	butter	45 mL
3 tablespoons	flour	45 mL
1 teaspoon	salt	5 mL
⅛ teaspoon	pepper	0.5 mL
dash	cayenne	dash
2 cups	milk	500 mL
1 cup	turkey broth	250 mL
2½ cups	cooked turkey, diced	625 mL
2 tablespoons	pimento, diced	30 mL
2½ cups	dry broad noodles	625 mL
1 tablespoon	butter, melted	15 mL
¼ cup	dry breadcrumbs	50 mL
¼ cup	almonds, blanched and slivered	50 mL

Sauté onions, celery, green pepper and mushrooms in butter for 5 minutes until soft. Stir in flour, salt, pepper and cayenne. Combine milk and turkey broth, and gradually add to flour mixture. Cook, stirring constantly, until smooth and thickened. Meanwhile, boil noodles in a large quantity of salted water. Drain thoroughly. Add turkey, pimento and noodles to vegetable mixture. Pour into a greased 9"x 13" (23 cm x 33 cm) baking dish. Combine melted butter, breadcrumbs and almonds. Sprinkle over top. Bake at 350°F (180°C) for 50 to 60 minutes, or until browned and bubbling.

Makes 6 servings.

TURKEY PASTRY ROLLS

FILLING

3 tablespoons	butter, melted	45 mL
3 tablespoons	flour	45 mL
1½ cups	milk	375 mL
2	egg yolks	2
2 cups	cooked turkey, diced	500 mL
1½ teaspoons	parsley flakes	7 mL
5 teaspoons	grated onion	25 mL
pinch	ground ginger	pinch
½ teaspoon	lemon juice	2 mL
¼ teaspoon	salt	1 mL
dash	pepper	dash

PASTRY

1½ cups	flour	375 mL
1 teaspoon	salt	5 mL
⅔ cup	butter	150 mL
	cold water and milk	

For filling, combine butter with flour in a saucepan. Slowly add milk, stirring constantly. Cook over medium heat until smooth and thickened. Remove from heat and add egg yolks, one at a time, blending well after each addition. Add turkey, parsley, onion, ginger, lemon juice, salt and pepper. Cook for 10 minutes, stirring frequently. Remove from heat and let cool. For pastry, combine flour and salt. Cut in butter with a pastry blender. Add enough water to make a soft dough. On a lightly floured surface, roll out dough ⅛" (3 mm) thick, and cut into 5" (12 cm) squares. Spread each square with filling and roll up in egg-roll fashion. Pinch ends together and place on a greased pan. Brush with milk. Bake at 400°F (200°C) for 30 minutes.
Makes 6 servings.

TURKEY PIE

1½ cups	milk	375 mL
1½ cups	broth	375 mL
½ cup	flour	125 mL
¾ cup	peas, canned or frozen and thawed	175 mL
½ cup	carrots, grated	125 mL
3 cups	cooked turkey, diced	750 mL
¼ cup	onion, chopped	50 mL
1½ teaspoons	paprika	7 mL
2	unbaked 9" (23 cm) pie shells	2

In a large saucepan, combine milk, broth and flour. Cook, stirring constantly, until smooth and thickened. Add peas, carrots, turkey, onions and paprika. Pour into pie shells and bake at 425°F (220°C) for 15 minutes. Reduce heat and continue baking at 350°F (180°C) for 30 minutes longer. Serve warm.

Makes 8-10 servings.

Turkey Pot Pie with Herbed Dumpling Crust

⅓ cup	butter	75 mL
½ pound	mushrooms, quartered	250 g
¼ cup	flour	50 mL
1 cup	chicken stock	250 mL
2 cups	milk	500 mL
½ teaspoon	dried thyme	2 mL
¼ teaspoon	Tabasco sauce	1 mL
4 cups	diced cooked turkey (or chicken or ham)	1 L
¼ cup	diced pimento	50 mL
1	package frozen peas and carrots, defrosted or (2½ cups cooked vegetables 625 mL)	1 (300 g)

HERB DUMPLING CRUST

1 cup	all purpose flour	250 mL
2 teaspoons	baking powder	10 mL
1 tablespoon	chopped fresh parsley (or ½ teaspoon (2 mL) dried)	15 mL
1 tablespoon	chopped fresh dill (or ½ teaspoon (2 mL) dried	15 mL
¼ teaspoon	salt	1 mL
⅓ cup	butter, cold	75 mL
½ cup	milk, cold	125 mL
2 tablespoons	milk	30 mL

Melt butter in a large saucepan. Add mushrooms. Cook a few minutes. Sprinkle with flour. Cook five minutes, do not brown. Whisk in stock and milk. Bring to a boil. Reduce heat. Add seasonings and salt and pepper to taste. Simmer gently, stirring occasionally, 10 minutes. Add turkey, pimento, carrots and peas to sauce. Preheat oven to 400°F (200°C). Butter a 3 quart (3 L) casserole dish. Prepare dumpling crust by combining flour with baking powder, parsley, dill, and salt. Cut in ⅓ cup (75 mL) butter until it is in tiny bits. Sprinkle mixture with milk. Gather together to form a dough. Roll dough on floured surface to fit the top of the casserole. Spoon turkey mixture into casserole. Place dough directly on top of turkey. Cut steam slits in dough. Brush with remaining 2 tablespoons (30 mL) milk. Bake 30 to 35 minutes.
Makes 6 servings.

TURKEY SWEET POTATO COMBO

3 cups	cooked turkey, diced	750 mL
3 cups	cooked turkey stuffing	750 mL
2	eggs, beaten	2
½ teaspoon	salt	2 mL
¼ teaspoon	pepper	1 mL
1 cup	milk	250 mL
4 tablespoons	soft butter	60 mL
1 (19 ounce)	tin sweet potatoes, drained and mashed	1 (540 mL)
	milk	
	nutmeg	

Combine turkey and stuffing and place in a greased 6-cup (1.5 L) casserole. Combine eggs, salt, pepper and milk. Pour over turkey mixture. Beat 3 tablespoons (45 mL) butter into mashed sweet potatoes. Add enough milk to make light and fluffy. Spread over turkey and dot with remaining tablespoon (15 mL) butter. Lightly sprinkle nutmeg on top. Bake at 375°F (190°C) for 45 minutes.

Makes 6 servings.

TURKEY TETRAZZINI

6 tablespoons	butter, melted	90 mL
4 tablespoons	flour	60 mL
1 teaspoon	salt	5 mL
¼ teaspoon	pepper	1 mL
dash	nutmeg	dash
2 ½ cups	milk	625 mL
2	green onions, chopped	2
¼ cup	Parmesan cheese, grated	50 mL
4 cups	cooked spaghetti	1 L
2 cups	cooked turkey, diced	500 mL
¼ cup	buttered breadcrumbs	50 mL

In a heavy saucepan, combine butter, flour, salt, pepper and nutmeg. Gradually add the milk, stirring constantly. Continue cooking over medium heat until smooth and thickened. Add green onions and cheese. Stir until cheese is melted. Fold in cooked spaghetti and turkey. Pour mixture into a greased 6-cup (1.5 L) casserole. Sprinkle crumbs over the top. Bake at 350°F (180°C) until bubbly and golden, about 30 minutes.

Makes 6 servings.

NOTES

ASPARAGUS CONTINENTAL

2 (12 ounce)	tins asparagus spears	2 (341 mL)
½ cup	mushrooms, sliced	125 mL
4 tablespoons	butter, melted	60 mL
4 tablespoons	flour	60 mL
1½ cups	milk	375 mL
2 tablespoons	pimento juice	30 mL
2	eggs, separated	2
¼ cup	pimento, diced	50 mL
½ teaspoon	prepared mustard	2 mL
½ cup	croutons	125 mL
½ cup	sharp Cheddar cheese, grated	125 mL

Drain asparagus spears, reserving ½ cup (125 mL) of the juice. Place asparagus spears on bottom of a 10″ (25 cm) flan dish. Sauté mushrooms in butter until tender, about 5 minutes. Stir in flour. Gradually add the milk and asparagus juice, stirring constantly, and continue to cook over medium heat until smooth and thickened. Add pimento juice and egg yolks and cook 1 minute longer. Remove from heat. Stir in diced pimento, and pour over asparagus. Beat egg whites until stiff but not dry. Fold in mustard, and spoon mixture over sauce. Sprinkle with croutons and cheese. Bake, uncovered, at 350°F (180°C) for 30 minutes.

Makes 8 servings.

ASPARAGUS CASHEW CASSEROLE

3	tablespoons	butter	45 mL
1	cup	chopped onion	250 mL
2½	cups	parboiled asparagus cut in	625 mL
		1½" (3.5 cm) lengths	
1	cup	shredded Canadian	250 mL
		Cheddar cheese	
2	tablespoons	butter	30 mL
2	tablespoons	all-purpose flour	30 mL
2	cups	milk	500 mL
½	teaspoon	salt	2 mL
pinch		pepper	pinch
⅓	cup	cashews	75 mL

Melt 3 tablespoons butter (45 mL) in saucepan. Sauté onion until tender. Place half of onions in a greased 2 quart (2 L) rectangular casserole. Cover with half of asparagus and half of cheese.

Repeat onion and asparagus layers. Reserve cheese. Melt 2 tablespoons butter (30 mL) in a medium saucepan. Blend in flour. Add milk slowly, stirring constantly until thickened and smooth. Season; pour over vegetables.

Bake in a 400°F (200°C) oven for 20 minutes.

Sprinkle with remaining cheese and nuts during last 10 minutes.

Makes 4 to 6 servings.

Cranberry Nut Stuffing,
page 367

BAKED WINTER SQUASH RING

2 pounds	butternut squash (or acorn, hubbard), cooked	1 kg
1 pound	carrots, cooked	500 g
3 tablespoons	brown sugar	45 mL
½ teaspoon	cinnamon	2 mL
¼ teaspoon	nutmeg	1 mL
¾ teaspoon	ground ginger	4 mL
5	eggs	5
1 cup	milk	250 mL
1 cup	fresh breadcrumbs – divided	250 mL
1 teaspoon	salt	5 mL
¼ teaspoon	black pepper	1 mL

Purée squash. Measure out 2 cups (500 mL). Purée carrots. Measure out 1 cup (250 mL) and add to squash. Beat in brown sugar, cinnamon, nutmeg and ginger. Beat eggs with milk; add to squash. Reserve 3 tablespoons (45 mL) breadcrumbs; add remaining breadcrumbs to squash with salt and pepper. Butter a 6 cup (1.5 L) ring mold. Sprinkle with reserved breadcrumbs. Pour squash mixture into the pan. Cover with buttered aluminum foil or waxed paper. Bake about one hour in a water bath (a larger pan half-filled with very hot water) or until a knife inserted comes out clean. Rest 10 minutes before unmolding. To unmold run a knife around edge and invert on a serving platter.

Note: Centre of ring can be filled with cooked vegetables.

Makes 6-8 servings.

BREADED CAULIFLOWER
WITH SOUR CREAM SAUCE

1	cauliflower, broken into flowerets	1
	salt	
2	egg yolks, beaten	2
¾ cup	sour cream	175 mL
¼ cup	whipping cream	50 mL
1 tablespoon	lemon juice	15 mL
dash	salt	dash
dash	sugar	dash
⅔ cup	dry breadcrumbs	150 mL
½ teaspoon	salt	2 mL
dash	pepper	dash
2	eggs	2
¼ cup	milk	50 mL
2 cups	vegetable oil for frying	500 mL

Wash the flowerets, and let stand in cold salted water for
1 hour. Rinse. Boil in salted water until cooked but still firm. Drain
well. Meanwhile, prepare the sauce by mixing egg yolks, sour
cream, whipping cream, lemon juice, dash of salt and sugar. Cook
in top part of double boiler for 5 minutes, whipping constantly.
Set sauce aside, but keep warm. Mix together breadcrumbs, ½
teaspoon (2 mL) salt, and pepper. In a separate bowl, mix together
the whole eggs and milk. Dip the drained cauliflower into egg
mixture, then into crumb mixture. Heat the oil to 365°F (185°C).
Carefully fry the cauliflower, a few at a time until golden brown.
Drain and keep warm. When complete, place in a heated serving
dish, and serve with sauce.
Makes 4-6 servings.

Broccoli a la Suisse

4 cups	coarsely-chopped broccoli	1 L
	water	
3 tablespoons	butter	45 mL
3 tablespoons	chopped onion	45 mL
2 tablespoons	flour	30 mL
1 teaspoon	salt	5 mL
1¼ cups	milk	300 mL
2 cups	shredded Canadian Swiss cheese	500 mL
2	eggs, beaten	2

Cook broccoli in a small amount of boiling water until crisp-tender. Drain well; set aside. Melt butter in a large saucepan. Sauté onion until tender. Blend in flour and salt. Gradually stir in milk. Cook over medium heat, stirring constantly, until mixture just comes to a boil and thickens. Remove from heat. Stir in broccoli and cheese until cheese melts slightly. Stir in eggs. Pour into a greased 1½-quart (1.5 L) shallow rectangular baking dish. Bake in preheated 325°F (160°C) oven 30 minutes or until firm to the touch.

Makes 6 servings.

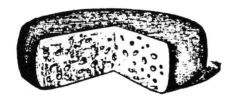

Broccoli cheese delight

1 pound	fresh broccoli	500 g
3	medium onions, quartered	3
¼ cup	butter, melted	50 mL
2 tablespoons	dry breadcrumbs	30 mL
2 tablespoons	flour	30 mL
1 cup	milk	250 mL
1 (4 ounce)	package cream cheese, cubed	1 (125 g)
½ cup	Cheddar cheese, grated	125 mL
dash	pepper	dash
¼ cup	dry breadcrumbs	50 mL

Cook broccoli and onions together in a small amount of water until tender, about 5 minutes. Drain. In a small saucepan, combine butter, flour and 2 tablespoons (30 mL) dry breadcrumbs. Slowly add milk and cook over medium heat, stirring constantly, until smooth and thickened. Add cheeses and pepper, and stir until melted. Arrange broccoli and onions in an 8"x 8" (20 cm x 20 cm) baking pan. Cover with sauce, and sprinkle with ¼ cup (50 mL) breadcrumbs. Refrigerate for one hour. Bake, covered, at 350°F (180°C) for 30 minutes.

Makes 6 servings.

Buffet Potato Casserole

6	large potatoes, peeled and cut up	6
½ cup	butter	125 mL
1½ cups	shredded Canadian Swiss cheese	375 mL
2 tablespoons	finely-chopped green onion	30 mL
1½ teaspoons	salt	7 mL
pinch	pepper	pinch
1¼ cups	milk	300 mL
¾ cup	fine dry bread crumbs	175 mL
⅓ cup	grated Canadian Parmesan cheese	75 mL
2 tablespoons	chopped parsley	30 mL
3 tablespoons	butter, melted	45 mL

Cook potatoes in a small amount of boiling salted water. Drain and break up into small pieces with a fork. Add ½ cup (125 mL) butter, Swiss cheese, onion, salt and pepper. Toss with a fork to combine. Spread mixture in a 2-quart (2 L) shallow rectangular baking dish. Pour milk over potato mixture. Combine bread crumbs, Parmesan cheese, parsley and 3 tablespoons (45 mL) melted butter. Sprinkle over potatoes. Bake in preheated 375°F (190°C) oven 20 to 25 minutes or until hot and topping has browned.

Makes 6 servings.

CABBAGE FRUIT SALAD

DRESSING

2 tablespoons	flour	30 mL
2 tablespoons	packed brown sugar	30 mL
1 teaspoon	salt	5 mL
1 teaspoon	dry mustard	5 mL
½ teaspoon	powdered ginger	2 mL
dash	cayenne	dash
1 cup	milk	250 mL
1	egg, beaten	1
2 tablespoons	white wine vinegar	30 mL
1 cup	creamy yogurt, unflavoured	250 mL
2 tablespoons	finely grated lemon rind	30 mL

For dressing, combine flour, sugar, salt, mustard, ginger and cayenne. Combine milk and egg. Gradually stir into flour mixture. Heat to boiling, stirring constantly and boil 1 minute. Stir in vinegar and chill. Stir in yogurt.

SALAD

4 cups	cabbage, coarsely shredded	1 L
2	seedless oranges, sectioned	2
1	red apple, diced	1
½ cup	chopped salted peanuts	125 mL

For salad, reserve a few orange sections for garnish. Cut remaining orange sections in half. Toss cabbage, oranges, apple and peanuts together. Just before serving, fold in 1 cup of dressing. Serve additional dressing separately. Garnish with reserved orange sections. (Leftover dressing may be stored up to 2 weeks in the refrigerator.)

Makes 6 servings.

CABBAGE RICE DUO

3 cups	cooked rice	750 mL
7	strips bacon	7
3 tablespoons	bacon fat	45 mL
½	medium cabbage, coarsely chopped	½
1 cup	onions, finely chopped	250 mL
1	clove garlic, minced	1
¾ cup	milk	175 mL

Cook bacon strips until done but not crisp. Reserve three slices for garnish. Crumble remainder and combine with cooked rice. Sauté cabbage, onions and garlic in bacon fat until soft, about 10 minutes. In a greased 6-cup (1.5 L) casserole, layer a third of the rice, then a third of the cabbage. Repeat layers twice more. Cut reserved bacon strips in half and use for garnish. Bake, uncovered, at 350°F (180°C) for 1 hour.

Makes 8 servings.

CAULIFLOWER CASSEROLE

1	medium cauliflower	1
4 tablespoons	butter, melted	60 mL
½ pound	fresh mushrooms, sliced	250 g
4 tablespoons	flour	60 mL
1 teaspoon	salt	5 mL
2 cups	milk	500 mL
¼ cup	Cheddar cheese, grated	50 mL
¼ cup	cracker crumbs	50 mL

Break cauliflower into small pieces and simmer in a small amount of water for 15 minutes. Drain well. Sauté mushrooms in butter until soft, about 5 minutes. Blend in flour and salt. Add milk gradually, stirring constantly, and continue to cook over medium heat until smooth and thickened. Arrange cauliflower pieces in baking dish and pour sauce over it. Combine cheese and crumbs. Sprinkle over top. Bake at 350°F (180°C) for 15-20 minutes, or until heated through.

Makes 6 servings.

CELERY A LA SUISSE

6 cups	celery, cut diagonally in 1″ (2.5 cm) pieces	1.5 L
4 tablespoons	butter, melted	60 mL
4	green onions, chopped stems and bulbs	4
4 tablespoons	flour	60 mL
2	packets instant chicken bouillon powder	2
2 cups	milk	500 mL
1 tablespoon	pimento, chopped	15 mL
½ cup	Swiss cheese, grated	125 mL

Simmer celery in a small amount of water for 5 minutes. Drain and set aside. Sauté onions in butter for 3 minutes until soft. Stir in flour and bouillon powder. Add milk gradually, and continue to cook, stirring constantly, over medium heat until smooth and thickened. Remove from heat and add pimento and cheese, stirring until cheese is melted. Stir in celery. Place in greased 8″x 8″ (20 cm x 20 cm) baking pan. Bake at 350°F (180°C) for 15 minutes until hot and bubbly.

Makes 6 servings.

CHEESE SCALLOPED CARROTS

12	medium carrots, peeled and sliced	12
4 tablespoons	butter, melted	60 mL
1	small onion, finely chopped	1
4 tablespoons	flour	60 mL
1 teaspoon	salt	5 mL
$\frac{1}{4}$ teaspoon	dry mustard	1 mL
$\frac{1}{8}$ teaspoon	pepper	0.5 mL
$\frac{1}{4}$ teaspoon	celery salt	1 mL
2 cups	milk	500 mL
2 cups	Mozzarella cheese, grated	500 mL
$\frac{1}{2}$ cup	dry breadcrumbs	125 mL
1 tablespoon	butter, melted	15 mL

Cook carrots in boiling water until just tender. Drain. Meanwhile sauté onion in 4 tablespoons (60 mL) butter until soft, about 5 minutes. Stir in flour, salt, mustard, pepper and celery salt. Gradually add milk and continue to cook, stirring constantly, over medium heat until smooth and thickened. Remove from heat, add cheese, and stir until melted. Stir in carrots and pour into a greased 6-cup (1.5 L) casserole. Combine crumbs and 1 tablespoon (15 mL) butter and sprinkle over carrots. Bake at 350°F (180°C) for 45 minutes.

Makes 6 servings.

CHEESY CABBAGE BAKE

1	small cabbage, coarsely shredded	1
2 tablespoons	butter, melted	30 mL
2 tablespoons	flour	30 mL
2 cups	milk	500 mL
$\frac{1}{4}$ teaspoon	celery salt	1 mL
$\frac{1}{4}$ teaspoon	pepper	1 mL
$\frac{1}{2}$ teaspoon	oregano	2 mL
$\frac{1}{8}$ teaspoon	nutmeg	0.5 mL
1 teaspoon	prepared mustard	5 mL
$\frac{1}{4}$ teaspoon	Worcestershire sauce	1 mL
6	drops Tabasco sauce	6
1 cup	Swiss cheese, grated	250 mL
$\frac{1}{2}$ cup	dry breadcrumbs	125 mL
1 tablespoon	butter, melted	15 mL

Cook cabbage in a small amount of water until tender. Drain and set aside. Combine 2 tablespoons (30 mL) butter with flour. Add milk gradually, and cook over medium heat, stirring constantly, until smooth and thickened. Add celery salt, pepper, oregano, nutmeg, mustard, Worcestershire sauce and Tabasco sauce. Bring to boil and simmer 2 minutes, stirring constantly. Remove from heat and add cheese. Stir until melted. Stir in the cooked cabbage. Turn into a greased 8″ square (20 cm²) baking pan. Combine breadcrumbs and 1 tablespoon butter (15 mL) and sprinkle over top of cabbage. Bake at 350°F (180°C) for 30 minutes or until hot and bubbly.

Makes 8 servings.

COMPANY BROCCOLI SOUFFLE

1 (10 ounce)	box frozen broccoli, cooked and drained	1 (300 g)
1½ cups	milk	375 mL
¾ teaspoon	salt	3 mL
¼ teaspoon	pepper	1 mL
¼ teaspoon	nutmeg	1 mL
2 tablespoons	soft butter	30 mL
3 tablespoons	flour	45 mL
6	eggs, separated	6

Purée broccoli in blender or food processor until smooth. Place in a saucepan. In blender, place milk, salt, pepper, nutmeg, butter, flour and egg yolks, and blend until smooth. Add to the broccoli. Cook over low heat, stirring constantly, until smooth. Cool slightly. Beat egg whites until stiff but not dry. Gently fold into broccoli mixture. Pour into greased 6-cup (1.5L) casserole. Bake at 350°F (180°C) for 30 minutes until browned and puffed.

Makes 6 servings.

CRANBERRY NUT STUFFING

½ cup	butter	125 mL
2	medium onions, chopped	2
2	celery stalks, chopped	2
6 cups	small bread cubes	1.5 L
½ cup	chopped nuts	125 mL
1½ teaspoons	poultry seasoning	7 mL
1 teaspoon	salt	5 mL
pinch	pepper	pinch
1½ cups	fresh cranberries*	375 mL
3	eggs	3
1 cup	milk	250 mL

Melt butter in a large frypan. Sauté onion and celery until tender. Add bread cubes, nuts, poultry seasoning, salt and pepper. Toss lightly to combine. Stir in cranberries. Beat eggs well; stir in milk. Pour over bread mixture. Mix lightly until all the bread is moistened. Spoon into a greased 6-cup (1.5 L) shallow rectangular casserole or loaf pan. Bake in preheated 325°F (160°C) oven 45 to 50 minutes.

Makes 6 servings.

* 1½ cups (375 mL) of frozen cranberries, thawed, may be substituted.

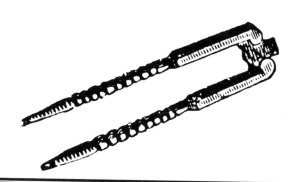

CREAMED CELERY WITH PECANS

4	cups	celery cut diagonally in ½″ (12 mm) pieces	1 L
2	tablespoons	butter	30 mL
2	tablespoons	all-purpose flour	30 mL
2	cups	milk	500 mL
1	teaspoon	salt	5 mL
¾	cup	pecan halves	175 mL
¼	cup	fresh bread crumbs	50 mL

Boil celery until tender in enough water to cover, then drain. Melt butter over medium heat, stir in flour and add milk slowly to make cream sauce, stirring until thick and smooth. Add salt and well-drained celery. Spoon into greased 1½ quart (1.5 L) casserole dish. Top with pecans and cover with bread crumbs. Bake in preheated 400°F (200°C) oven for 15 minutes.
Makes 4 to 6 servings.

CREAMY VEGETABLE CASSEROLE

10 cups	combination of carrots, broccoli, cauliflower and onions	2.5 L
2 tablespoons	butter, melted	30 mL
2 tablespoons	flour	30 mL
1 1/2 cups	milk	375 mL
1/4 teaspoon	salt	1 mL
dash	pepper	dash
1 (4 ounce)	package cream cheese, cubed	1 (125 g)
1/4 cup	Cheddar cheese, grated	50 mL
1/2 cup	dry breadcrumbs	125 mL
1 tablespoon	butter, melted	15 mL
1/4 cup	Parmesan cheese, grated	50 mL

Separate cauliflower flowerets. Cut broccoli heads and stems into bite-sized portions. Slice carrots and onions. Boil the vegetables gently about 10 minutes, until tender. Drain well. In a saucepan over medium heat, combine 2 tablespoons (30 mL) butter and flour. Add milk gradually, stirring constantly, and continue to heat until smooth and thickened. Add salt, pepper, cream cheese and Cheddar cheese. Stir until melted and smooth. Stir in vegetables and place in 2 greased 6-cup (1.5 L) casseroles. Combine breadcrumbs with 1 tablespoon (15 mL) melted butter and Parmesan cheese and sprinkle over top of casseroles. Bake, uncovered, at 350°F (180°C) for 35 minutes.

Makes 12 servings.

CREAMY ONIONS AND POTATOES

12	large potatoes, peeled and sliced very thinly	12
4	medium onions, sliced	4
1	clove garlic, minced	1
2 tablespoons	butter, melted	30 mL
1½ cups	milk	375 mL
⅓ cup	flour	75 mL
2½ teaspoons	salt	12 mL
½ teaspoon	pepper	2 mL
¼ cup	fresh parsley, chopped	50 mL

Parboil potatoes for 2 minutes in salted water. Drain well and place in a greased 9"x 13" (23 cm x 33 cm) baking pan. Sauté onions and garlic in butter until soft, but not brown, about 5 minutes and place over potatoes. Combine milk, flour, salt, pepper and parsley. Pour over top of vegetables. Bake, uncovered, at 375°F (190°C) for 1 hour.

Makes 12 servings.

CREAMY BROCCOLI AND ZUCCHINI

2 cups	fresh broccoli, chopped	500 mL
2 cups	chopped zucchini, unpeeled	500 mL
½ cup	chopped onion	125 mL
1	clove garlic, minced	1
2 tablespoons	butter, melted	30 mL
3 tablespoons	flour	45 mL
½ teaspoon	salt	2 mL
2 tablespoons	chopped parsley	30 mL
½ teaspoon	oregano	2 mL
¾ cup	milk	175 mL
1½ cups	creamed cottage cheese	375 mL
½ pound	noodles, cooked, drained and buttered	250 g

Cook broccoli and zucchini until tender. Drain well. In a frypan, sauté onion and garlic in butter until soft, about 5 minutes. Blend in flour, salt, parsley and oregano. Add milk, stirring constantly, and continue cooking over medium heat until smooth and thickened. Add broccoli, zucchini and cottage cheese and blend well. Serve over hot noodles.

Makes 8 servings.

DEEP FRIED ONION RINGS

5 or 6	medium onions, peeled	5 or 6
1 cup	all-purpose flour	250 mL
1 teaspoon	baking powder	5 mL
½ teaspoon	dry parsley flakes	2 mL
½ teaspoon	salt	2 mL
¼ teaspoon	pepper	1 mL
1 cup	milk	250 mL
1	egg, beaten	1
	oil for deep fat frying	

Slice onions in ¼-inch (6 mm) thick slices and separate into rings. Set aside. Combine flour, baking powder, parsley flakes, salt and pepper in a large bowl. Add milk and egg. Beat until smooth. Carefully stir onion rings into batter until well coated. Heat oil to 375°F (190°C). With a fork, pick up one onion ring at a time. Fry in deep fat, a few at a time, until golden brown on both sides. Drain. Serve immediately.

Makes 8 servings.

DEEP DISH VEGETABLE PIE

6 cups	assorted vegetables, such as broccoli, carrots, onions, cauliflower, celery and cabbage, coarsely chopped	1.5 L
3 tablespoons	butter, melted	45 mL
3 tablespoons	flour	45 mL
1 teaspoon	salt	5 mL
¼ teaspoon	pepper	1 mL
½ teaspoon	curry powder	2 mL
1 cup	chicken bouillon	250 mL
1 cup	milk	250 mL
1 cup	Cheddar cheese, grated	250 mL

BISCUIT TOPPING

2 cups	flour	500 mL
4 teaspoons	baking powder	20 mL
1 teaspoon	salt	5 mL
¼ cup	shortening	50 mL
1 cup	old Cheddar cheese, grated	250 mL
1 cup	milk	250 mL

Parboil vegetables in a small amount of water until just tender (about 10 minutes). Meanwhile, add flour, salt, pepper and curry powder to melted butter. Combine milk and bouillon. Gradually add the liquid to flour mixture, and continue to cook over medium heat, stirring constantly, until smooth and thickened. Add cheese and stir until melted. Add drained vegetables. Pour into a 10" square (25 cm²) baking pan. For biscuit topping, mix flour, baking powder and salt together. Cut in shortening and cheese until mixture is the consistency of coarse crumbs. Add milk all at once, and stir vigorously until combined. Turn dough out on floured board, and knead lightly for 1 minute. Roll out approximately ½" (1 cm) thick, so that pastry will fit top of pan. Place over vegetables and make a few slashes in dough. Bake at 400°F (200°C) for 30 minutes, or until dough is lightly browned.

Makes 8 servings.

ELEGANT CARROT RING

8	medium carrots	8
2	small onions	2
½ teaspoon	salt	2 mL
⅛ teaspoon	pepper	0.5 mL
3	eggs, beaten	3
½ cup	flour	125 mL
2 cups	milk	500 mL
1½ cups	hot, buttered green peas	375 mL
	slivered almonds	

Cook carrots and onions in a small amount of water until tender. Drain and mash. Add salt and pepper. Combine eggs and flour together until well blended. Slowly add milk, and beat until smooth. Add mashed vegetables and mix well. Pour into a very well greased 6 cup (1.5 L) ring mold. Bake at 375°F (190°C) for 45 minutes or until set. Unmold onto heated serving platter. Fill centre with hot, buttered peas. Garnish with slivered almonds.

Makes 8 servings.

Fresh Asparagus Mimosa

2 pounds	fresh trimmed asparagus	1 kg
	water	

Place a collapsible steamer basket in a large saucepan. Add sufficient water to cover bottom of saucepan but not enough to touch bottom of steamer basket. Bring water to a boil. Place asparagus in basket. Cover pan and cook 3 to 8 minutes, depending on thickness of spears, or until asparagus is crisp-tender. Lift steamer basket from pan and serve immediately with MIMOSA SAUCE.

Makes 4 servings.

MIMOSA SAUCE

⅓ cup	mayonnaise or salad dressing	75 mL
3 tablespoons	flour	45 mL
1¼ teaspoons	salt	6 mL
1¾ cups	milk	425 mL
1½ teaspoons	grated lemon rind	7 mL
2 tablespoons	lemon juice	30 mL
1	hard-cooked egg, separated	1

Combine mayonnaise, flour and salt in a saucepan. Gradually stir in milk. Cook over medium heat, stirring constantly, until mixture just comes to a boil and thickens. Add lemon rind and juice. Sliver egg white and fold into sauce. Sieve egg yolk. Pour sauce over asparagus; garnish with egg yolk.

Makes 2⅓ cups/575 mL.

GARDEN LOAF

2 tablespoons	butter, melted	30 mL
½	green pepper, chopped	½
1	large onion, chopped	1
1	stalk celery, chopped	1
2	medium zucchini, chopped	2
1 tablespoon	soy sauce	15 mL
2	medium carrots, grated	2
3	eggs, beaten	3
½ cup	yogurt	125 mL
¼ cup	sour cream	50 mL
¾ cup	milk	175 mL
½ teaspoon	salt	2 mL
¼ teaspoon	pepper	1 mL
dash	cayenne	dash
¼ teaspoon	thyme	1 mL
1 cup	Cheddar cheese, grated	250 mL

Sauté green pepper, onion, celery and zucchini in butter until soft, about 10 minutes. Add soy sauce and let stand a few minutes to marinate. Add grated carrots. In a separate bowl, combine eggs, yogurt, sour cream, milk, salt, pepper, cayenne and thyme. Add half of the grated cheese and the sautéed vegetables. Pour into a well greased 9" x 5" (23 cm x 13 cm) loaf pan. Sprinkle remaining cheese on top. Bake at 375°F (190°C) for 40 minutes or until knife inserted in centre comes out clean. Let stand 20 minutes before turning out of pan.

Makes 8 servings.

GREEN BEANS WITH WATER CHESTNUTS

3 tablespoons	butter, melted	45 mL
½ cup	celery, chopped	125 mL
3 tablespoons	flour	45 mL
2 teaspoons	celery salt	10 mL
⅛ teaspoon	pepper	0.5 mL
1½ cups	milk	375 mL
2 (10 ounce)	boxes frozen green beans, thawed	2 (300 g)
1 (8 ounce)	can water chestnuts, drained and sliced	1 (227 mL)
¼ cup	Parmesan cheese, grated	50 mL
1 cup	canned French fried onions	250 mL

Sauté celery in butter until tender. Blend in flour, celery salt and pepper. Gradually add milk, stirring constantly, and continue to heat until smooth and thickened. Combine green beans and water chestnuts. Place in a greased 8"x 8" (20 cm x 20 cm) baking pan. Pour sauce over the vegetables and top with the Parmesan cheese. Bake, covered, at 375°F (190°C) for 45 minutes. Uncover, and sprinkle onions on top. Bake 10 minutes longer, or until onions are crisp.

Makes 8 servings.

Homemade Potato Salad

DRESSING

1 teaspoon	salt	5 mL
2 teaspoons	dry mustard	10 mL
2 tablespoons	brown sugar	30 mL
2 tablespoons	all purpose flour	30 mL
pinch	cayenne pepper	pinch
2 cups	milk	500 mL
3	egg yolks	3
3 tablespoons	butter	45 mL
⅓ cup	vinegar	75 mL
1 teaspoon	white horseradish	5 mL
1 tablespoon	Dijon mustard (optional)	15 mL

SALAD

3 pounds	new potatoes or red potatoes	1.5 kg
4	green onions, thinly sliced	4
1	large red onion, thinly sliced	1
¾ cup	black olives, drained	175 mL
	salt and pepper	

Combine salt, mustard, brown sugar, flour, and cayenne in the top of a double boiler (or heavy saucepan). Add milk and egg yolks. Stir until smooth. Cook over boiling water (or medium heat), stirring constantly until thickened. Stir in butter and vinegar. Chill thoroughly. Scrub potatoes (or peel). Cut potatoes into 2" (5 cm) pieces. Place in cold water, bring to a boil, cook until tender. Drain well. Combine potatoes with green onions, red onions and olives. Stir horseradish and mustard into dressing. Season to taste. Add all or as much as you wish to potatoes. Combine well. Season to taste. Note: Refrigerated, this keeps well for up to 1 week.

Makes 6 to 8 servings.

Mixed Vegetables Italian Style

2 tablespoons	butter, melted	30 mL
1	medium onion, sliced	1
2	green peppers, diced	2
¼	cauliflower	¼
2	cloves garlic, minced	2
3	firm tomatoes, cut in wedges	3
2 tablespoons	butter, melted	30 mL
2 tablespoons	flour	30 mL
1 cup	milk	250 mL
½ teaspoon	oregano	2 mL
½ cup	Parmesan cheese, grated	125 mL
½ teaspoon	salt	2 mL
¼ teaspoon	pepper	1 mL

Sauté the onion, green pepper, cauliflower and garlic in 2 tablespoons (30 mL) butter until the onion is soft, about 5 minutes. Add tomatoes, and stir 1 minute longer. Set aside. In a medium pot, combine 2 tablespoons (30 mL) butter with flour. Gradually add milk, stirring constantly, and cook over medium heat until smooth and thickened. Add oregano, half of the cheese, salt and pepper and stir until cheese is melted. Stir the cooked vegetables into the sauce and place in a greased 6 cup (1.5 L) casserole. Sprinkle remaining cheese over the top. Bake at 350°F (180°C) for 30 minutes or until hot and bubbly.

Makes 6 servings.

MUSHROOM CREAMED PEAS

2 (10 ounce)	boxes frozen green peas	2 (284 g)
¼ cup	butter, melted	50 mL
2	green onions, chopped	2
½ pound	fresh mushrooms, sliced	250 g
2 tablespoons	flour	30 mL
¾ cup	milk	175 mL
1 teaspoon	sugar	5 mL
⅛ teaspoon	thyme	0.5 mL
⅛ teaspoon	nutmeg	0.5 mL
½ teaspoon	savory	2 mL
1 teaspoon	salt	5 mL
¼ teaspoon	pepper	1 mL

Cook green peas as directed on box. In a large skillet, sauté onions and mushrooms in butter until soft, about 5 minutes. Sprinkle with flour. Gradually add the milk, stirring constantly, and continuing to cook over medium heat until smooth and thickened. Add sugar, thyme, nutmeg, savory, salt and pepper, and simmer 2 minutes. Add peas and reheat to serving temperature.

Makes 8 servings.

Mushrooms Supreme

2 cups	sliced mushrooms	500 mL
¼ cup	onions, finely chopped	50 mL
2 tablespoons	butter, melted	30 mL
1 tablespoon	whole wheat flour	15 mL
½ cup	whole wheat flour	125 mL
3 tablespoons	butter	45 mL
2 cups	milk	500 mL
½ teaspoon	salt	2 mL
dash	pepper	dash
¼ teaspoon	nutmeg	1 mL
4	eggs, beaten	4
1½ cups	Cheddar cheese, grated	375 mL

Sauté mushrooms and onions in 2 tablespoons (30 mL) butter for 5 minutes until soft. Sprinkle with 1 tablespoon (15 mL) flour, stirring to coat evenly. Set aside. Combine ½ cup (125 mL) flour with 3 tablespoons (45 mL) butter in a large saucepan. Add milk slowly until evenly blended. Cook over medium heat, stirring constantly, until smooth and thickened. Remove from heat. Add salt, pepper, nutmeg, eggs and 1 cup (250 mL) of the cheese. Stir until cheese is melted. Put half of the sauce into a greased 6-cup (1.5 L) casserole. Spread with mushroom mixture. Cover with remaining sauce. Sprinkle remaining cheese over top. Bake at 375°F (190°C) for 45 minutes, or until puffy, firm and golden. Serve immediately. This is a rich, tasty accompaniment to a main course.

Makes 12 servings.

PARMESAN BROCCOLI

2 (10 ounce)	boxes frozen broccoli	2 (300 g)
¼ cup	onion, chopped	50 mL
3 tablespoons	butter, melted	45 mL
3 tablespoons	flour	45 mL
¼ teaspoon	salt	1 mL
¼ teaspoon	pepper	1 mL
1½ cups	milk	375 mL
⅓ cup	Parmesan cheese, grated	75 mL
¼ cup	dry breadcrumbs	50 mL
2 tablespoons	Parmesan cheese, grated	30 mL

Cook broccoli according to package directions. Drain well and place in an 8″ square (20 cm²) baking pan. Sauté onion in butter until soft. Sprinkle in flour, salt and pepper. Gradually add milk, stirring constantly, and continue cooking over medium heat until smooth and thickened. Remove from heat, add ⅓ cup (75 mL) cheese and stir until melted. Pour sauce over broccoli. Combine 2 tablespoons (30 mL) cheese and breadcrumbs and sprinkle over top. Bake at 375°F (190°C) for 15 minutes until browned.

Makes 6 servings.

PARMESAN SAUCED CAULIFLOWER

1	large cauliflower	1
	water	
3	slices bacon, chopped	3
2 tablespoons	finely-chopped green onion	30 mL
2 tablespoons	flour	30 mL
¾ teaspoon	salt	3 mL
2 cups	milk	500 mL
¾ cup	grated Canadian Parmesan cheese	175 mL

Cook trimmed cauliflower, whole, in boiling salted water until tender, about 20 minutes; drain well. Cook bacon in a saucepan until crisp. Drain, reserving 1½ tablespoons (25 mL) drippings; set bacon aside. Sauté onion in reserved drippings until tender. Blend in flour and salt. Gradually stir in milk. Cook over medium heat, stirring constantly, until mixture just comes to a boil and thickens. Remove from heat; add cheese and stir until well combined. Place cauliflower on a serving dish; cut almost through into 8 wedges. Pour sauce over all and garnish with reserved bacon.

Makes 8 servings.

Prestige scalloped potatoes

5 pounds	potatoes, peeled and sliced	2.5 kg
2	stalks celery, sliced	2
1 (1½ ounce)	package onion soup mix	1 (45 g)
4 tablespoons	flour	60 mL
2 cups	milk	500 mL
2 teaspoons	Angostura bitters	10 mL
1 cup	toasted croutons	250 mL

In a large saucepan, boil potatoes and celery in a small amount of water until done, about 10 minutes. Drain, reserving 1½ cups (375 mL) of the cooking water. To the water, add onion soup mix and flour, whisking well to avoid lumps. Bring mixture to the boil, stirring constantly. Add milk and bitters. Continue to cook over medium heat, stirring constantly, until thickened. In a 9"x 13" (23 cm x 33 cm) baking dish, layer a third of the vegetables and a third of the sauce. Repeat layers twice more, ending with the sauce. Sprinkle croutons over the top. Bake, uncovered, at 350°F (180°C) for 15 minutes, or until hot and bubbly.

Makes 12 servings.

Ranch Style Potatoes

6 cups	potatoes, peeled and sliced	1.5 L
4 tablespoons	butter, melted	60 mL
4 tablespoons	flour	60 mL
2 cups	milk	500 mL
1 teaspoon	salt	5 mL
¼ teaspoon	pepper	1 mL
4	chopped green onions, white and green parts	4
1 cup	sharp Cheddar cheese, grated	250 mL
½ cup	barbecue sauce	125 mL
1	green pepper, cut in rings	1

Combine butter and flour. Add milk gradually, stirring constantly, and cook over medium heat until smooth and thickened. Add salt, pepper, onions and cheese, stirring until cheese is melted. In a greased 9"x 13" (23 cm x 33 cm) baking dish, alternate layers of potatoes and cheese sauce starting with potatoes. Top with barbecue sauce and green pepper rings. Bake at 350°F (180°C) for 40-45 minutes, or until potatoes are cooked through.

Makes 8 servings.

SCALLOPED POTATOES WITH CHEESE AND HERBS

3 tablespoons	butter	45 mL
2	onions, finely chopped	2
2	cloves garlic, minced	2
3 cups	milk	750 mL
¼ teaspoon	dried rosemary	1 mL
¼ teaspoon	dried basil	1 mL
¼ teaspoon	dried oregano	1 mL
1 teaspoon	salt	5 mL
¼ teaspoon	black pepper	1 mL
2 cups	grated Swiss cheese	500 mL
	(approx. 8 ounce/250 g)	
1 cup	fresh breadcrumbs	250 mL
5	medium potatoes	5
	(approx. 2 pounds/1 kg)	

Melt butter in skillet; add onions and garlic. Cook until tender without browning. Reserve. Heat milk, rosemary, basil, oregano, salt and pepper. Reserve. Combine cheese with breadcrumbs. Peel potatoes; slice thinly. Place one third of potatoes in bottom of a buttered 2 quart (2 L) casserole. Spread half the onions over potatoes and sprinkle with a third of the cheese mixture. Place another third of potatoes on top; the remaining onions on top of that and another third of the cheese. Add the remaining potatoes, pour milk mixture over; sprinkle with remaining cheese. Bake at 350°F (180°C) for 1 hour and 35 minutes or until potatoes are tender when pierced with a knife. Rest 10 minutes before serving.
     Serves 6 to 8.

Seaside Potatoes

5	medium potatoes, peeled, about 1½ pounds (750 g)	5
1	small onion, thinly sliced	1
2	eggs, hard-cooked and sliced	2
1 (12 ounce)	tin asparagus, drained	1 (341 mL)
1 cup	baby shrimp	250 mL
4 tablespoons	sour cream	60 mL
3 tablespoons	butter	45 mL
3 tablespoons	flour	45 mL
½ teaspoon	salt	2 mL
¼ teaspoon	pepper	1 mL
2 cups	milk	500 mL
¼ cup	buttered breadcrumbs	50 mL

Thinly slice potatoes and place half of them in bottom of greased 8-cup (2 L) casserole. Make a cream sauce by melting butter in a saucepan, and stirring in flour, salt and pepper. Gradually add milk, stirring constantly over medium heat, until smooth and thickened. Pour some of sauce over potatoes. Arrange layers of onion, egg, asparagus and shrimp over potatoes, covering each with some of the sauce. Top with remaining potatoes, and cover with remaining sauce. Sprinkle buttered crumbs over all. Bake, covered, at 350°F (180°C) for 1 hour and 15 minutes, or until potatoes are soft when pierced with a fork.

Makes 8 servings.

Sour Cream Baked Potatoes

8	medium potatoes	8
1 cup	sour cream	250 mL
¼ teaspoon	ground cumin	1 mL
½ teaspoon	salt	2 mL
dash	pepper	dash
1 tablespoon	butter	15 mL
	milk, as necessary	
	paprika	

Scrub potatoes well but do not peel. Prick with a fork in several spots in each potato. Bake at 425°F (220°C) for 40-50 minutes, or until soft when gently squeezed. Set aside until cool enough to handle. Meanwhile, combine sour cream, cumin, salt, and pepper. Cut a thin slice off the top of each potato and discard. Scoop out the centre of the potatoes and add to the sour cream mixture. Add butter, and beat until fluffy, adding milk as necessary. Spoon the mixture back into the potato shells and sprinkle with paprika. Bake at 375°F (190°C) for 20-25 minutes until heated through.

Makes 8 servings.

SPINACH BALLS IN TOMATO SAUCE

1 (10 ounce)	box frozen, chopped spinach	1 (300 g)
1½ cups	dry breadcrumbs	375 mL
2 teaspoons	baking powder	10 mL
1 teaspoon	basil	5 mL
¼ teaspoon	nutmeg	1 mL
1 teaspoon	salt	5 mL
1½ cups	small curd cottage cheese	375 mL
½ cup	Parmesan cheese, grated	125 mL
2	eggs, beaten	2
1	clove garlic, crushed	1
1 (28 ounce)	can stewed tomatoes, well drained	1 (796 mL)
1 (5½ ounce)	can tomato paste	1 (156 mL)
1	medium onion	1
1	clove garlic	1
1 teaspoon	salt	5 mL
½ cup	water	125 mL
1 cup	milk	250 mL
½ teaspoon	oregano	2 mL
1	small bay leaf	1
¼ cup	Parmesan cheese, grated	50 mL

Cook spinach as directed. Drain and press out as much water as possible. In a medium bowl, combine crumbs, baking powder, basil, nutmeg and 1 teaspoon salt. Add cottage cheese, ½ cup (125 mL) Parmesan cheese, eggs, crushed garlic clove, drained tomatoes (reserve liquid for other uses) and spinach. Mix well. For the balls, shape the mixture with wet hands into 32 balls, about 1½" (4 cm) in diameter. In a large pot, bring 3" (8 cm) of water to the boil. Reduce heat to simmer and drop in the balls a few at a time. Poach, uncovered, until cooked through, about 7 minutes. As they are done lift out with a slotted spoon and place in an 8"x 8" (20 cm x 20 cm) baking pan. For the sauce, purée tomato paste, onion and garlic clove in blender. Pour into a small saucepan and add 1 teaspoon salt, water, milk, oregano and bay leaf. Heat to boiling, reduce heat, and simmer for 10 minutes. Remove bay leaf, and pour sauce over spinach balls in baking pan. Sprinkle ¼ cup (50 mL) Parmesan cheese over all. Bake, uncovered, at 350°F (180°C) for 30 minutes.
Makes 8 servings.

SPINACH RING

4 (10 ounce)	boxes frozen chopped spinach	4 (300 g)
1½ cups	cooked rice	375 mL
1¼ cups	milk	300 mL
2 tablespoons	butter	30 mL
2 teaspoons	salt	10 mL
¼ teaspoon	nutmeg	1 mL
3	eggs, beaten	3

Cook spinach according to directions. Drain well. Toss with cooked rice. Scald milk and add butter, salt, nutmeg. Let cool slightly. Stir in beaten eggs. Add spinach and rice mixture. Mix well. Pour into well-greased 10" (25 cm) ring mold. Place pan in water and bake at 350°F (180°C) for 50-60 minutes until set. Turn out onto lipped platter and serve with curry sauce.

CURRY SAUCE

4 tablespoons	butter, melted	60 mL
1	small apple, peeled and diced	1
¼	onion, minced	¼
4 tablespoons	flour	60 mL
1-3 teaspoons	curry powder	5-15 mL
2 cups	milk	500 mL
1 teaspoon	salt	5 mL
1 teaspoon	sugar	5 mL
1 teaspoon	lemon juice	5 mL
½ teaspoon	Worcestershire sauce	2 mL

Sauté apple and onion in butter until soft, about 5 minutes. Blend in flour and curry powder. Add milk gradually, stirring constantly over low heat until smooth and thickened. Add salt, sugar, lemon juice and Worcestershire sauce. Cook an additional 2 minutes. Pour over spinach ring.

Makes 10 servings.

SPINACH SOUFFLE

1 (10 ounce)	package fresh spinach, finely chopped	1 (284 g)
4 tablespoons	butter, melted	60 mL
4 tablespoons	flour	60 mL
1 cup	milk	250 mL
3	eggs, separated	3
½ teaspoon	salt	2 mL
dash	pepper	dash

Chop spinach very finely in a food processor, or by hand. In a saucepan, combine butter and flour. Add milk gradually, stirring constantly, and cook over medium heat until smooth and thickened. Remove from heat. Beat yolks slightly; add hot mixture gradually, beating vigorously. Add salt and pepper. Beat egg whites until stiff, but not dry. Fold egg whites and spinach into the sauce mixture. Pour into well greased 8"x 8" (20 cm x 20 cm) baking pan. Place within another pan of hot water and bake at 325°F (160°C) for 35 minutes, or until set.

Makes 6 servings.

SQUASH SOUFFLE

3 ½ cups	diced, frozen squash	875 mL
1 cup	milk	250 mL
2 tablespoons	butter	30 mL
1 cup	cracker crumbs, coarsely crushed	250 mL
2 tablespoons	pimento, finely chopped	30 mL
½ teaspoon	salt	2 mL
1 teaspoon	grated onion	5 mL
dash	pepper	dash
dash	nutmeg	dash
2	eggs, beaten	2
	pimento strips	

Cook squash in water until tender. Drain and mash. In a large saucepan, heat milk and butter. Add ¾ cup (175 mL) of the cracker crumbs and stir well. Add squash, pimento, salt, onion, pepper and nutmeg. Stir in beaten eggs. Place in a well greased 8"x 8" (20 cm x 20 cm) baking pan, and top with the remaining cracker crumbs. Bake at 350°F (180°C) for 1 hour, or until a knife inserted in the centre comes out clean. Garnish with pimento strips.

Makes 6 servings.

Tomato Bacon Wraps

8		bacon slices	8
8		medium tomatoes	8
4		bread slices	4
1	cup	milk	250 mL
3		green onions, finely chopped	3
1	cup	grated Canadian Cheddar cheese	250 mL
1		egg, beaten	1
1	teaspoon	oregano	5 mL
½	teaspoon	salt	2 mL
¼	teaspoon	freshly ground pepper	1 mL

Pre-heat oven to 350°F (180°C).

Partially cook bacon slices; set aside. Cut tops from tomatoes; scoop out pulp and retain, discarding juice and seeds. Turn tomatoes upside down on a paper towel to drain while preparing filling.

Remove crusts from bread; crumble bread and combine with milk; soak 5 minutes. Stir in remaining ingredients, including chopped tomato pulp. Season the inside of each tomato cup with a pinch of oregano, salt and pepper; fill with savory mixture. Wrap a bacon slice around each tomato; secure with tooth picks. Bake in oven until filling is set, 15-20 minutes.

Makes 4 servings.

TURNIP APPLE BAKE

1	large turnip	1
1 tablespoon	butter	15 mL
1 cup	milk	250 mL
1½ cups	apples, peeled and sliced	375 mL
3-4 tablespoons	brown sugar	45-60 mL
½ teaspoon	cinnamon	2 mL

TOPPING

¼ cup	flour	50 mL
¼ cup	brown sugar	50 mL
2 tablespoons	butter	30 mL

Peel, dice and cook turnip in water until tender. Drain and mash, adding 1 tablespoon (15 mL) butter and milk. Toss apples with ¼ cup (50 mL) brown sugar and cinnamon. In a greased 6 cup (1.5 L) casserole, arrange alternate layers of turnip and apples, beginning and ending with turnip. Mix topping ingredients together until crumbly. Sprinkle over top of casserole. Bake at 350°F (180°C) for 1 hour. Serve hot.

Makes 8 servings.

ZUCCHINI CASSEROLE

6	small zucchini, diced	6
1 pound	ground beef	500 g
½ cup	onion, chopped	125 mL
½ cup	dry breadcrumbs	125 mL
¾ teaspoon	salt	3 mL
⅛ teaspoon	pepper	0.5 mL
¼ teaspoon	garlic powder	1 mL
1 tablespoon	fresh parsley, chopped	15 mL
4 tablespoons	butter, melted	60 mL
4 tablespoons	flour	60 mL
2 cups	milk	500 mL
1 cup	sharp Cheddar cheese, grated	250 mL
1 tablespoon	Parmesan cheese, grated	15 mL
½ cup	dry breadcrumbs	125 mL
2 tablespoons	butter, melted	30 mL

Cook zucchini in water until tender. Drain well. Cook beef and chopped onions together until beef is browned. Remove excess fat. Stir in ½ cup (125 mL) breadcrumbs, salt, pepper, garlic and parsley. Remove from heat. In a saucepan, blend 4 tablespoons (60 mL) butter and flour. Add milk gradually, stirring constantly, and continue to cook over medium heat until smooth and thickened. Add Cheddar cheese, and stir until melted. Add milk mixture to meat. In a greased 6-cup (1.5 L) casserole, place half of zucchini, followed by half of meat mixture. Then layer remainder of zucchini, and remainder of meat. Combine ½ cup (125 mL) breadcrumbs, Parmesan cheese and 2 tablespoons (30 mL) butter. Sprinkle on top of casserole. Bake, uncovered, at 350°F (180°C) for 45 minutes.
Makes 6 servings.

CHEDDAR SAUCE SUPREME

2	tablespoons	butter	30 mL
2	tablespoons	flour	30 mL
1		chicken bouillon cube	1
pinch		dry mustard	pinch
1½ cups		milk	375 mL
1½ cups		shredded Canadian Cheddar cheese	375 mL

Melt butter in a medium saucepan. Blend in flour, bouillon cube, and dry mustard. Gradually stir in milk. Cook over medium heat, stirring constantly until thick and mixture comes to a boil. Reduce heat to low; add cheese and stir until melted.

Makes 2 cups/500 mL.

CLASSIC WHITE SAUCE

THIN

1	tablespoon	butter	15 mL
1	tablespoon	flour	15 mL
¼	teaspoon	salt	1 mL
1	cup	milk	250 mL

MEDIUM

2	tablespoons	butter	30 mL
2	tablespoons	flour	30 mL
¼	teaspoon	salt	1 mL
1	cup	milk	250 mL

THICK

3	tablespoons	butter	45 mL
3	tablespoons	flour	45 mL
¼	teaspoon	salt	1 mL
1	cup	milk	250 mL

Melt butter in a saucepan. Blend in flour and salt. Gradually stir in milk. Cook over medium heat until thickened and mixture comes to a boil.

Makes 1 cup/250 mL.

COUNTRY KITCHEN PANCAKE SAUCE

2	cups	lightly packed brown sugar	500 mL
1	cup	corn syrup	250 mL
½	cup	butter	125 mL
1	cup	milk	250 mL
2	teaspoons	vanilla	10 mL

Mix brown sugar, corn syrup, butter and milk in a medium saucepan. Cook over low heat, stirring constantly until smooth and sauce is heated through. Stir in vanilla. Serve warm.
Makes 3½ cups/875 mL.

CREAMY ONION SAUCE

3	tablespoons	butter	45 mL
2		onions, chopped	2
2	tablespoons	flour	30 mL
2		chicken bouillon cubes, crushed	2
1½	cups	milk	375 mL

Melt butter in a medium saucepan; sauté onions for 10 minutes or until tender. Blend in flour and bouillon cubes. Gradually stir in milk. Cook over medium heat, stirring constantly until thick and mixture comes to a boil.

Makes 2 cups/500 mL.

CURRY SAUCE

¼	cup	butter	50 mL
2	tablespoons	finely chopped onion	30 mL
2	tablespoons	finely chopped celery	30 mL
3	tablespoons	all-purpose flour	45 mL
1	teaspoon	curry powder	5 mL
1		chicken bouillon cube, crushed	1
½	teaspoon	salt	2 mL
2	cups	milk	500 mL

Melt butter in a large saucepan; sauté onion and celery until tender. Blend in flour, curry powder, bouillon cube and salt. Gradually stir in milk. Cook over medium heat, stirring constantly until thick and mixture comes to a boil.

Makes 2¼ cups/550 mL.

Hearty spaghetti sauce

3 tablespoons	olive or other vegetable oil	45 mL
3	cloves garlic, finely chopped	3
2	onions, finely chopped	2
2	ribs celery, finely chopped	2
2	carrots, finely chopped	2
1½ pounds	ground beef	750 g
1 cup	milk	250 mL
2 (28 ounce)	tins tomatoes with juices	2 (796 mL)
¼ teaspoon	dried hot red chili peppers (more or less to taste)	1 mL
½ teaspoon	dried basil	2 mL
½ teaspoon	dried oregano	2 mL
1 tablespoon	chopped fresh parsley (or 1 teaspoon/5 mL dried)	15 mL
1 teaspoon	salt or more to taste	5 mL
¼ teaspoon	pepper or more to taste	1 mL

Heat oil in a deep large skillet or Dutch oven. Add garlic, onions, celery and carrots. Cook until vegetables are tender (about 8 minutes). Add meat and brown, discard excess fat, if any. Add milk. Cook, covered, 20 minutes over medium low heat. Break up tomatoes. Add to meat mixture with juices. Add remaining ingredients. Cook uncovered 1 hour. Taste and add seasoning if necessary.
Makes 6 cups/1.5L.

Mornay Sauce

3 tablespoons	butter	45 mL
3 tablespoons	flour	45 mL
2	chicken bouillon cubes, crushed	2
$\frac{1}{2}$ teaspoon	dry mustard	2 mL
2 cups	milk	500 mL
$\frac{1}{2}$ cup	shredded Swiss cheese	125 mL
2 tablespoons	grated Parmesan cheese	30 mL

Melt butter in a medium saucepan. Blend in flour, bouillon cubes, and mustard. Gradually stir in milk. Cook over medium heat, stirring constantly until thick and mixture comes to a boil. Reduce heat to low. Add cheeses and stir until melted.

Makes about 2 cups (500mL).

Excellent over fish, eggs and vegetable dishes. If dish is to be browned in the oven or under the broiler, sprinkle top with additional cheese first.

Mustard Sauce

2	tablespoons	butter	30 mL
3	tablespoons	flour	45 mL
¼	teaspoon	garlic salt	1 mL
¼	teaspoon	salt	1 mL
1½	cups	milk	325 mL
3	tablespoons	prepared mustard	45 mL

Melt butter in a medium saucepan. Blend in flour, garlic salt and salt. Gradually stir in milk. Cook over medium heat, stirring constantly until thick and mixture comes to a boil. Stir in mustard. Heat through.

Makes 1¾ cups/425 mL.

SAVOURY MUSHROOM SAUCE

3	tablespoons	butter	45 mL
3	cups	sliced fresh mushrooms	750 mL
2	tablespoons	finely chopped onion	30 mL
3	tablespoons	all-purpose flour	45 mL
1		chicken bouillon cube	1
¼	teaspoon	salt	1 mL
1¾	cups	milk	475 mL

Melt butter in a medium saucepan. Sauté mushrooms and onions until tender and mushroom liquid has evaporated. Blend in flour, chicken bouillon cube and salt. Gradually stir in milk. Cook over medium heat, stirring constantly until thick and mixture comes to a boil.

Makes 2½ cups/625 mL.

Smoke-House Bacon Sauce

6	slices bacon, cooked	6
3 tablespoons	butter	45 mL
2 tablespoons	finely chopped onion	30 mL
3 tablespoons	all-purpose flour	45 mL
¼ teaspoon	salt	1 mL
2¼ cups	milk	300 mL
¾ cup	grated Parmesan cheese	175 mL
¼ cup	chopped parsley	50 mL

Crumble bacon, set aside. In a medium saucepan, melt butter; sauté onion until tender. Blend in flour and salt. Gradually stir in milk. Cook over medium heat, stirring constantly until thick and mixture comes to a boil. Add Parmesan cheese and crumbled bacon. Heat through. Remove from heat and stir in parsley.

Makes 2⅔ cups/650 mL.

REGAL CHOCOLATE SAUCE

1	(12-ounce)	package semi-sweet chocolate pieces	1 (340 g)
⅓	cup	butter	75 mL
3	tablespoons	corn syrup	45 mL
1	cup	milk	250 mL
1	teaspoon	vanilla	5 mL

Combine chocolate, butter and corn syrup in a saucepan. Stir over low heat until chocolate is melted and mixture is thoroughly combined. Gradually stir in milk. Add vanilla. Cool to room temperature. Spoon over ice cream to serve. Refrigerate to store.

Makes 2½ cups/625 mL.

FRUIT SALAD DRESSING

¼	cup	flour	50 mL
1	teaspoon	salt	5 mL
1½	teaspoons	dry mustard	7 mL
3	tablespoons	sugar	45 mL
1½	cups	buttermilk	375 mL
2		eggs, beaten	2
1	tablespoon	butter	15 mL
2	tablespoons	vinegar	30 mL
¼	teaspoon	basil	1 mL

Mix dry ingredients together in a saucepan. Blend in a little buttermilk and stir until smooth. Add remaining buttermilk and eggs. Cook over low heat, stirring constantly, until thickened. Do not boil. Remove from heat. Add butter, vinegar and basil. Cool. Serve with a tossed fruit salad of apples, oranges, grapes and bananas.

Makes 2 cups/500 mL.

TANGY TOMATO DRESSING

2	cups	buttermilk	500 mL
1	(5½ ounce)	can tomato paste	1 (156 mL)
2	teaspoons	lemon juice	10 mL
⅓	cup	finely chopped green pepper	125 mL
2	teaspoons	grated onion	10 mL
		salt and pepper to taste	

Combine buttermilk and tomato paste; mix well. Stir in lemon juice, green pepper, onion, salt and pepper to taste. Pour over mixed greens and toss well.

Makes about 3 cups/750 mL.

OATMEAL BARS

½ cup	shortening	125 mL
¾ cup	brown sugar	175 mL
1 cup	all purpose flour	250 mL
½ teaspoon	baking powder	2 mL
½ teaspoon	salt	2 mL
¾ cup	milk	175 mL
1 teaspoon	vanilla extract	5 mL
1 cup	quick cooking oats	250 mL
¾ cup	seedless raisins	175 mL

Cream shortening with sugar until well blended. Combine flour, baking powder and salt. Stir into shortening mixture, alternating with milk, until well mixed. Add vanilla, oats and raisins. Spoon into a greased 8" (1L) square baking pan. Bake in a 350°F (180°C) oven for 30-35 minutes until top springs back when touched. Cool and cut into 2" squares.

Makes 16 squares.

BUTTERMILK BISCUITS

2	cups	all-purpose flour	500 mL
1	tablespoon	baking powder	15 mL
½	teaspoon	salt	2 mL
¼	teaspoon	baking soda	1 mL
6	tablespoons	butter	90 mL
1	cup	buttermilk	250 mL

Stir together flour, baking powder, salt and baking soda. Cut in butter until mixture resembles coarse crumbs. Add buttermilk all at once; stir until dough clings together and forms a ball. Knead dough 10-12 strokes on lightly floured surface. Pat or roll to form a rectangle about ½" (12 mm) thick. Cut into squares with a knife or into rounds with a 2" (5 cm) biscuit cutter. Place on cookie sheet; brush tops with additional buttermilk. Bake in preheated 450°F (230°C) oven 15-18 minutes.

Makes 1 dozen.

Variation:
Cheese Biscuits: Place biscuits about ¼" (6 mm) apart on cookie sheet. Sprinkle tops of biscuits with 1 cup (250 mL) shredded Cheddar cheese after brushing with buttermilk. Bake as above.

WHOLE WHEAT SESAME BISCUITS

1 cup	all purpose flour	250 mL
1 cup	whole wheat flour	250 mL
2 tablespoons	wheat germ	30 mL
4 teaspoons	baking powder	20 mL
1 teaspoon	salt	5 mL
1/3 cup	butter	75 mL
3/4 cup	milk	175 mL
1/3 cup	sesame seeds	75 mL

Preheat oven to 425°F (220°C).

In a large bowl stir together flours, wheatgerm, baking powder and salt. Cut in butter and blend until mixture resembles coarse crumbs. Add milk to flour mixture, tossing lightly with a fork until combined. Turn dough onto a lightly floured board and knead gently 20 times. Roll out to 1/2" (1.5 cm) thickness and cut into 2 1/2" (6 cm) rounds. Brush surfaces lightly with milk and press biscuits into sesame seeds. Place on an ungreased baking sheet. Bake 12-15 minutes.

Makes about 15 biscuits.

MINI CHEESE BISCUITS

2 cups	all-purpose flour	500 mL
4 teaspoons	baking powder	20 mL
1 tablespoon	sugar	15 mL
½ teaspoon	salt	2 mL
½ cup	shortening	125 mL
¾ cup	shredded Canadian Cheddar cheese	175 mL
1	egg	1
⅔ cup	milk	150 mL

Stir together flour, baking powder, sugar and salt. Cut in shortening until mixture resembles coarse crumbs. Stir in cheese. Beat together egg and milk; add to flour mixture, mixing lightly with a fork until just combined. Turn out dough onto floured board and knead gently 20 times. Roll to ½-inch (12 mm) thickness. Cut with floured 2-inch (5 cm) round cutter. Place on ungreased cookie sheet. Bake in preheated 450°F (230°C) oven 8 to 10 minutes. Serve warm with butter.

Makes about 2½ dozen.

PUFFY CHEESE BISCUITS

1 cup	milk	250 mL
⅓ cup	cold butter, cut into bits	75 mL
½ teaspoon	salt	2 mL
1 teaspoon	dry mustard	5 mL
¼ teaspoon	black pepper	1 mL
	pinch of cayenne pepper	
1 cup	all-purpose flour	250 mL
4	eggs	4
1 tablespoon	milk	15 mL
1 cup	grated Swiss or Cheddar cheese (approx. 4 ounce/125 g)	250 mL
3 tablespoons	grated Parmesan cheese	45 mL
2 tablespoons	sesame seeds	30 mL

Combine milk, butter, salt, mustard, pepper and cayenne in a medium-sized saucepan; bring to a boil. Remove from heat; beat in flour all at once and stir until mixture forms a ball of dough. Return to medium heat; cook a few minutes longer scraping mixture along the bottom of pot to dry dough slightly. Transfer dough to a bowl; cool 5 minutes. Beat in three eggs, one at a time. Mixture will be slippery. Beat the fourth egg lightly and reserve 2 tablespoons (30 mL) egg with 1 tablespoon (15 mL) milk. Beat remaining egg into dough. Add both cheeses; combine well. Butter a cookie sheet; dust lightly with flour; trace out an 8" (22 cm) circle. Spoon batter in mounds around the outside edge of circle; mounds should barely touch each other. Brush tops with egg-milk mixture; sprinkle with sesame seeds. Bake in a pre-heated oven at 425°F (210°C) for 10 minutes. Reduce heat to 350°F (180°C); continue to bake 45 to 55 minutes longer or until biscuits are firm and golden. Serve warm.
Makes 8 to 10 biscuits.

Scottish Raisin Scones

1	tablespoon	vinegar	15 mL
1	cup	milk	250 mL
2	cups	unsifted all-purpose flour	500 mL
3	tablespoons	sugar	45 mL
1	teaspoon	salt	5 mL
½	teaspoon	baking soda	2 mL
⅓	cup	shortening	75 mL
½	cup	seedless raisins	125 mL
1		egg yolk, beaten	1
		sugar	

Stir vinegar into milk; set aside. Combine flour, sugar, salt and baking soda in a mixing bowl; stir well to blend. Cut in shortening until mixture resembles coarse crumbs. Stir in raisins. Add milk mixture to dry ingredients all at once and stir with a fork until all ingredients are moistened. Turn out on a lightly floured board and knead gently about 20 times. Place dough on an ungreased cookie sheet. Pat or roll out to a ½″ (12 mm) thick circle. Cut into 8 wedges but do not separate. Brush with egg yolk and sprinkle sugar on top. Bake in pre-heated 450°F (230°C) oven 12 to 15 minutes or until done.
Makes 8 servings.

Herbed Parmesan Batter Bread

1 cup	milk	250 mL
¼ cup	butter	50 mL
¼ cup	sugar	50 mL
1½ tablespoons	oregano leaves	25 mL
1 teaspoon	salt	5 mL
1 teaspoon	onion salt	5 mL
½ teaspoon	celery salt	2 mL
½ cup	lukewarm water	125 mL
1 teaspoon	sugar	5 mL
2	envelopes dry granular yeast	2
1	egg, slightly beaten	1
½ cup	grated Canadian Parmesan cheese	125 mL
3¾ to 4 cups	all-purpose flour	925 to 1000 mL

Heat milk, ¼ cup (50 mL) butter, ¼ cup (50 mL) sugar, oregano, salt, onion salt and celery salt until butter melts. Cool to lukewarm. Pour lukewarm water into a large bowl. Stir in 1 teaspoon (5 mL) sugar until dissolved. Sprinkle yeast over mixture. Let stand 10 minutes, then stir well. Stir in lukewarm milk-butter mixture, egg and ½ cup (125 mL) Parmesan cheese. Add 3 cups (750 mL) of the flour. Beat until smooth and elastic. Gradually work in sufficient additional flour to make a soft dough, ¾ cup (175 mL) to 1 cup (250 mL) more. Turn dough into a large greased bowl. Cover and let rise in a warm place until double in bulk (about 1¼ hours). Stir down batter with a wooden spoon and beat vigorously for 30 seconds. Turn into two greased 1-quart (1 L) round casseroles. Cover and let rise until almost double in bulk (about ½ hour). Bake in preheated 350°F (180°C) oven 40 to 45 minutes. Remove from pans at once and cool on wire rack.

Makes 2 loaves.

KULICH EASTER BREAD

1 cup	milk	250 mL
½ cup	lukewarm water	125 mL
2 teaspoons	sugar	10 mL
2	envelopes dry yeast	2
6 to 6½ cups	unsifted all-purpose flour	1500 to 1625 mL
5	eggs	5
1 teaspoon	salt	5 mL
½ cup	sugar	125 mL
½ cup	butter, melted	125 mL
2 teaspoons	vanilla	10 mL
1½ tablespoons	grated lemon rind	25 mL
2 cups	seedless raisins	500 mL

Scald milk; cool to lukewarm. Pour luke-warm water into a large bowl. Stir in 2 teaspoons (10 mL) sugar. Sprinkle with yeast. Let stand 10 minutes; stir well. Stir in milk and 1 cup (250 mL) of the flour. Beat until smooth. Cover with damp cloth. Let rise in a warm place until light and spongy, about 1 hour.

Beat eggs; gradually beat in salt and ½ cup (125 mL) sugar and continue beating until light. Beat in butter, vanilla and lemon rind. Add egg mixture and raisins to yeast mixture and mix well. Gradually work in sufficient additional flour to make a soft dough, 5 to 5½ cups (1250 to 1375 mL) more. Turn dough onto floured board and knead until smooth and elastic, about 10 minutes. Place in greased bowl and brush top with melted butter. Cover and let rise in a warm place until double in bulk; about 45 minutes.

Punch dough down; turn onto lightly floured board and knead until smooth. Divide dough into 4 equal portions. Shape each into a ball and place in 4 well greased 1 pound (500 g) coffee cans. Brush tops with melted butter. Cover and let rise in a warm place until triple in bulk and dough reaches tops of cans, about 25 to 30 minutes. Bake in preheated 350°F (180°C) oven 30 to 35 minutes. Remove from cans and cool loaves on wire racks. Turn loaves occasionally while cooling. Frost tops.

For the Frosting: combine 2 cups (500 mL) sifted icing sugar and 2 tablespoons (30 mL) milk. Pour over loaves. Decorate as desired.

Makes 4 loaves.

EGG TWIST BREAD

6-7 cups	all purpose flour	1.5 L
		1.75 L
¼ cup	sugar	50 mL
2 teaspoons	salt	10 mL
2 packages	rapid mix yeast	2
1½ cups	milk	375 mL
½ cup	butter, cut up	125 mL
4	eggs	4
1 teaspoon	water	5 mL
2 tablespoons	poppy seeds	30 mL
2 tablespoons	finely chopped onion	30 mL

Combine 2 cups (500 mL) flour, sugar, salt and yeast. Heat butter and milk until warm. Gradually add milk mixture to dry ingredients and beat one minute at medium speed with mixer. Add 1 cup (250 mL) flour, 3 eggs and 1 egg white. Beat 1 minute more at high speed. Add additional flour to make a soft dough. Knead on floured surface until smooth and elastic (8-10 minutes). Place in covered buttered bowl. Let rise (about 1 hour) until double in bulk. Pound dough down. Divide in half. Divide each half into 2 pieces, one about ⅓ of the dough, the other ⅔. Divide larger piece into 3 pieces, and shape each into 11″ (28 cm) ropes. Braid ropes together. Pinch ends to seal. Do the same with smaller piece of dough making a 9″ (23 cm) rope. Braid. Place smaller braid on top of larger one. Place in buttered 9″x 5″ (2 L) loaf pan. Form a second loaf with remaining dough. Beat together egg yolk and water. Brush loaves, sprinkle with poppy seed and onion. Let rise until double in bulk. Bake about 25-30 minutes in preheated 375°F (190°C) oven. Remove from pans and cool on racks.

Makes 2 loaves.

NUTS AND SEEDS BREAD

1½ cups	all purpose flour	375 mL
½ cup	whole wheat flour	125 mL
1 teaspoon	baking powder	5 mL
1 teaspoon	baking soda	5 mL
½ teaspoon	salt	2 mL
1 cup	lightly packed brown sugar	250 mL
2 tablespoons	wheat germ	30 mL
2 tablespoons	sesame seeds	30 mL
2 tablespoons	poppy seeds	30 mL
½ cup	chopped nuts	125 mL
1	egg, beaten	1
1 cup	buttermilk	250 mL
¼ cup	vegetable oil	50 mL

Combine dry ingredients, seeds and nuts. Mix well. Combine egg, buttermilk and oil. Add to dry ingredients, stirring just until blended. Pour into a lightly greased 9"x 5" (2 L) loaf pan. Bake in a 350°F (180°C) oven about one hour, until nicely browned and cooked through.

Makes 1 large loaf.

Savoury Cheddar Bread

2 cups	all purpose flour	500 mL
4 teaspoons	baking powder	20 mL
1 tablespoon	sugar	15 mL
½ teaspoon	dried oregano	2 mL
½ teaspoon	onion salt	2 mL
¼ teaspoon	dry mustard	1 mL
1¼ cups	grated old Canadian Cheddar Cheese	300 mL
1	egg, beaten	1
1 cup	milk	250 mL
2 tablespoons	butter, melted	30 mL

Preheat oven to 350°F (180°C).

In a large bowl combine flour, baking powder, sugar, oregano, onion salt, mustard and cheese. Mix egg with milk and melted butter. Add egg mixture all at once to dry ingredients; stir just until moistened. Spread batter into a greased 8½"x 4½"x 2¾" (1.5 L) loaf pan. Bake 45 minutes or until done. Cool 10 minutes on a rack and remove from pan.

Makes 1 loaf.

SIMPLY SEASONED CASSEROLE BREAD

3⅓ cups	variety baking/biscuit mix	825 mL
¾ teaspoon	dill weed	3 mL
½ teaspoon	celery seed	2 mL
⅓ cup	soft butter	75 mL
½ cup	finely-chopped onion	125 mL
1	egg	1
1 cup	milk	250 mL
1 tablespoon	butter, melted	15 mL
2 tablespoons	grated Canadian Parmesan cheese	30 mL

Combine baking mix, dill weed and celery seed in a medium bowl. Cut in ⅓ cup (75 mL) butter until mixture resembles coarse crumbs. Stir in onion. In a small bowl, beat egg; gradually stir in milk. Add to crumbled mixture; stir just until ingredients are moistened. Turn batter into a greased 1½-quart (1.5 L) round casserole. Cut through batter with a knife forming a circle about 1 inch (2.5 cm) from edge of casserole. Bake in preheated 350°F (180°C) oven 30 to 35 minutes or until done. Brush top of loaf with 1 tablespoon (15 mL) melted butter and sprinkle with cheese. Cool 10 minutes. Serve warm.

Makes one round loaf.

Swiss Cheese Bacon Bread

3 ½ cups	variety baking/biscuit mix	875 mL
¼ cup	sugar	50 mL
1 ½ cups	grated Swiss cheese	375 mL
6	slices bacon, crisp cooked and crumbled	6
1 ⅓ cups	milk	325 mL
1	egg, beaten	1

Preheat oven to 350°F (180°C).

Combine baking mix and sugar in a large bowl. Add cheese and bacon. Blend milk and egg together and stir into dry ingredients mixing until just blended. Spoon batter into a lightly buttered 9"x 5"x 3" (2 L) loaf pan. Bake for approximately 45 minutes or until a wooden pick inserted near the center comes out clean. Cool 5 minutes before slicing.

Makes 1 loaf.

CRANBERRY NUT BREAD

½	cup	butter, softened	125 mL
½	cup	sugar	125 mL
1		egg	1
2	teaspoons	grated orange peel	10 mL
2½	cups	all-purpose flour	625 mL
2½	teaspoons	baking powder	12 mL
¾	teaspoon	salt	3 mL
1	cup	milk	250 mL
2	cups	chopped cranberries	500 mL
½	cup	chopped nuts	125 mL

Beat together butter and sugar until fluffy. Beat in egg and orange peel. Stir in flour mixed with baking powder and salt alternately with milk. Batter will be stiff. Mix in cranberries and nuts. Turn into greased and floured 9" x 5" x 3" (2 L) loaf pan. Bake in 350°F (180°C) oven 1 hour or until pick inserted in centre comes out dry. Cool 10 minutes in pan. Invert and cool on wire rack.

Makes 1 loaf.

CRISP-CRUSTED SODA-BREAD

1 cup	all purpose flour	250 mL
2 tablespoons	brown sugar	30 mL
1 teaspoon	baking soda	5 mL
½ teaspoon	salt	2 mL
¼ cup	butter	50 mL
2 cups	whole wheat or graham flour	500 mL
⅓ cup	rolled oats	75 mL
1½ cups	buttermilk	375 mL

Preheat oven to 375°F (190°C).

In a large bowl, mix together white flour, sugar, soda and salt. Cut in 2 tablespoons (30 mL) of the butter to form coarse crumbs. Stir in whole wheat, or graham, flour and oats. Melt remaining butter and add to dry ingredients with buttermilk, tossing all together to form a soft dough. Turn out on a lightly floured board and knead 10 times. Form dough into a ball, place on a greased pie plate and cut a deep cross on the top. Bake 40-50 minutes or until a toothpick inserted in the centre comes out clean. Cool briefly. Enjoy while still warm, spread with sweet butter.

Makes 1 loaf.

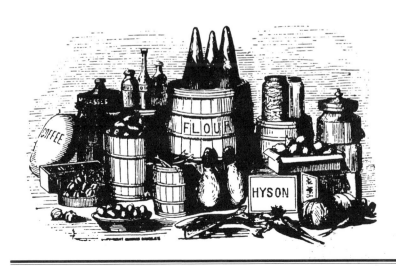

CARROT TEA LOAF

2 cups	all purpose flour	500 mL
¼ cup	brown sugar	50 mL
2 teaspoons	baking powder	10 mL
½ teaspoon	salt	2 mL
½ teaspoon	cinnamon	2 mL
½ teaspoon	nutmeg	2 mL
1 teaspoon	grated orange rind	5 mL
1	egg, beaten	1
1 cup	milk	250 mL
¼ cup	molasses	50 mL
1 cup	grated carrot	250 mL
¼ cup	melted butter	50 mL
½ cup	seedless raisins	125 mL
½ cup	chopped walnuts	125 mL

Combine flour, brown sugar, baking powder, salt, cinnamon, nutmeg and orange rind in a large bowl. Toss well to mix. Whisk egg, milk, molasses together. Stir in butter and carrot. Stir wet ingredients into dry ingredients until just mixed. Add nuts and raisins. Stir to distribute evenly. Spoon into an 8½"x 4½" (1.5 L) greased loaf pan. Bake in 375°F (190°C) oven until cooked through, about 50 minutes. Cool before slicing.

Makes 1 small loaf.

LEMON NUT BREAD

2¼ cups	all-purpose flour	550 mL
1 tablespoon	baking powder	15 mL
½ teaspoon	salt	2 mL
¾ cup	finely-chopped nuts	175 mL
1	egg	1
1¼ cups	milk	300 mL
⅔ cup	sugar	150 mL
⅓ cup	butter, melted	75 mL
1 tablespoon	grated lemon rind	15 mL
½ teaspoon	vanilla	2 mL
2 tablespoons	sugar	30 mL
2 tablespoons	lemon juice	30 mL

In a large bowl stir together flour, baking powder, salt and nuts. Beat egg well in a small bowl; stir in milk, ⅔ cup (150 mL) sugar, butter, lemon rind and vanilla. Add all at once to dry ingredients, stirring just until moistened. Turn batter into a greased 9 x 5 x 3-inch (2 L) loaf pan. Bake in preheated 350°F (180°C) oven 60 to 65 minutes or until done. Cool 10 minutes. Remove from pan. Combine 2 tablespoons (30 mL) sugar and lemon juice. Brush over surface of loaf. Cool completely.

Makes 1 loaf.

Apple Streusel Coffeecake

4 cups	chopped apples	1 L
1 cup	water	250 mL
2 tablespoons	lemon juice	30 mL
1¾ cup	sugar	425 mL
⅓ cup	cornstarch	75 mL
3 cups	all purpose flour	750 mL
1 tablespoon	baking powder	15 mL
1 teaspoon	salt	5 mL
1 teaspoon	ground cinnamon	5 mL
¼ teaspoon	ground nutmeg	1 mL
1 cup	butter	250 mL
2	eggs, lightly beaten	2
1 cup	milk	250 mL
1 teaspoon	vanilla extract	5 mL

TOPPING

½ cup	sugar	125 mL
½ cup	all-purpose flour	125 mL
¼ cup	butter	50 mL
½.cup	chopped walnuts	125 mL

Combine fruit and water in saucepan. Simmer until fruit is tender. Stir in lemon juice. Mix together 1 cup (250 mL) sugar and cornstarch. Blend into fruit. Cook and stir until thickened and bubbly. Cool. Combine flour, ¾ cup (175 mL) sugar, baking powder and spices. Cut in butter until mixture is crumbly. Blend eggs, milk and vanilla. Add to flour mixture and mix well. Spread half the batter in a 9"x 13"x 2" (3L) baking pan. Spread fruit over batter. Spoon remaining batter in small mounds over fruit mixture, spreading out as much as possible. Prepare topping by combining sugar and flour. Cut in butter until mixture is crumbly. Add walnuts. Place on top of batter. Bake in 350°F (180°C) oven for 45-50 minutes. Cool and cut in squares, to serve.

Makes 1 cake.

Cinnamon Buttermilk Coffeecake

2⅓ cups	sifted all-purpose flour, divided	575 mL
2 cups	firmly packed light brown sugar	500 mL
½ cup	butter	125 mL
1 teaspoon	baking soda	5 mL
1 teaspoon	cinnamon	5 mL
1	egg	1
1 cup	buttermilk	250 mL
½ cup	chopped nuts	125 mL

Mix 2 cups (500 mL) flour and brown sugar. Cut butter in until mixture resembles coarse meal. Set aside ¾ cup (175 mL) of this mixture to be used for topping. To mixture, add remaining flour, baking soda and cinnamon. Mix well, then add egg and buttermilk. Mix only until dry ingredients are well moistened. Pour into a greased 9″ (2.5 L) pan. Mix the ¾ cup (175 mL) sugar mixture that was set aside and the chopped nuts. Sprinkle over the top.

Bake in preheated 350°F (180°C) oven 50-55 minutes. Let cool in the pan and cut.

Makes 9 servings.

MAKE IT EASIER FOR YOURSELF: If brown sugar is lumpy, press through course sieve; or heat in a slow oven; or crush lumps with a rolling pin. Measure it just before you use it. Otherwise it will cake or lump.

CHEDDAR STREUSEL COFFEECAKE

1¾ cup	all-purpose flour	425 mL
⅓ cup	sugar	75 mL
3½ teaspoons	baking powder	17 mL
¾ teaspoon	salt	3 mL
⅓ cup	butter, chilled	75 mL
1 cup	shredded Canadian Cheddar cheese	250 mL
1	egg	1
¾ cup	milk	175 mL
½ teaspoon	vanilla	2 mL

Topping

½ cup	packed brown sugar	125 mL
⅓ cup	chopped nuts	75 mL
1 teaspoon	ground cinnamon	5 mL

In a large bowl stir together flour, sugar, baking powder and salt. With pastry blender cut in butter; stir in cheese. Beat egg well; stir in milk and vanilla. Add all at once to dry ingredients; stir just until moistened. Set aside.

In a small bowl, combine brown sugar, nuts and cinnamon. Set aside.

Spread half of the batter in greased 8" (2 L) square cake pan; top with half of the streusel topping. Repeat layering. Bake at 375°F (190°C) for 25-30 minutes. Serve warm with butter.

Makes 1 coffeecake.

Honey Pecan Coffeecake

1	cup	milk	250 mL
1	tablespoon	lemon juice	15 mL
1	cup	butter, softened	250 mL
1	cup	liquid honey	250 mL
2		eggs	2
1	teaspoon	vanilla	5 mL
2	cups	flour	500 mL
1¼	teaspoon	baking soda	6 mL
¼	teaspoon	salt	1 mL
1	teaspoon	cinnamon	5 mL

STREUSEL TOPPING

1	cup	chopped pecans	250 mL
½	cup	brown sugar	125 mL
1	teaspoon	cinnamon	5 mL
¼	cup	butter	50 mL

Pre-heat oven to 350°F (180°C).

Grease and flour a cake pan, either 8″ x 8″ square or 9″ round (2 L). Stir lemon juice into milk; set aside. Cream butter with honey until light and fluffy; beat in eggs; add vanilla. Sift together flour, baking soda, salt and cinnamon; add to creamed mixture alternately with milk mixture, beginning and ending with dry ingredients.

Combine streusel ingredients. Pour half the batter into pre-pared pan; sprinkle with half the streusel mixture; add the remaining batter; top with remaining streusel. Bake in oven about one hour or until done. Decorate with extra pecans, if desired.

Makes 1 cake.

Swedish Tea Ring

1 cup	warm milk	250 mL
2 packages	rapid mix yeast	2
1 cup	butter, softened	250 mL
½ cup	sugar	125 mL
2	eggs, lightly beaten	2
1 teaspoon	salt	5 mL
½ teaspoon	ground nutmeg	2 mL
5-6 cups	all purpose flour	1.25 L-1.5 L
1 (19 ounce)	can crushed pineapple	1 (540 mL)
¾ cup	brown sugar	175 mL
½ cup	raisins	125 mL
2 tablespoons	cornstarch	30 mL
¼ teaspoon	cinnamon	1 mL
1	egg yolk, beaten	1
2 tablespoons	milk	30 mL
	sugar glaze	

Place milk in large bowl. Add yeast. Stir to dissolve. Blend in butter, sugar, eggs, salt, nutmeg and 2 cups (500 mL) flour. Beat until smooth. Add enough flour to make a soft dough. Knead dough on floured surface until smooth and satiny.

Place in buttered bowl. Cover and let rise until double in bulk (1½ hours). Meanwhile for filling combine pineapple, brown sugar, raisins, cornstarch and cinnamon in a medium size saucepan. Cook over medium heat stirring constantly until thickened and clear. Boil 2 minutes. Cool. Punch dough down. Divide into 3 equal parts. Roll into 20"x 5" (50.8 cm x 12.7 cm) rectangles. Spoon ⅓ of the filling to an inch (2.5 cm) of edges. Roll up lengthwise. Pinch edges and seam to seal. Place on buttered cookie sheets curving to form 3 separate rings. Pinch ends together. Cut diagonal slits 1" (2.5 cm) apart cutting ⅔ of the way through. Pull out sections slightly alternating from right and left. Brush with a mixture of egg yolk and milk. Cover and let rise until double (about 30 minutes). Bake in preheated 375°F (190°C) oven until golden, about 18 minutes. Drizzle with sugar glaze.

continued on next page

Swedish Tea Ring (continued)

SUGAR GLAZE

2 cups	icing sugar	500 mL
3 tablespoons	milk	45 mL
½ teaspoon	vanilla	5 mL

Combine ingredients. Mix well. Drizzle over cakes.

BLUEBERRY STREUSEL COFFEECAKE

1¾ cups	all-purpose flour	425 mL
⅓ cup	granulated sugar	75 mL
3½ teaspoons	baking powder	17 mL
¾ teaspoon	salt	3 mL
⅓ cup	chilled butter	75 mL
1	egg	1
¾ cup	milk	175 mL
2 teaspoons	grated lemon rind	10 mL
½ teaspoon	vanilla	2 mL
1 cup	fresh blueberries*	250 mL
¼ cup	firmly-packed brown sugar	50 mL
2 tablespoons	chopped pecans	30 mL
½ teaspoon	ground cinnamon	2 mL

In a large bowl stir together flour, granulated sugar, baking powder and salt. With a pastry blender, cut butter in finely. Beat egg well. Stir in milk, lemon rind, vanilla and blueberries; add all at once to dry ingredients, stirring just until moistened. Spread batter evenly in a greased 8-inch (2 L) square cake pan. Combine brown sugar, nuts and cinnamon. Sprinkle evenly over batter. Bake in preheated 375°F (190°C) oven 25 to 30 minutes or until done. Serve warm with butter.

Makes 1 square coffeecake.

* 1 cup frozen blueberries, thawed and well-drained may be substituted.

MILK AND HONEY CAKE

1½ cups	all-purpose flour	375 mL
2½ teaspoons	baking powder	12 mL
½ teaspoon	salt	2 mL
⅓ cup	butter	75 mL
⅓ cup	sugar	75 mL
1 teaspoon	vanilla	5 mL
1	egg	1
¼ cup	liquid honey	50 mL
⅔ cup	milk	150 mL

Combine flour, baking powder and salt. Cream butter; gradually beat in sugar and vanilla. Beat in egg and honey. Add dry ingredients to creamed mixture alternately with milk, combining lightly after each addition. Turn batter into a greased 8-inch (2 L) square cake pan. Bake in preheated 350°F (180°C) oven 30 to 35 minutes or until done. Cool on wire rack. Frost with HONEY COCONUT TOPPING.

Makes one 8-inch/2 L cake.

HONEY COCONUT TOPPING

1 cup	flaked or shredded coconut	250 mL
½ cup	liquid honey	125 mL
1 tablespoon	butter, melted	15 mL

Combine coconut, honey and butter. Spread over top of cake. Broil until coconut is lightly browned.

Frosts one 8-inch/2 L cake.

GRANDMA'S CHOCOLATE CAKE

2 ½ cups	all purpose flour	625 mL
1 teaspoon	salt	5 mL
⅔ cup	cocoa	150 mL
2 teaspoons	baking soda	10 mL
1 cup	hot water	250 mL
1 cup	butter	250 mL
2 cups	sugar	500 mL
2	eggs	2
1 teaspoon	vanilla	5 mL
1 cup	buttermilk	250 mL

Preheat oven to 350°F (180°C).

Sift together flour and salt and set aside. Combine cocoa, soda and hot water and set aside. Cream butter and gradually add sugar, beating until mixture is light and fluffy. Beat in eggs. Combine vanilla and buttermilk and add to mixture alternately with dry ingredients, beginning and ending with flour. Quickly blend in cocoa mixture; mix well. Pour batter into two greased and floured 9" (1.5 L) cake tins or a 9" x 13" (3 L) cake pan. Bake 30 minutes (50 minutes for large cake) or until toothpick inserted in centre comes out clean. Let rest in pans 10 minutes then turn out onto racks to cool. Fill and frost with Easy Chocolate Icing.

Makes one cake.

EASY CHOCOLATE ICING:

¾ cup	cocoa	175 mL
⅓ cup	butter, melted	75 mL
3 cups	sifted icing sugar	750 mL
	pinch salt	
⅓ cup	milk	75 mL
1 teaspoon	vanilla	5 mL

Combine cocoa and butter and mix to form a smooth paste. Add to remaining ingredients and blend well. Let stand, stirring occasionally until icing reaches spreading consistency.

Makes about 2 cups (500 mL).

POPPY SEED CAKE

¾ cup	poppy seeds	150 mL
1 cup	milk	250 mL
1 teaspoon	vanilla extract	5 mL
¾ cup	butter	175 mL
1½ cups	sugar	375 mL
2 cups	flour	500 mL
2½ teaspoons	baking powder	12 mL
½ teaspoon	salt	2 mL
4	egg whites	4

Soak poppy seeds in milk, 15 minutes. Add vanilla. Cream butter with sugar. Beat until light. Combine flour, baking powder and salt. Add to butter mixture alternating with milk and poppy seeds. Beat egg whites until stiff. Fold into batter. Pour into a greased 9"x 5" (2 L) loaf pan. Bake in a 350°F (180°C) oven for 50-60 minutes until a toothpick inserted in centre comes out clean.
Makes one large loaf.

Rhubarb cake

2 cups	all purpose flour	500 mL
1 teaspoon	baking powder	5 mL
1 teaspoon	baking soda	5 mL
½ teaspoon	salt	2 mL
½ cup	butter	125 mL
1½ cups	sugar	375 mL
1 teaspoon	vanilla	5 mL
1	egg	1
1 cup	milk	250 mL
2 cups	fresh diced rhubarb	500 mL
½ cup	brown sugar	125 mL
1 teaspoon	cinnamon	5 mL

Preheat oven to 325°F (160°C).

Sift together flour, baking powder, soda and salt. Cream butter and sugar until light; beat in vanilla and egg and blend in milk. Stir sifted dry ingredients into batter; add rhubarb. Pour mixture into a greased and floured 9"x 13" (3 L) baking pan. Combine brown sugar and cinnamon and sprinkle over top; bake 45-50 minutes. Serve warm with cream.

Optional: Add ½ cup (125 mL) chopped dates, raisins or nuts. Makes 1 cake.

TRIPLE TREAT LEMON CAKE

1 (520 g)	package lemon cake mix	1 (520 g)
1 (4-ounce)	package lemon pudding and pie filling	1 (113 g)
2 cups	milk	500 mL
2	eggs	2
½ cup	sugar	125 mL
¼ cup	lemon juice	50 mL

Combine cake mix, pudding and pie filling, milk and eggs in a large mixer bowl. Blend ingredients at low speed. Beat at medium speed of electric mixer 3 minutes. Pour into a greased 12-cup (3 L) tube pan. Bake in preheated 350°F (180°C) oven 55 to 60 minutes or until toothpick inserted in centre comes out clean. Cool 10 minutes on wire rack. Remove cake from pan. Combine sugar and lemon juice. Stir until most of the sugar has dissolved. Brush sugar mixture evenly over top and sides of warm cake. Cool before serving.

Makes 1 ring cake.

Pineapple Tote Cake

2	cups	flour	500 mL
1½	cups	sugar	375 mL
1	teaspoon	baking soda	5 mL
2	teaspoons	cinnamon	10 mL
½	teaspoon	salt	2 mL
3		eggs	3
½	cup	oil	125 mL
1	tablespoon	lemon juice	15 mL
¾	cup	milk	175 mL
1	teaspoon	vanilla	5 mL
1	cup	crushed pineapple, well drained	250 mL
2	cups	shredded carrots	500 mL
1	cup	flaked coconut	250 mL
1	cup	finely chopped almonds	250 mL

Grease 9″ x 13″ x 2″ (3.5 L) pan and dust lightly with flour. Preheat oven to 350°F (180°C).

Add lemon juice to milk in measuring cup and set aside to sour. Sift together flour, sugar, soda, cinnamon and salt. Beat together eggs, oil, sour milk and vanilla. Add sifted dry ingredients and mix well. Stir in pineapple, carrots, coconut and almonds. Pour batter into prepared pan and bake for 1 hour, or until cake springs back when lightly touched. Let cake rest in pan on cooling rack for 10 minutes, then prick carefully all over with fork, cake tester or toothpick. Slowly pour hot glaze over cake. Note: this cake may be baked in a greased and floured 10″ (25 cm) Bundt pan.

GLAZE

⅓	cup	sugar	75 mL
¼	teaspoon	baking soda	2 mL
¼	cup	milk	50 mL
¼	cup	butter	50 mL
1	teaspoon	corn syrup	5 mL
½	teaspoon	vanilla	2 mL

In a small saucepan combine sugar, soda, milk, butter and corn syrup. Bring to a boil over medium heat, boil 5 minutes. Remove from heat and add vanilla.

Makes 1 cake.

ORANGE DATE CAKE

4 cups	all purpose flour	1 L
1 teaspoon	baking soda	5 mL
¼ teaspoon	salt	1 mL
1⅓ cups	chopped dates	325 mL
1 cup	chopped pecans	250 mL
1 cup	butter	250 mL
2 cups	sugar	500 mL
4	eggs	4
1½ cups	buttermilk	375 mL
2 tablespoons	grated orange rind	30 mL

Preheat oven to 325°F (160°C).

Toss ¼ cup (50 mL) flour with chopped dates and pecans. Sift together remaining flour with soda and salt. In a large bowl cream butter; gradually add sugar and beat until light and fluffy. Beat in eggs one at a time. Add flour mixture alternately with buttermilk. Stir in dates, nuts and orange rind. Turn into a greased and floured 10″ (3 L) tube pan. Bake 1¼-1½ hours or until a toothpick inserted in the centre comes out clean. Prick hot cake and spoon ORANGE GLAZE over top. Cool in pan.

Makes 1 cake.

ORANGE GLAZE

½ cup	sugar	125 mL
½ cup	orange juice	125 mL
2 tablespoons	lemon juice	30 mL
2 teaspoons	grated orange rind	10 mL

Combine ingredients. Heat and stir until sugar is dissolved. Spoon over hot cake.

Makes ½ cup (125 mL).

Coconut Meringue Cake

1 cup	all purpose flour	250 mL
1 teaspoon	baking powder	5 mL
$\frac{1}{3}$ cup	butter	75 mL
$1\frac{1}{4}$ cup	sugar	300 mL
2	egg yolks	2
$\frac{1}{2}$ cup	milk	125 mL
3	egg whites	3
$\frac{1}{2}$ cup	shredded coconut	125 mL
	Vanilla Filling	

Preheat oven to 350°F (180°C).

Sift flour and baking powder. Cream butter and $\frac{1}{2}$ cup (125 mL) sugar until light. Beat in egg yolks. Add dry ingredients, to creamed mixture alternately with milk, beginning and ending with flour. Spread mixture into two greased and floured 8″ (1.2 L) cake pans. Beat egg whites. Add remaining sugar gradually and beat until firm peaks form. Spread meringue on top of each cake batter and sprinkle with coconut. Bake 20-25 minutes or until done. Cool. Place one cake layer, coconut side down, on a cake plate. Spread VANILLA FILLING over cake layer and place other layer on top coconut side up. Serve at room temperature.

Makes 1 cake.

VANILLA FILLING

1	egg yolk	1
$\frac{1}{3}$ cup	sugar	75 mL
1 tablespoon	cornstarch	15 mL
1 cup	milk	250 mL
1 teaspoon	vanilla	5 mL

Combine egg yolk, sugar, cornstarch, milk and vanilla in a small saucepan. Bring to a boil, stirring constantly and cook until mixture is smooth and thick. Cool.

Makes 1 cup (250 mL).

ALMOND SYRUP CAKE

½ cup	ground almonds	125 mL
6	eggs, separated	6
2 cups	sugar	500 mL
3 cups	cake and pastry flour	750 mL
3 ½ teaspoons	baking powder	17 mL
1 cup	vegetable oil	250 mL
1 cup	milk	250 mL
1 teaspoon	almond extract	5 mL

Preheat oven to 350°F (180°C).

Grease a 12 cup (3 L) tube pan and sprinkle ground almonds on bottom; set aside. Beat egg whites until stiff; set aside. Sift sugar, flour and baking powder into a large bowl. Combine egg yolks, oil, milk and almond extract. Add milk mixture to dry ingredients, and combine well. Fold in beaten egg whites. Pour into prepared pan and bake about one hour or until toothpick inserted in centre comes out clean. Allow cake to cool in pan for 10 minutes then turn out and set in a deep baking dish upside down. Pour Almond Syrup over warm cake and leave to cool. About a half hour before serving set cake right side up on a serving plate. This will allow excess syrup to drain back through cake. Excellent served with whipped cream and fresh berries.

Makes 1 cake.

ALMOND SYRUP

½ cup	sugar	125 mL
1 cup	water	250 mL
½ teaspoon	almond extract	2 mL

Boil sugar and water together 2-3 minutes. Flavour with almond extract and set aside.

Makes 1 cup (250 mL).

BUTTERSCOTCH SUNDAE CAKE

1 cup	butter	250 mL
1 cup	brown sugar	250 mL
4	eggs, separated	4
2⅔ cups	sifted cake and pastry flour	650 mL
1 tablespoon	baking powder	15 mL
½ teaspoon	salt	2 mL
1 cup	milk	250 mL
1½ teaspoon	vanilla	7 mL
	sliced bananas	
	pecans	

Preheat oven to 350°F (180°C). Generously grease and flour three 8" (20 cm) or two 9" (23 cm) cake tins. Cream butter. Gradually add sugar and beat until light and fluffy. Add egg yolks one at a time, beating well after each addition. Sift together flour, baking powder and salt. Combine milk and vanilla. Add sifted dry ingredients to creamed mixture, alternately with milk mixture, beginning and ending with dry ingredients. Beat egg whites until stiff and fold gently into batter. Pour into prepared pans. Bake 25-30 minutes, or until toothpick inserted in centre comes out clean. Let rest in pans 5 minutes then turn out on racks to cool. Place Butterscotch Filling between layers adding slices of bananas if desired. Frost sides and top of cake with Boiled Icing and drizzle Butterscotch Glaze over top allowing it to run down the sides. Arrange pecans on top.

Makes one cake.

continued on next page

BUTTERSCOTCH SUNDAE CAKE (continued)

BUTTERSCOTCH FILLING

½ cup	brown sugar	125 mL
3 tablespoons	flour	45 mL
	pinch salt	
1 cup	milk	250 mL
2	egg yolks, beaten	2
2 tablespoons	butter	30 mL
1 teaspoon	vanilla	5 mL

Combine sugar, flour and salt in saucepan. Blend in milk. Set over moderate heat and bring to a boil stirring constantly while sauce thickens. Cook 1-2 minutes. Spoon half hot sauce into egg yolks, then return mixture to heat and cook briefly. Remove from heat and add butter and vanilla. Chill before using between cake layers.

Makes 1 cup (250 mL).

BOILED ICING

½ cup	sugar	125 mL
2 tablespoons	water	30 mL
¼ cup	corn syrup	50 mL
2	egg whites	2
1 teaspoon	vanilla	5 mL

Combine sugar, water and corn syrup in small saucepan. Bring to boil and boil hard, without stirring. Beat egg whites until stiff. Pour hot syrup slowly into egg whites, beating constantly until stiff peaks form. Blend in vanilla.

Makes about 1½ cups (375 mL).

continued on next page

BUTTERSCOTCH SUNDAE CAKE (continued)

BUTTERSCOTCH GLAZE

¼ cup	brown sugar	50 mL
3 tablespoons	butter	45 mL
2 tablespoons	water	30 mL

Combine all ingredients in small saucepan. Bring to a boil and boil hard 1½ minutes without stirring. Cool slightly. Drizzle over frosting on cake.

Makes about ¼ cup/50 mL.

LEMON ICE-BOX CAKE

1 tablespoon	cornstarch	15 mL
½ cup	sugar	125 mL
2	eggs, separated	2
1 cup	milk, hot	250 mL
1 teaspoon	butter	5 mL
2 tablespoons	lemon juice	30 mL
2 teaspoons	grated lemon rind	10 mL
1	Angle Food Cake	1
1 cup	whipping cream	250 mL
	blanched flaked almonds, toasted	

Combine cornstarch, sugar and egg yolks in a medium-sized bowl. Stir in hot milk. Return to saucepan and stir constantly over medium heat until mixture comes to a boil and becomes smooth and thick. Remove from heat, stir in butter and lemon juice and rind. Beat egg whites until stiff and fold into custard. Line a large bowl with wax paper. Break angel food cake into small cubes and layer in bowl with lemon mixture – beginning and ending with cake. Cover and chill for 24 hours. Turn out onto serving plate. Decorate with whipped cream and toasted almonds.

Makes one cake.

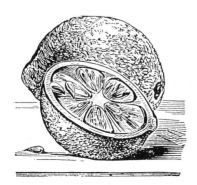

Peach Upside Down Cake

½ cup	butter	125 mL
1½ cups	brown sugar	375 mL
4	fresh peaches peeled, and sliced (or use canned)	4
½ teaspoon	nutmeg	2 mL
2	eggs, well beaten	2
1 cup	milk	250 mL
1 teaspoon	vanilla extract	5 mL
2 cups	flour	500 mL
2 teaspoons	baking powder	10 mL

Melt ¼ cup (50 mL) butter in a large cast iron frying pan or a 10″ (25 cm) deep dish pie plate. Stir in ½ cup (125 mL) brown sugar. Melt to blend. Arrange fruit on top in a decorative pattern. Sprinkle with nutmeg and set aside. Cream remaining butter and sugar together. Add beaten eggs, milk and vanilla. Combine flour and baking powder. Stir into wet ingredients. Pour batter over fruit. Bake in preheated 350°F (180°C) oven for 30-35 minutes until cake is cooked through. Leave in pan 5 minutes. Then turn out onto a platter.

Makes 8 servings.

ORANGE RAISIN CAKE

2 cups	sifted all purpose flour	500 mL
1 teaspoon	soda	5 mL
½ teaspoon	salt	2 mL
½ cup	butter	125 mL
1 cup	sugar	250 mL
2	eggs	2
1	orange, juice and rind	1
1 cup	raisins	250 mL
1 cup	milk	250 mL

Preheat oven to 350°F (180°C). Sift together flour, soda and salt. Beat butter and sugar until light; blend in eggs, one at a time. Run raisins and orange rind through a grinder or chop in a food processor. (Reserve orange juice for glaze.) Blend dry ingredients into creamed mixture alternately with milk and stir in ground raisins and orange rind. Spread mixture in a greased and floured 13"x 9" (3.5 L) pan. Bake 30-35 minutes or until toothpick inserted in centre comes out clean. Leave cake in pan and spread with ORANGE GLAZE while still warm.

Makes 1 cake.

ORANGE GLAZE

	juice of 1 orange	
¼ cup	sugar	50 mL
½ teaspoon	cinnamon	2 mL
¼ cup	finely chopped walnuts	50 mL

In a small saucepan combine orange juice and sugar. Stir over medium heat until sugar melts; add remaining ingredients. Makes ½ cup/125 mL.

DOUBLE FUDGE CHOCOLATE CAKE

4 ounces	unsweetened chocolate, chopped coarsely	112 g
⅓ cup	cocoa	75 mL
1 cup	boiling water	250 mL
1 cup	butter	250 mL
2¼ cups	sugar	550 mL
3	eggs	3
1 teaspoon	pure vanilla extract	5 mL
2¼ cups	all purpose flour	550 mL
2 teaspoons	baking powder	10 mL
1 teaspoon	baking soda	5 mL
1 cup	milk	250 mL

ICING

8 ounces	semi-sweet chocolate, chopped	225 g
¼ cup	cocoa	50 mL
¾ cup	milk	175 mL
½ cup	butter	125 mL
3 cups	icing sugar, sifted	750 mL
1½ teaspoons	pure vanilla extract	7 mL
4 ounces	cream cheese	125 g

Preheat oven to 350°F (180°C). Butter two 9″ (23 cm) cake pans. Line with parchment or waxed paper rounds. Butter again. Combine chopped chocolate and cocoa. Pour boiling water over mixture. Stir to melt. Cool slightly. Cream butter. Beat in sugar. Beat in eggs one at a time. Beat in vanilla and melted chocolate. Sift flour, with baking powder and baking soda. Add to butter mixture alternately with milk, beginning and ending with dry ingredients. Mix ingredients only until blended. Divide batter between pans. Bake 45 to 50 minutes or until centre springs back when lightly pressed. Allow to cool 10 minutes. Invert onto cooling racks. To prepare icing place chocolate, cocoa, milk and butter in top of double boiler. Heat gently over simmering water just until chocolate melts. Mixture should be smooth. Cool slightly. Beat in icing sugar and vanilla. Chill until spreadable. Combine ½ cup (125 mL) of the icing with the cream cheese. Beat until smooth. Reserve for the decoration. Using the dark icing sandwich the two layers of cake together. Frost the top and sides. Decorate as desired with the cream cheese chocolate frosting.

Makes 9″ (23 cm) layer cake.

COUNTRY KITCHEN CHOCOLATE CAKE

2⅔ cups	all purpose flour	650 mL
½ cup	cocoa	125 mL
1 teaspoon	salt	5 mL
1 cup	sour cream	250 mL
1 tablespoon	lemon juice	15 mL
2 teaspoons	baking soda	10 mL
2	eggs, separated	2
2 cups	sugar	500 mL
1 tablespoon	vanilla	15 mL
¼ cup	vegetable oil	50 mL
2 cups	milk	500 mL
½ cup	chopped pecans (optional)	125 mL

Preheat oven to 350°F (180°C). Sift together flour, cocoa and salt. Combine sour cream with lemon juice and baking soda. In a large bowl beat egg yolks with sugar, vanilla, sour cream mixture and oil. Add 1 cup (250 mL) milk and all the dry ingredients and mix well. Beat in remaining milk. Whisk egg whites until stiff and fold into mixture. Pour batter into a greased 12 cup (3 L) tube pan. Bake about 70 minutes or until toothpick inserted in centre comes out clean. Cool 10 minutes on wire rack. Remove cake from pan. Spread CHOCOLATE FROSTING on cake while still warm and sprinkle with chopped pecans.

Makes 1 ring cake.

CHOCOLATE FROSTING

2 tablespoons	cocoa	30 mL
1½ tablespoons	butter	22 mL
3 tablespoons	half and half cream	45 mL
	pinch salt	
1½ cups	icing sugar	375 mL
½ cup	semi-sweet chocolate chips	125 mL
10	large marshmallows cut in quarters	10
1 teaspoon	vanilla	5 mL

continued on next page

COUNTRY KITCHEN CHOCOLATE CAKE (continued)

In a small saucepan combine cocoa, butter, cream, salt and ½ cup (125 mL) icing sugar. Stir constantly over medium heat until bubbling. Add chips and marshmallows and stir until melted. Off heat stir in vanilla and remaining icing sugar. Beat well and set aside, stirring occasionally. Spread on warm cake.

Makes 2 cups (500 mL).

Strawberries 'n' Cream Cake

2 cups	all purpose flour	500 mL
2 teaspoons	baking powder	10 mL
½ teaspoon	salt	2 mL
⅔ cup	butter	150 mL
1¾ cup	sugar	425 mL
3	eggs	3
1 teaspoon	almond extract	5 mL
1⅓ cup	milk	325 mL
2 cups	whipping cream	500 mL
4 cups	fresh strawberries	1 L

Preheat oven to 350°F (180°C).

Combine dry ingredients; set aside. Cream butter and 1½ cups (375 mL) sugar until light. Beat in eggs, one at a time. Mix almond with milk. Add dry ingredients to mixture alternately with milk, beginning and ending with dry ingredients. Spread mixture in two greased and floured 8″ (1.2 L) cake pans. Bake 20-25 minutes, or until toothpick inserted in centre comes out clean. Let layers rest in pan 5 minutes then turn out onto racks to cool. Shortly before serving beat cream until stiff with remaining sugar. Slice half the berries, reserving the perfect whole ones. Cover one cake layer with whipped cream, arrange sliced berries on top and then the other cake layer. Pile cream on top and decorate with whole berries.

Makes one cake.

PEACHES 'N' CREAM SHORTCAKE

4 cups	sliced peaches	1 L
1 tablespoon	lemon juice	15 mL
½ cup plus	sugar	125 mL;
5 tablespoons	sugar	75 mL
4 cups	variety baking mix	1 L
½ cup	milk	125 mL
1	egg beaten	1
¼ cup	melted butter	50 mL
1 teaspoon	grated lemon rind	5 mL
1 cup	whipping cream	250 mL
1 cup	crushed peanut brittle	250 mL

Combine peaches with lemon juice and ½ cup (125 mL) sugar. Toss well. Let stand 30 minutes. Place biscuit mix in bowl with 2 tablespoons (30 mL) sugar. Whisk together milk, egg and butter. Add to biscuit mix. Stir lightly with a fork to make a soft dough. Divide dough into 8 pieces and pat into ¼ inch (6 mm) rounds on a greased cookie sheet. Mix one tablespoon (15 mL) sugar with lemon rind. Sprinkle over biscuits. Bake in 400°F (200°C) oven for 10-15 minutes or until puffed and golden brown. Remove from oven. Set on wire rack to cool 15 minutes. Whip cream with remaining sugar until stiff. Halve biscuits. Pile with peaches, whipped cream and crushed peanut brittle. Cover with second half and top with more peaches, cream and peanut brittle.

Makes 8 servings.

ORANGE RUM CAKE

1 cup	butter, softened	250 mL
1 cup	sugar	250 mL
1 tablespoon	grated orange rind	15 mL
1 tablespoon	grated lemon rind	15 mL
2	eggs	2
2 ½ cups	all purpose flour	625 mL
2 teaspoons	baking powder	10 mL
1 teaspoon	soda	5 mL
½ teaspoon	salt	2 mL
1 cup	buttermilk	250 mL
1 cup	finely chopped walnuts	250 mL

Cream butter and sugar together. Beat until light. Add fruit rinds and eggs, one at a time, mixing well. Combine dry ingredients and add to egg mixture, alternating with buttermilk, beating after each addition until smooth. Fold in nuts. Pour into greased 10" (3 L) tube pan. Bake in 350°F (180°C) oven about one hour. Remove cake from oven. Pour half the Citrus Syrup over cake. Let soak in. Pour over remaining syrup. Turn cake out after it has cooled.

Makes one 10"/3 L cake.

CITRUS SYRUP

½ cup	freshly squeezed orange juice	125 mL
¼ cup	lemon juice	50 mL
1 cup	sugar	250 mL
2 tablespoons	rum	30 mL

Strain fruit juices into saucepan. Add sugar and rum. Bring to a boil. Cook until sugar is dissolved. Cool.

RASPBERRY DELIGHT CAKE

⅔ cup	shortening	150 mL
1½ cups	sugar	375 mL
3	eggs, beaten	3
1 teaspoon	vanilla extract	5 mL
2 cups	all purpose flour	500 mL
1 teaspoon	salt	5 mL
2 teaspoons	baking powder	10 mL
1⅓ cups	milk	325 mL
1 teaspoon	grated lemon rind	5 mL
2 cups	whipping cream	500 mL
4 cups	fresh raspberries	1 L

Cream together shortening, 1¼ cups (300 mL) sugar, eggs, and vanilla. Combine flour, salt and baking powder. Mix dry ingredients with creamed mixture alternating with milk. Stir in lemon rind. Spoon into 2 well greased 9″ (23 cm) cake pans. Bake in 350°F (180°C) oven for 30-40 minutes until toothpick inserted in centre comes out clean. Turn out onto wire rack to cool. Cut cakes in half. Whip cream until almost stiff. Add remaining sugar, and continue beating until stiff. Spread half the cream on one cake layer, top with raspberries. Repeat for remaining layers. Let rest, refrigerated 30 minutes before serving.

Makes 8 servings.

LAYERED RASPBERRY SQUARES

1 (250 g)	package soft cream cheese	1 (250 g)
½ cup	sugar	125 mL
1 teaspoon	vanilla	5 mL
1	envelope unflavoured gelatin	1
¼ cup	water	50 mL
1 cup	milk	250 mL
1 (15-ounce)	package frozen raspberries, thawed	1 (425 g)
1 (3-ounce)	package raspberry jelly powder	1 (85 g)
1 cup	boiling water	250 mL

Beat cream cheese and sugar until light and smooth. Blend in vanilla. Sprinkle gelatin over ¼ cup (50 mL) water. Let stand 5 minutes to soften. Combine milk and gelatin in a saucepan and heat until gelatin dissolves. Gradually beat milk mixture into cream cheese mixture. Pour into 9x9x2-inch (2.5 L) pan. Chill until firm. Drain raspberries, reserving ¾ cup (175 mL) syrup. Dissolve jelly powder in 1 cup (250 mL) boiling water. Stir in raspberries and reserved syrup. Chill until slightly thickened. Spoon over cream cheese layer. Chill until set. Cut into squares to serve.

Makes one 9x9x2-inch/2.5 L pan.

Happy Face Cupcakes

1¼ cups	all-purpose flour	300 mL
¾ cup	unsweetened cocoa	175 mL
1 tablespoon	baking powder	15 mL
½ teaspoon	salt	2 mL
½ cup	butter	125 mL
1⅓ cups	sugar	325 mL
3	eggs	3
⅔ cup	milk	150 mL
1 teaspoon	vanilla	5 mL

Sift together flour, cocoa, baking powder and salt. Cream butter; gradually beat in sugar. Add eggs one at a time, beating well after each addition. Add sifted dry ingredients to creamed mixture alternately with milk and vanilla, combining lightly after each addition. Divide batter evenly among 22 large paper baking cups set in muffin tins. Bake in preheated 375°F (190°C) oven 20 to 25 minutes or until done. Cool completely. Frost with CHEESY FROSTING and decorate as desired.
   Makes 22 cupcakes.

CHEESY FROSTING

1 (250 g)	package soft cream cheese	1 (250 g)
3 tablespoons	honey	45 mL
1 teaspoon	vanilla	5 mL

Beat cream cheese until light. Beat in honey and vanilla. Frost each cupcake. Make 'happy faces' with peanuts, raisins, dried banana chips, grapes, apple slices, etc.
   Makes 1⅔ cups/400 mL.

CHOCOLATE CREAM CHEESE CUPCAKES

FILLING

8 ounces	cream cheese	250 g
¼ cup	sugar	75 mL
1	egg, beaten	1

CUP CAKES

1½ cups	all purpose flour	375 mL
¾ cup	sugar	175 mL
¼ teaspoon	salt	1 mL
¼ cup	cocoa	50 mL
1 teaspoon	baking soda	5 mL
1 teaspoon	cinnamon	5 mL
1 cup	milk	250 mL
⅓ cup	oil	75 mL
1 teaspoon	vanilla extract	5 mL
1 teaspoon	grated orange rind	5 mL
1 tablespoon	vinegar	15 mL
	chopped nuts	

To prepare filling beat cream cheese, sugar and egg together until smooth. Set aside. In a large bowl combine flour, sugar, salt, cocoa, soda and cinnamon. In a separate bowl blend together milk, oil, vanilla, orange rind and vinegar. Gradually whisk wet ingredients into dry, beating until smooth. Fill lightly greased muffin tins about ½ full. Top with a tablespoon (15 mL) of filling. Sprinkle with chopped nuts if desired. Bake in 350°F (180°C) oven until cooked through.

Makes 12 cupcakes.

Brown Sugar Nuggets

3¾ cups	all-purpose flour	925 mL
1 teaspoon	baking soda	5 mL
1 teaspoon	salt	5 mL
1 cup	shortening	250 mL
2 cups	firmly-packed brown sugar	500 mL
2	eggs	2
2 teaspoons	lemon juice	10 mL
½ cup	milk	125 mL
	whole almonds or pecans	

Combine flour, baking soda and salt. Cream shortening; gradually beat in sugar. Add eggs and beat until light and fluffy. Add lemon juice. Blend in milk. Stir in dry ingredients. Chill dough 1 hour. Drop by rounded teaspoons onto lightly greased cookie sheets. Top each mound with a whole nut. Bake in preheated 400°F (200°C) oven 8 to 10 minutes. Makes about 50 cookies.

Whirly-Twirly Pinwheel Cookies

1½ cups	all-purpose flour	375 mL
½ teaspoon	baking powder	3 mL
¼ teaspoon	salt	2 mL
½ cup	shortening	125 mL
½ cup	sugar	125 mL
1	egg yolk	1
3 tablespoons	milk	50 mL
1	square unsweetened chocolate, melted	1

Combine flour, baking powder and salt. Cream shortening; gradually beat in sugar. Beat in egg yolk and milk. Stir in dry ingredients. Divide dough in half and add melted chocolate to one half. Roll out plain dough on lightly floured surface to about ¼-inch (6 mm) thickness. Roll chocolate dough to same size and place on top of plain dough. Roll the two doughs together like a jelly roll. Wrap roll in foil or plastic wrap and chill until firm, several hours or overnight. Cut dough into slices and place on greased cookie sheets. Bake in preheated 400°F (200°C) oven 10 to 12 minutes. Makes about 30 cookies.

PEACH KUCHEN

2 cups	variety baking/biscuit mix	500 mL
2 tablespoons	sugar	30 mL
¼ cup	chilled butter	50 mL
1 (19-ounce)	can sliced peaches, drained	1 (540 mL)
1	egg, slightly beaten	1
⅓ cup	sugar	75 mL
¼ teaspoon	ground cinnamon	1 mL
1 cup	milk	250 mL
	ground nutmeg	

Combine baking mix and 2 tablespoons (30 mL) sugar. Cut in butter until mixture resembles coarse crumbs. Press evenly over bottom and half way up sides of an 8x8x2-inch (2 L) baking dish. Bake in preheated 350°F (180°C) oven 15 minutes. Arrange peach slices in rows over crust. Combine egg, ⅓ cup (75 mL) sugar and cinnamon. Gradually beat in milk. Pour over fruit. Sprinkle with nutmeg. Return to oven and bake 30 to 35 minutes longer or until custard is set. Serve warm.
     Makes 6 servings.

Peach dumplings

3 cups	all purpose flour	750 mL
3 teaspoons	baking powder	15 mL
1 teaspoon	salt	5 mL
6 tablespoons	sugar	90 mL
½ cup	butter	125 mL
3 tablespoons	butter	45 mL
1 cup	milk	250 mL
6	fresh peaches	6
¼ cup	brown sugar	50 mL
¼ cup	finely chopped nuts	50 mL
¼ teaspoon	cinnamon	2 mL
3	egg yolks	3
1¼ cups	milk	300 mL

Combine flour, baking powder, salt and 3 tablespoons (45 mL) sugar. Cut in ½ cup (125 mL) butter until mixture resembles cornmeal. Add milk gradually and blend until mixture is moistened and will hold together. Blend together remaining butter, brown sugar, nuts, and cinnamon. Set aside. Peel peaches, halve and remove pits. Roll dough out to ¼ inch (.6 cm) thickness on a lightly floured surface. Cut into 4 inch (10 cm) squares. Place a peach half on each square. Place 1 teaspoon nut mixture in cavity. Wrap dough around fruit. Pinch corners to seal. Place on lightly greased baking sheet. Bake in 375°F (190°C) oven for 25-30 minutes until crisp and lightly browned. While dumplings are baking prepare sauce. Whisk egg yolks with remaining sugar until lemony and smooth. Heat milk until hot. Add a little hot milk to eggs, then return to pot. Stir and cook until gently thickened. Serve warm or chilled with dumplings.

Makes 12 servings.

BRANBERRY MUFFINS

¾ cup	whole wheat flour	175 mL
¾ cup	bran	175 mL
½ cup	wheat germ	125 mL
¾ cup	lightly packed brown sugar	175 mL
1 teaspoon	baking soda	5 mL
½ teaspoon	salt	2 mL
1 teaspoon	grated orange rind	5 mL
1 cup	blueberries, fresh or frozen	250 mL
½ cup	raisins or currants	125 mL
1	egg	1
¾ cup	buttermilk or sour milk	175 mL
¼ cup	oil	50 mL

In a large mixing bowl, mix dry ingredients. Stir in orange rind, blueberries and raisins.

In a small bowl, beat together egg, buttermilk and oil. Add to dry ingredients and mix until well blended.

Spoon into greased muffin tins.

Bake in 400°F (200°C) oven for 20-25 minutes.

Makes 12 large or 18 medium muffins.

CARROT WHEAT MUFFINS

1	cup	whole wheat flour	250 mL
1	cup	all-purpose flour	250 mL
1	tablespoon	baking powder	15 mL
½	teaspoon	salt	2 mL
¼	cup	brown sugar	50 mL
½	teaspoon	cinnamon	2 mL
⅛	teaspoon	allspice or ground cloves	0.5 mL
1	cup	coarsely grated carrots	250 mL
1	teaspoon	grated orange rind	5 mL
1		egg	1
1	cup	milk	250 mL
¼	cup	molasses	50 mL
¼	cup	butter, melted	50 mL
½	cup	raisins	125 mL
¼	cup	chopped nuts	50 mL

Measure dry ingredients into a large mixing bowl. Stir in carrots and orange rind. In a small bowl, beat egg with milk, molasses and melted butter. Add to dry mixture all at once, stirring just to moisten. Fold in raisins and nuts. Fill greased muffin tins ¾ full. Bake in a 400°F (200°C) oven for 25 minutes.
Makes 12 large muffins.

CHOCOLATE CHIP MUFFINS

1¾ cups	all purpose flour	425 mL
½ cup	whole wheat flour	125 mL
2½ teaspoons	baking powder	12 mL
1	ripe banana, mashed (approximately ½ cup/125 mL)	1
½ cup	brown sugar, packed	125 mL
2	eggs	2
1 cup	milk	250 mL
⅓ cup	butter, melted	75 mL
1 cup	chocolate chips	250 mL

Preheat oven to 400°F (200°C). Butter 12 large muffin pans well or line with paper muffin cups. Sift together flours, and baking powder. Reserve. Combine banana with brown sugar until smooth. Beat in eggs, milk and melted butter. Stir in dry ingredients only until blended. Quickly stir in chocolate chips. Do not overbeat. Spoon mixture into prepared muffin cups. (An ice cream scoop works well for this.) Bake 25 minutes or until browned slightly. Cool on racks.

Makes 12 large muffins.

Apple 'n' Oats Muffins

1 cup	milk	250 mL
1 tablespoon	lemon juice	15 mL
1 cup	quick cooking rolled oats	250 mL
1 cup	all purpose flour	250 mL
1 teaspoon	baking powder	5 mL
½ teaspoon	baking soda	2 mL
1 teaspoon	cinnamon	5 mL
1 teaspoon	nutmeg	5 mL
½ teaspoon	salt	2 mL
½ cup	brown sugar	125 mL
1	egg, beaten	1
¼ cup	melted butter	50 mL
½ cup	raisins	125 mL
½ cup	grated apple	125 mL
½ cup	chopped walnuts	125 mL

Set oven to 400°F (200°C). Combine lemon juice and milk and set aside 5 minutes. Stir in oats. In a large bowl toss together flour, baking powder, baking soda, cinnamon, nutmeg and salt. Stir brown sugar and egg into oat milk mixture. Add to dry ingredients with melted butter, stirring swiftly just to combine. Fold in raisins, apple and nuts. Spoon into lightly greased 2½″ (6 cm) muffin tins and bake in a preheated oven about 20 minutes or until done.

Makes one dozen.

MAPLE NUT MUFFINS

½	cup	maple syrup	125 mL
2	tablespoons	melted butter	30 mL
¼	cup	chopped walnuts	50 mL
2	cups	all-purpose flour	500 mL
1	tablespoon	baking powder	15 mL
1	teaspoon	salt	5 mL
1	cup	milk	250 mL
3	tablespoons	maple syrup	45 mL
¼	cup	salad oil	50 mL
1		egg	1

Grease 12 large muffin cups. Put 2 teaspoons (10 mL) maple syrup, ½ teaspoon (2 mL) melted butter and 1 teaspoon (5 mL) chopped nuts into each muffin cup. Set aside. Sift flour, baking powder and salt into mixing bowl. Mix remaining ingredients together. Add to flour mixture stirring with a fork, just to blend. Spoon into muffin cups, filling about ⅔ full.

Bake at 425°F (220°C) for 20 minutes. Invert on rack set on waxed paper. Leave 2-3 minutes. Remove pan. Serve warm or cold.

Makes 12 muffins.

Halloween Pumpkin Muffins

1⅔ cups	all-purpose flour	400 mL
½ cup	sugar	125 mL
1 tablespoon	baking powder	15 mL
½ teaspoon	salt	2 mL
½ teaspoon	ground cinnamon	2 mL
½ teaspoon	ground nutmeg	2 mL
1	egg	1
¼ cup	cooking oil	50 mL
½ cup	canned pumpkin	125 mL
⅔ cup	milk	150 mL
¼ cup	orange marmalade	50 mL

In a large bowl, stir together flour, sugar, baking powder, salt, cinnamon and nutmeg. Beat egg well in a small bowl; beat in oil, pumpkin and milk. Add all at once to dry ingredients, stirring just until moistened. Divide mixture evenly among 12 greased medium muffin cups. Place about 1 teaspoon (5 mL) marmalade on top of each muffin. Bake in preheated 400°F (200°C) oven 20 to 25 minutes or until done. Serve warm.

Makes 1 dozen.

LEMON BLUEBERRY MUFFINS

1 cup	butter, at room temperature	250 mL
1½ cups	sugar	375 mL
4	eggs	4
1 teaspoon	pure vanilla extract	5 mL
2 tablespoons	lemon juice	30 mL
2 tablespoons	grated fresh lemon rind	30 mL
2½ cups	all purpose flour	625 mL
½ cup	yellow cornmeal	125 mL
2 teaspoons	baking powder	10 mL
1 teaspoon	baking soda	5 mL
1 cup	milk	250 mL
1½ cups	fresh blueberries, patted dry	375 mL

Cream butter and sugar until light. Add eggs one at a time; beat well after each addition. Beat in vanilla, lemon juice and lemon rind. Do not worry if mixture looks curdled. Reserve 2 tablespoons (30 mL) flour for the blueberries. Combine remaining flour, cornmeal, baking powder and baking soda. Add dry ingredients to butter mixture alternately with milk, beginning and ending with dry ingredients. Toss reserved flour with blueberries and stir into batter gently. Spoon batter into buttered muffin pans (or line with paper cups); batter should come almost to top of pans. Bake in a preheated 350°F (180°C) oven for 25 to 30 minutes or until just turning golden. Remove from pans and place on wire racks to cool.

Makes 18 large or 24 medium muffins.

PEANUT BUTTER RAISIN MUFFINS

2 cups	all purpose flour	500 mL
¼ cup	brown sugar	50 mL
1 tablespoon	baking powder	15 mL
½ teaspoon	salt	2 mL
½ teaspoon	allspice	2 mL
2	eggs, beaten	2
1 cup	milk	250 mL
½ cup	peanut butter	125 mL
¼ cup	melted butter	50 mL
¾ cup	raisins	175 mL

Combine flour, sugar, baking powder, salt and allspice in a large bowl. In another bowl mix eggs with milk, peanut butter and butter. Add to dry ingredients and mix until just combined. Fold in raisins. Spoon batter into 12 greased 2 ½ inch (6.2 cm) muffin tins. Bake in 400°F (200°C) oven 20-25 minutes. Remove from pan and serve warm.

PEANUT BUTTER AND JELLY MUFFINS

1¾ cups	all-purpose flour	425 mL
3 tablespoons	sugar	45 mL
2½ teaspoons	baking powder	12 mL
¾ teaspoon	salt	3 mL
⅓ cup	smooth peanut butter	75 mL
3 tablespoons	butter	45 mL
1	egg	1
¾ cup	milk	175 mL
	grape jelly	

Combine flour, sugar, baking powder and salt. Cut in peanut butter and butter until mixture resembles coarse crumbs. Beat egg; stir in milk. Add all at once to dry ingredients, stirring just until moistened. Divide batter evenly among 12 greased medium muffin cups. Press about ½ teaspoon (2 mL) grape jelly into the top of each muffin. Bake in preheated 400°F (200°C) oven 20 to 25 minutes or until golden brown. Serve warm or cool.

Makes 1 dozen.

Pumpkin Raisin Muffins

1½ cups	whole wheat flour	375 mL
½ cup	sugar	125 mL
2 teaspoons	baking powder	10 mL
½ teaspoon	salt	2 mL
½ teaspoon	cinnamon	2 mL
½ teaspoon	nutmeg	2 mL
1	egg	1
½ cup	milk	125 mL
½ cup	mashed, cooked or canned pumpkin	125 mL
¼ cup	butter, melted	50 mL
½ cup	seedless raisins	125 mL

In a large bowl, stir together flour, sugar, baking powder, salt, cinnamon and nutmeg. Beat egg in a small bowl; stir in milk, pumpkin and butter. Add all at once to dry ingredients, stirring just until moistened. Stir in raisins. Divide mixture evenly among 12 greased medium muffin cups. Bake in pre-heated 400°F (200°C) oven 20 to 25 minutes or until done. Serve warm or cool.

Makes 1 dozen.

SAVOURY CHEESE MUFFINS

2	cups	all-purpose flour	500 mL
1	tablespoon	baking powder	15 mL
½	teaspoon	salt	2 mL
½	cup	softened butter	125 mL
¼	cup	sugar	50 mL
2		eggs	2
1	cup	milk	250 mL
1	cup	grated Canadian Cheddar cheese	250 mL
½	cup	chopped smoked ham or pepperoni	125 mL
1	teaspoon	basil	5 mL

Pre-heat oven to 350°F (180°C).

Grease and flour twelve large muffin cups. Sift together flour, baking powder and salt. Beat butter with sugar until creamy; add eggs; beat well. Gradually add flour mixture and milk to butter mixture, alternating dry and liquid ingredients. Quickly fold in cheese, meat and basil. Spoon batter into prepared pan and bake in oven for 25-30 minutes until done. Remove to a rack. Delicious served at once or reheated and buttered.

Variation: Batter may be baked in a 9" x 5" (2 L) greased loaf pan for 50-60 minutes. Serve sliced and toasted with butter.

Makes 12 muffins.

DATE 'N' NUT MUFFINS

2 cups	all-purpose flour	500 mL
1 tablespoon	baking powder	15 mL
1 teaspoon	salt	5 mL
⅓ cup	chopped nuts	75 mL
1	egg, well beaten	1
1¼ cups	milk	300 mL
1 cup	finely-chopped dates	250 mL
½ cup	firmly-packed brown sugar	125 mL
¼ cup	butter, melted	50 mL

In a large bowl stir together flour, baking powder, salt and nuts. In a small bowl thoroughly combine egg, milk, dates, sugar and butter; add all at once to dry ingredients, stirring just until moistened. Divide mixture evenly among 12 greased medium muffin cups. Bake in preheated 400°F (200°C) oven 22 to 25 minutes. Serve warm.

Makes 1 dozen.

GOLDEN PENNY PANCAKES

¾ cup	all-purpose flour	175 mL
1½ teaspoons	baking powder	7 mL
1½ teaspoons	sugar	7 mL
¼ teaspoon	salt	1 mL
¾ cup	shredded Canadian	175 mL
	Cheddar cheese	
1	egg	1
¾ cup	milk	175 mL
2 tablespoons	butter, melted	30 mL
	applesauce	

In a large bowl combine flour, baking powder, sugar and salt. Add and mix in cheese. In a medium bowl beat egg well. Stir in milk and butter. Add to flour mixture; stir until well combined. Using about 1 heaping tablespoon (15 mL) batter for each pancake, pour onto a preheated greased griddle or frypan. Cook 2 or 3 minutes or until underside is golden and surface is bubbly. Turn; cook until done. Serve with warm applesauce.

Makes about 30 mini pancakes.

APPLE PANCAKE

¾ cup	all purpose flour	175 mL
½ teaspoon	salt	2 mL
3	eggs, lightly beaten	3
¾ cup	milk	175 mL
½ cup	thinly sliced peeled apples	125 mL
2 tablespoons	butter	30 mL

Preheat oven to 450°F (230°C). Combine flour and salt in a large bowl. Add eggs and milk and beat until smooth. Stir in apple slices. Melt butter in a large heavy frying pan. When butter is bubbling, pour in batter and set pan in oven for 15 minutes. Lower heat to 350°F (180°C) and continue baking for about 10 minutes more or until pancake is light brown and crisp. Spread with SPICY APPLE FILLING and serve hot cut into wedges as for a pie.

Makes 1 large pancake – 6 servings.

SPICY APPLE FILLING

2 tablespoons	butter	30 mL
8-10	tart apples, sliced ¼″ (.5 cm)	8-10
¼ cup	sugar	50 mL
1 teaspoon	cinnamon	5 mL

Melt butter in a large frying pan. Add apple slices and cook over medium heat until tender, about 10-15 minutes. Stir in sugar and cinnamon.

Makes about 4 cups/1 L.

BERRY PUFF PANCAKES

4	eggs, separated	4
1	cup	milk	250 mL
3	tablespoons	butter, melted	45 mL
1	cup	all-purpose flour	250 mL
¼	teaspoon	salt	1 mL
1	tablespoon	honey	15 mL
2	cups	fresh berries	500 mL
		icing sugar	

Combine egg yolks, milk, 1 tablespoon (15 mL) butter, flour, salt and honey. Beat egg whites until stiff but not dry; fold into milk mixture. Heat 1 tablespoon (15 mL) butter in a 9″ or 10″ (23-25 cm) frypan. Pour half of the mixture into pan; cook over low heat 3 to 5 minutes, or until bubbles appear on the surface. Spoon 1 cup (250 mL) berries over top. Cover handle of frypan with foil; place under hot broiler until pancake is puffed and golden brown, about 2 to 3 minutes. Remove from pan and keep warm; repeat with remaining mixture. Sprinkle pancakes with icing sugar and serve in wedges. Accompany with maple syrup, fresh cream, bacon or sausages.
Note: Raspberries, blueberries or any other kind of berry or pitted cherries may be used.
    Makes 2 pancakes – 2 to 4 servings.

Pina colada pancakes

2 cups	all-purpose flour	500 mL
2 tablespoons	baking powder	30 mL
⅓ cup	sugar	75 mL
½ teaspoon	salt	2 mL
1 (14-ounce)	can crushed pineapple	1 (398 mL)
	water	
2	eggs	2
1 cup	milk	250 mL
¼ cup	butter, melted	50 mL
1 cup	toasted, flaked coconut	250 mL

In a large bowl, combine flour, baking powder, sugar and salt. Drain pineapple and reserve syrup; set pineapple aside. Add water to syrup to make 1 cup (250 mL) liquid. In a medium bowl, beat eggs well. Stir in syrup-water mixture, milk and butter. Add to flour mixture; stir only until combined (batter will be lumpy). Using about ¼ cup (50 mL) batter for each pancake, pour onto preheated greased griddle or frypan. Sprinkle each pancake with 1 tablespoon (15 mL) drained pineapple. Cook 2 or 3 minutes or until underside is golden and surface is bubbly. Turn and cook until brown. Pour CREAMY RUM SAUCE over each serving and sprinkle with toasted coconut.
Makes 16 pancakes.

CREAMY RUM SAUCE

1	package (4 serving size) vanilla pudding and pie filling	1
2¾ cups	milk	675 mL
¼ cup	light rum	50 mL
1 teaspoon	vanilla	5 mL

Combine pudding mix and milk in a saucepan. Cook according to package directions. Stir in rum and vanilla. Serve warm.
Makes 3 cups/750 mL.

MAPLE BUTTERED PANCAKES

1½ cups	all-purpose flour	375 mL
¾ teaspoon	baking powder	4 mL
¾ teaspoon	baking soda	4 mL
¾ teaspoon	salt	4 mL
¾ cup	plain yogurt	175 mL
1 cup	milk	250 mL
2	eggs, beaten	2
¼ cup	butter, melted	50 mL

In large bowl, mix together flour, baking powder, baking soda, and salt. In separate bowl combine yogurt, milk, eggs, and butter; add all at once to dry ingredients, stirring just until combined.

Cook pancakes in hot, lightly greased skillet or on griddle until bubbles appear on surface. Turn and brown underside.

Serve with hot maple butter.

Makes 12 – 5″ (12.5 cm) pancakes.

HOT MAPLE BUTTER

1 cup	maple syrup	250 mL
½ cup	butter	125 mL

Combine maple syrup and butter in a saucepan. Cook, stirring constantly, until butter melts and sauce is heated through.

Makes 1½ cups (375 mL).

OATMEAL PANCAKES

1½	cups	rolled oats	375 mL
2	cups	milk	500 mL
½	cup	whole wheat flour	125 mL
½	cup	all-purpose flour	125 mL
1	tablespoon	brown sugar	15 mL
1	tablespoon	baking powder	15 mL
1	teaspoon	salt	5 mL
½	teaspoon	cinnamon	2 mL
2		eggs, beaten	2
¼	cup	butter, melted	50 mL

In a large mixing bowl, blend rolled oats and milk; let stand 5 minutes. Stir together flours, sugar, baking powder, salt and cinnamon. Add dry ingredients, eggs and melted butter to oats, stirring until combined. Pour ¼ cup (50 mL) batter for each pancake onto a hot, lightly greased griddle. Cook each pancake until edges become dry and surface is covered with bubbles. Turn and cook second side until golden brown.

Makes about 16-18 medium pancakes.

CORN PANCAKES

⅓ cup	whole wheat flour	75 mL
⅔ cup	all purpose flour	150 mL
1 tablespoon	baking powder	15 mL
2 teaspoons	sugar	10 mL
½ teaspoon	celery salt	2 mL
1	egg, beaten	1
2 tablespoons	oil	30 mL
1 cup	milk	250 mL
1 cup	canned kernel corn, drained	250 mL

Combine whole wheat flour, all purpose flour, baking powder, sugar and salt. In a separate bowl, combine egg, oil and milk. Pour liquid into dry ingredients and stir just until moistened. Stir in corn kernels. On a hot, greased griddle, fry pancakes 3″ (8 cm) in diameter. Serve with applesauce as an accompaniment to pork.
Makes 4 servings.

Cinnamon Popovers

3	eggs	3
1 cup	milk	250 mL
1 cup	all-purpose flour	250 mL
3 tablespoons	soft butter	45 mL
1 teaspoon	ground cinnamon	5 mL
¼ teaspoon	salt	1 mL

In a blender container combine eggs, milk, flour, butter, cinnamon and salt. Cover and blend at high speed for 15 seconds. Scrape down any flour that has collected on the sides of the container. Continue blending at high speed until thoroughly combined. Fill six well-greased 6-ounce (150 mL) custard cups half full with batter. Bake in preheated 400°F (200°C) oven 40 minutes or until puffed and golden. Remove from pan and serve immediately with HONEY BUTTER.

Makes 6 popovers.

HONEY BUTTER

½ cup	butter	125 mL
⅓ cup	liquid honey	75 mL

Cream butter until light and fluffy. Gradually beat in honey. Spoon into a small pot to serve.

Makes about 1 cup/250 mL.

ORANGE FRENCH TOAST

8 slices	French bread cut 1″ (2.5 cm) thick	8
4	eggs	4
$\frac{1}{8}$ teaspoon	ground nutmeg	0.5 mL
2 teaspoons	sugar	10 mL
1 teaspoon	grated orange rind	5 mL
1 cup	milk	250 mL
$\frac{1}{2}$ cup	finely chopped nuts	125 mL

Place slices of bread in shallow dish. Beat eggs. Add nutmeg, sugar, orange rind and milk. Blend well. Pour over bread. Let stand 2 minutes then turn over. Leave for 1 hour. Place bread on greased cookie sheet and bake in 375°F (190°C) oven for 15-20 minutes until lightly browned, turning once. Serve with Orange Honey Syrup.

Makes 4-8 servings.

ORANGE HONEY SYRUP

1 (6$\frac{1}{4}$ ounce)	can frozen orange juice	1 (178 mL)
$\frac{1}{4}$ cup	honey	50 mL
2 teaspoons	butter	10 mL
1 teaspoon	vanilla extract	5 mL

Combine juice with honey, heat slowly stirring constantly to boiling. Simmer, stirring once or twice until well thickened. Stir in butter and vanilla.

SPICE ISLANDS FRENCH TOAST

4	eggs	4
1 cup	milk	250 mL
2 tablespoons	sugar	30 mL
½ teaspoon	vanilla	2 mL
6	¾-inch/18 mm thick slices day-old French bread	6
2 tablespoons	butter	30 mL

Beat eggs lightly to combine. Beat in milk, sugar and vanilla. Pour over bread slices in a large pan. Turn slices of bread over once; cover and refrigerate several hours or preferrably over night. Melt butter in a large frypan or electric skillet. Saute slices of bread until golden on both sides. Serve with Cinnamon-Nutmeg Butter and maple syrup.

Makes 6 servings.

CINNAMON-NUTMEG BUTTER

½ cup	soft butter	125 mL
¼ cup	sifted icing sugar	50 mL
½ teaspoon	ground cinnamon	2 mL
¼ teaspoon	ground nutmeg	1 mL

Cream butter; beat in sugar, cinnamon and nutmeg. Makes about ½ cup/125 mL.

Carrot Wheat Muffins,
page 462

BANANA FRENCH TOAST

1 cup	milk	250 mL
2	eggs	2
½	medium banana	½
1 tablespoon	honey	15 mL
2 tablespoons	frozen orange juice concentrate, thawed	30 mL
6	slices stale whole wheat bread	6
2 tablespoons	butter	30 mL
	sliced bananas	
	maple syrup	

Combine milk, eggs, banana, honey and orange juice concentrate in blender container. Cover and blend at high speed until smooth. Pour over bread slices in a large pan. Turn slices of bread over once; cover and refrigerate several hours or preferably overnight. Melt butter in a large frypan or electric skillet. Sauté slices of bread until golden on both sides. Serve with sliced bananas and maple syrup.

Makes 6 servings.

NOTES

Peanut Butter Pudding

½ cup	sugar	125 mL
2 tablespoons	corn starch	30 mL
¼ teaspoon	salt	1 mL
2½ cups	milk	625 mL
½ cup	peanut butter	125 mL
1 teaspoon	vanilla	5 mL
	chopped salted peanuts	

Combine sugar, corn starch and salt in a saucepan. Gradually stir in milk. Cook over medium heat, stirring constantly, until mixture comes to a boil and thickens. Reduce heat and cook 2 minutes longer, stirring occasionally. Stir in peanut butter and vanilla. Cover surface and cool. Chill well before serving. Garnish with peanuts if desired.

Makes 6 servings.

BAKED LEMON SPONGE PUDDING

1 cup	milk	250 mL
¾ cup	sugar	175 mL
3 tablespoons	flour	45 mL
¼ teaspoon	salt	1 mL
1	lemon	1
2	eggs, separated	2
¼ cup	sugar	50 mL

Place milk, ¾ cup (175 mL) sugar, flour and salt in blender container. Cut thin outer rind of the lemon into strips. Add to blender container. Cover and blend at high speed until smooth. Add juice of the lemon and egg yolks; blend at high speed until well combined. Beat egg whites until frothy. Gradually beat in the ¼ cup (50 mL) sugar and continue to beat until stiff peaks form. Pour blended mixture over egg whites; fold to combine. Pour into 1-quart (1 L) casserole. Set in pan of hot water. Bake in preheated 325°F (160°C) oven 50 to 60 minutes. Serve immediately.

Makes 4 or 5 servings.

CHOCOLATE RICE PUDDING

3 cups	milk	750 mL
¾ cup	uncooked long grain rice	175 mL
⅔ cup	sugar	150 mL
¼ cup	cocoa	50 mL
½ teaspoon	salt	2 mL
2 tablespoons	butter	30 mL
1 teaspoon	vanilla	5 mL

Scald milk in top part of double broiler. Combine rice, sugar, cocoa and salt. Stir into milk. Cover and cook over hot water, stirring occasionally, until rice is tender, about 1¼ hours. Remove from heat. Stir in butter and vanilla. Cool. Serve warm.

Makes about 3½ cups/875 mL.

APRICOT BREAD PUDDING

½ cup	raisins	125 mL
5	dried apricots, cut into quarters	5
6	slices bread (preferably egg bread)	6
⅓ cup	butter (preferably unsalted), at room temperature	75 mL
6	eggs	6
¾ cup	sugar	175 mL
3 cups	milk, hot	750 mL
2 teaspoons	pure vanilla extract	10 mL
½ cup	apricot jam or jelly	125 mL

Place raisins and apricots in a bowl; cover with boiling water. Soften 10 minutes; drain well. Sprinkle half the fruit in the bottom of a buttered 8″ square (2 L) baking dish. Butter bread and arrange, buttered side up, on top of the fruit, overlapping if necessary. Sprinkle with remaining fruit. Combine eggs and sugar; blend well. Beat in milk and vanilla. Pour over bread. Place in a water bath (a larger pan half-filled with very hot water). Bake at 350°F (180°C) for about one hour until a knife inserted into the centre comes out clean. Heat jam or jelly and brush over the bread; rest 10 minutes before serving. Serve warm or cold.

Serves 4 to 6.

LEMON PUDDING

¾ cup	sugar	175 mL
¼ cup	flour	50 mL
¼ cup	melted butter	50 mL
	pinch salt	
2	eggs, separated	2
¼ cup	lemon juice	50 mL
2 teaspoons	grated lemon rind	10 mL
1 cup	milk	250 mL
½ teaspoon	baking powder	2 mL

Preheat oven to 350°F (180°C).

Beat together sugar, flour, butter, salt and egg yolks. Add lemon juice and grated rind. Slowly add milk and beat until smooth. Whisk egg whites until stiff but not dry; beat in baking powder. Fold egg whites into lemon/flour mixture. Pour batter into a lightly buttered 4 cup (1L) baking dish. Bake 40 minutes or until pudding is puffed and cooked through. Serve hot or cold. Delicious served with a raspberry sauce.

Makes 4-6 servings.

PEARS CONDE RICE PUDDING

3 ½ cups	milk	875 mL
2 tablespoons	milk	30 mL
3 tablespoons	butter	45 mL
½ cup	sugar	125 mL
¼ teaspoon	salt	1 mL
1 teaspoon	vanilla extract	5 mL
	pinch nutmeg	
1 ⅓ cups	quick cooking rice	325 mL
2	egg yolks	2
1 (28 ounce)	can pear halves	1 (796 mL)
¼ cup	brandy	50 mL
	chopped nuts	
	whipping cream	

Combine 3 ½ cups (875 mL) milk, butter, sugar, salt, vanilla, nutmeg and rice in a saucepan. Bring to a boil. Cook, over medium heat 20 minutes stirring frequently. Blend 2 egg yolks with remaining milk. Beat briefly. Add a small amount of hot rice mixture and stir well. Return to saucepan. Stir well. Remove from heat and chill. While pudding is cooling combine pear halves and juice with brandy in a saucepan. Bring to just a boil. Remove from heat and chill. Stir 3 tablespoons (45 mL) pear syrup into chilled rice mixture. Pile into serving dishes. Arrange pear halves around rice. Garnish with chopped nuts. Top with whipped cream if desired.
Makes 4-6 servings.

Tipsy Bread Pudding

10	slices raisin bread	10
¼ cup	butter, melted	50 mL
½ teaspoon	ground nutmeg	2 mL
½ cup	chopped nuts	125 mL
4	eggs, slightly beaten	4
½ cup	sugar	125 mL
1½ cups	milk	375 mL
½ cup	light rum	125 mL
1 teaspoon	vanilla	5 mL
	light cream	

Cut each slice of bread into 4 pieces. Combine butter and nutmeg; drizzle over bread and toss lightly. Add nuts. Arrange bread and nuts in a greased 2½-quart (2.5 L) round baking dish. Combine eggs, sugar, milk, rum and vanilla. Pour over bread. Set baking dish in pan of hot water. Bake in preheated 350°F (180°C) oven 60 minutes or until knife inserted near centre comes out clean. Serve warm or cold with cream.

Makes 6 to 8 servings.

EASY STEAMED CHRISTMAS PUDDINGS

1½ cups	vanilla wafer crumbs	375 mL
½ cup	all-purpose flour	125 mL
½ teaspoon	baking soda	2 mL
½ teaspoon	ground cinnamon	2 mL
½ teaspoon	ground nutmeg	2 mL
¼ teaspoon	salt	1 mL
¼ cup	shortening	50 mL
½ cup	sugar	125 mL
1	egg	1
⅔ cup	milk	150 mL
½ cup	chopped, blanched almonds	125 mL
½ cup	chopped, drained maraschino cherries	125 mL

Stir together wafer crumbs, flour, baking soda, cinnamon, nutmeg and salt. Cream shortening; beat in sugar. Beat in egg. Add dry ingredients to creamed mixture alternately with milk, combining lightly after each addition. Stir in almonds and cherries. Spoon into six greased 6-ounce (150 mL) custard cups; cover with foil. Pour water into an electric frying pan to depth of ¾-inch (18 mm). Bring to a boil. Place custard cups in frying pan. Cover and steam puddings on very low heat setting 250°F (120°C) 45 to 50 minutes. Remove from water; let stand 10 minutes. Unmold and serve with CUSTARD SAUCE.
Makes 6 servings.

CUSTARD SAUCE

1½ cups	milk	375 mL
3	eggs	3
⅓ cup	sugar	75 mL
½ teaspoon	vanilla	2 mL

Scald milk. Beat eggs until light; beat in sugar. Gradually stir hot milk into eggs. Cook over medium heat, stirring constantly, until mixture will coat a metal spoon. Do not boil. Add vanilla. Cover surface with plastic wrap; cool.
Makes 2 cups/500 mL.

APPLE RAISIN COTTAGE PUDDING

1 (19-ounce)	can apple pie filling	1 (540 mL)
½ cup	raisins	125 mL
1⅓ cups	all-purpose flour	325 mL
¾ cup	sugar	175 mL
1 tablespoon	baking powder	15 mL
½ teaspoon	salt	2 mL
¼ cup	chilled butter	50 mL
¾ cup	milk	175 mL
1	egg, slightly beaten	1

Combine apple pie filling and raisins in an 8x8x2-inch (2 L) baking dish. Combine flour, sugar, baking powder and salt. Cut in butter until mixture resembles coarse crumbs. Beat together milk and egg. Add to butter mixture all at once, stirring just to moisten. Spread batter over apple mixture. Bake in preheated 350°F (180°C) oven 45 to 50 minutes. Serve warm with Brown Sugar Sauce.

Makes 6 to 8 servings.

BROWN SUGAR SAUCE

1 cup	firmly-packed brown sugar	250 mL
½ cup	corn syrup	125 mL
¼ cup	butter	50 mL
½ cup	milk	125 mL
1 teaspoon	vanilla	5 mL

Combine sugar, corn syrup, butter and milk in a saucepan. Cook over low heat, stirring constantly, until smoothly combined and heated through. Add vanilla. Serve warm.

Makes about 1¾ cups/425 mL.

CREAMY RHUBARB-APRICOT PUDDING

4	eggs, separated	4
¾ cup	sugar	175 mL
⅓ cup	flour	75 mL
2 cups	milk	500 mL
1 teaspoon	vanilla extract	5 mL
¼ teaspoon	grated nutmeg	1 mL
1½ cups	fresh rhubarb, cut in ¾" (2 cm) pieces	375 mL
½ cup	dried apricots	125 mL
1 tablespoon	grated orange rind	15 mL
	water	

Beat egg yolks with ½ cup (125 mL) sugar until light and creamy. Fold in flour. Cook milk until hot. Add a little hot milk to egg mixture. Whisk into milk in saucepan. Cook stirring constantly until well thickened. Stir in vanilla and nutmeg. Cover with waxed paper and chill well. Combine rhubarb and 1 tablespoon (15 mL) water in a saucepan. Cover and cook over low heat until rhubarb is soft but not mushy. Blend remaining sugar and cornstarch and stir into rhubarb. Cook, stirring constantly until mixture boils and thickens. Cool. Place apricots in small saucepan, cover with cold water (about ½ cup, 125 mL) and cook over low heat until soft (approximately 15 minutes). Drain, cool and blend until smooth. Add to rhubarb along with 1 tablespoon (15 mL) finely grated orange rind. Blend fruit with cream ½ hour before serving.

Makes 4 servings.

CRAZY RAISIN RICE PUDDING

1	package (4 serving size) vanilla pudding and pie filling	1
2½ cups	milk	625 mL
¾ cup	raisins	175 mL
2½ cups	cooked rice	625 mL
½ teaspoon	vanilla	2 mL
	ground cinnamon	
	ground nutmeg	

Combine pudding mix and milk in a saucepan. Add raisins. Cook according to package directions. Cool slightly. Stir in cooked rice and vanilla; season to taste with cinnamon and nutmeg. Serve warm or cooled.

Makes 8 servings.

Rice and Apricot Dessert

½ cup	rice	125 mL
1¼ cups	water	300 mL
2½ cups	milk	625 mL
¼ cup	sugar	50 mL
1 teaspoon	vanilla	5 mL
2 cups	dried apricots, sliced	500 mL
2 tablespoons	rum, or brandy	30 mL
	or lemon juice	
⅔ cup	whipping cream	150 mL

Cook rice in boiling water for five minutes; strain. Combine parboiled rice with milk, sugar and vanilla in the top of a double boiler. Set over hot water, cover and leave to cook until rice is soft, and milk is absorbed, stirring occasionally. Cover and cool. Meanwhile simmer apricots in water to cover until just soft. Flavour with a dash of rum, brandy or lemon juice. Set aside. Stir 1 cup (250 mL) fruit into rice. Beat cream until stiff and fold into rice/fruit mixture. Spoon into a 4 cup (1L) mold and chill. Whirl remaining apricots in a blender to form a purée to serve with the dessert.

Makes 6 servings.

VARIATION: STRAWBERRY RICE PARFAITS

Prepare dessert as above but substitute 2 cups (500 mL) fresh sliced strawberries, sprinkled with sugar and marinated in a little liqueur, for apricots. Arrange layers of creamy rice and strawberries in parfait glasses and top with additional whipped cream and whole berries.

Makes 6 servings.

BANANA SCOTCH-ER-OO PUDDING

1	package (4 serving size) butterscotch pudding and pie filling	1
2	eggs, separated	2
2½ cups	milk	625 mL
2	large ripe bananas, sliced	2
¼ cup	sugar	50 mL

Combine pudding mix, egg yolks and milk in a saucepan. Cook according to package directions. Cover surface with plastic wrap; cool. Alternately layer pudding and banana slices in a 1-quart (1 L) round casserole, ending with pudding. Beat egg whites until frothy. Gradually beat in sugar. Continue to beat until stiff peaks form. Spread meringue over pudding; seal edges well. Bake in preheated 350°F (180°C) oven 8 to 10 minutes or until meringue is lightly browned. Serve warm.

Makes 6 to 8 servings.

DANISH RUM PUDDING WITH RASPBERRY SAUCE

1	envelope unflavoured gelatin	1
1½ cups	milk	375 mL
½ teaspoon	salt	2 mL
¾ cup	sugar	175 mL
4	eggs, separated	4
2 tablespoons	dark rum	30 mL
1 cup	cream	250 mL

Sprinkle gelatin over cold milk. Add salt and ½ cup (125 mL) sugar. Heat on top of double boiler to boiling point. Pour over slightly beaten egg yolks. Return to heat and cook, stirring constantly until mixture coats back of spoon. Add rum. Chill until mixture reaches consistency of unbeaten egg whites. Beat egg whites with remaining sugar until stiff. Fold into rum mixture. Beat cream until stiff and fold gently but thoroughly into custard. Turn into 6 cup (1.5 L) mold. Chill until set. Serve with Raspberry Sauce.

RASPBERRY SAUCE

2 cups	frozen raspberries, partially thawed	500 mL
2 tablespoons	sugar	30 mL
1 tablespoon	lemon juice	15 mL

Combine ingredients in blender container. Cover and blend briefly. Strain and chill.

Makes 6 servings.

HOLIDAY DESSERT PUDDING

2	envelopes unflavoured gelatin	2
¾ cup	sugar	175 mL
¼ teaspoon	salt	1 mL
2 cups	milk	500 mL
2	eggs, separated	2
1 teaspoon	grated lemon rind	5 mL
¼ teaspoon	nutmeg	1 mL
1 tablespoon	rum (optional)	15 mL
1 teaspoon	vanilla extract	5 mL
¼ cup	blanched slivered almonds	50 mL
½ cup	candied cherries, quartered	125 mL
1 cup	whipping cream	250 mL

Combine gelatin, ½ cup (125 mL) sugar and salt on top of a double boiler. Add milk and beaten egg yolks. Whisk until well blended. Cook over simmering water stirring constantly until mixture is lightly thickened. Stir in nutmeg, rum and vanilla. Strain into bowl. Chill until well thickened and cooled. Beat egg whites until almost stiff. Gradually add remaining sugar and beat until stiff. Fold into custard. Beat whipping cream until stiff. Fold in custard with nuts and cherries. Spoon into 1½ quart (1.5 L) mold. Chill until set, preferably overnight. Unmold. Serve with additional whipped cream.
Makes 8 servings.

Sweet Noodle Pudding Deluxe

1-12 ounce	package regular egg noodles	1 (340 g)
4	eggs	4
¾ cup	sugar	175 mL;
2 tablespoons	sugar	30 mL
1 cup	sour cream	250 mL
2½ cups	milk	625 mL
½ cup	seedless raisins	125 mL
1 teaspoon	cinnamon	5 mL
1 tablespoon	melted butter	15 mL
2 cups	crushed cornflakes	500 mL

Cook noodles in boiling water until just tender. Drain. Beat eggs. Gradually add ¾ cup (175 mL) sugar, sour cream and milk. Stir in noodles and raisins. Pour into a lightly greased 10-inch square (3 L) pan. Combine cornflakes, cinnamon, remaining sugar and butter. Sprinkle on top of pudding. Bake uncovered for 1 hour at 300°F (150°C). Raise heat to 350°F (180°C) and continue cooking 40 minutes more until set. Serve, cut into squares, warm or cooled to room temperature.

Makes 1-10 inch/3 L pan.

CREAMY RICE PUDDING

½ cup	rice (preferably short grain)	125 mL
1 cup	boiling water	250 mL
⅓ cup	sugar	75 mL
1 teaspoon	cornstarch	5 mL
	pinch of salt	
4 cups	milk	1 L
1 cup	18% table cream	250 mL
	nutmeg to taste	
½ cup	raisins	125 mL
2	egg yolks	2
1 teaspoon	pure vanilla extract	5 mL
2 tablespoons	butter	30 mL
1 tablespoon	cinnamon	15 mL

Combine rice with boiling water in a medium sized saucepan. Cover. Simmer gently 15 minutes or until water is absorbed. Combine sugar with cornstarch and salt. Whisk in 1 cup (250 mL) milk. Stir until smooth. Add sugar mixture along with remaining milk and cream to the saucepan with the rice. Combine well. Add nutmeg and raisins. Stirring steadily, bring to a boil. Cover, reduce heat to the barest simmer, stir occasionally. Cook 1 to 1½ hours or until mixture is no longer liquidy but very creamy. Beat egg yolks. Remove pudding from heat. Whisk a little of the pudding into yolks. Add yolk mixture to rest of pudding. Cook 1 minute. Remove from heat again, stir in vanilla and butter. Transfer to an attractive serving bowl and sprinkle top with cinnamon. Serve hot or cold.

Makes 6 servings.

Pudding Pompadour

3 cups	milk	750 mL
3	strips orange rind	3
3 tablespoons	cornstarch	45 mL
½ cup	sugar	125 mL
	pinch salt	
2	egg yolks, beaten	2
1 teaspoon	vanilla	5 mL
	chocolate meringue	

In top of double boiler, heat 2½ cups (625 mL) milk with orange rind; remove rind. Blend remaining milk with cornstarch, sugar and salt. Gradually beat in hot milk. Set over hot water and bring to boil, stirring constantly, until mixture is smooth and thick. Cook 10 minutes. Beat half of hot sauce into egg yolks then return to heat and cook 2 minutes more. Add vanilla. Pour into 4 cup (1 L) baking dish, cover with Chocolate Meringue and bake in a preheated 300°F (150°C) oven until meringue is slightly browned. Chill.

Makes 4-6 servings.

CHOCOLATE MERINGUE

1 ounce	semi-sweet chocolate, grated	28 g
1 tablespoon	milk	15 mL
2	egg whites	2
3 tablespoons	sugar	45 mL

Combine grated chocolate with milk in a bowl and set in hot water until chocolate is melted. Cool slightly. Beat egg whites until stiff. Gradually beat in sugar until stiff peaks form. Add melted chocolate.

FESTIVE RASPBERRY TRIFLE

1	package (4 serving size) vanilla pudding and pie filling	1
3¼ cups	milk	800 mL
1 teaspoon	vanilla	5 mL
1 (15-ounce)	package frozen raspberries, thawed	1 (425 g)
1 (285 g)	package mini raspberry jelly rolls	1 (285 g)
⅓ cup	sherry (optional) whipped cream	75 mL

Combine pudding and milk in a saucepan. Cook according to package directions. Stir in vanilla. Cover surface with plastic wrap; cool completely. Drain raspberries, reserving syrup. Slice each jelly roll into 5 pieces. Line bottom and sides of a 1½ quart (1.5 L) glass serving bowl with all but 5 of the slices. Brush some of the reserved syrup over the slices (part of the syrup may be replaced with sherry). Place half the raspberries on top and cover with remaining five jelly roll slices. Brush slices with raspberry syrup or sherry. Top with remaining raspberries. Stir pudding until smooth; pour over fruit and rolls. Chill several hours. Garnish with whipped cream.

Makes 8 servings.

Fresh Fruit Trifle

1 ½ cups	sliced strawberries	375 mL
1 ½ cups	raspberries	375 mL
4	kiwi fruit, peeled and sliced	4
½ cup plus	sugar	125 mL
3 tablespoons	sugar	45 mL
4	egg yolks	4
⅓ cup	all purpose flour	75 mL
2 cups	milk	500 mL
1 teaspoon	grated lemon rind	5 mL
1 teaspoon	vanilla extract	5 mL
1 ½ pounds	pound cake	750 g
8 teaspoons	sherry	40 mL
1 cup	whipping cream	250 mL
	fresh fruit	

Toss strawberries with 1 tablespoon (15 mL) sugar. Toss raspberries with 1 tablespoon (15 mL) sugar. Peel and slice kiwi fruit. Toss with 1 tablespoon (15 mL) sugar. Refrigerate until ready to use. Beat egg yolks with remaining sugar until light and creamy. Fold in flour. Heat milk. Blend a little hot milk into egg mixture. Return to pot and cook and stir over medium heat until thick. Stir in lemon rind and vanilla. Cover with waxed paper and chill. To assemble trifle, slice pound cake into ½" (6 mm) slices. Place on bottom of quart (1 L) deep dish. Drizzle with 2 teaspoons (10 mL) sherry. Spread with ⅓ of the pastry cream. Top with strawberries. Repeat layers using raspberries in one, kiwi in another. Top with pound cake. Drizzle with remaining sherry. Cover and let stand refrigerated overnight (or several hours). Whip cream and pile on top. Garnish with fresh fruit.

Makes 8 servings.

Peach parfaits

1	envelope unflavoured gelatin	1
¼ cup	water	50 mL
1½ cups	milk	375 mL
⅓ cup	sugar	75 mL
1 tablespoon	lemon juice	15 mL
1 teaspoon	sherry (optional)	5 mL
4	fresh peaches,	4
	peeled and sliced	
	whipped cream	

Sprinkle gelatin over water in a small saucepan and stir over low heat until gelatin dissolves. Stir in sugar and milk. Flavour with lemon juice and sherry. Chill until mixture reaches consistency of unbeaten egg whites. Arrange layers of mixture and peaches in tall glasses. Chill until firm. Serve decorated with whipped cream.

Makes 4 servings.

Super Creamy Pecan Parfaits

1 cup	milk	250 mL
2 cups	softened butter pecan ice cream	500 mL
1	package (4 serving size) butter pecan instant pudding whipped cream toasted pecans	1

Combine milk and ice cream in a large bowl. Stir until ice cream is melted. Add pudding mix and beat with a rotary beater 2 minutes or until pudding mix is completely dissolved. Chill until mixture starts to thicken. Spoon mixture into four parfait glasses and chill until set. Garnish with whipped cream and toasted pecans if desired.

Makes 4 servings.

CAPPUCCINO PARFAITS

1	envelope unflavoured gelatin	1
1½ cups	milk	375 mL
¼ cup	sugar	50 mL
2 teaspoons	instant coffee crystals	10 mL
2 cups	coffee ice cream	500 mL
¼ cup	coffee liqueur	50 mL
	whipped cream	
	ground cinnamon	

Sprinkle gelatin over milk. Let stand 10 minutes to soften. Cook over low heat, stirring constantly, until gelatin has dissolved. Add sugar and coffee crystals; stir until dissolved. Add and break up ice cream; stir until melted. Add liqueur. Pour into four 8-ounce (250 mL) parfait glasses. Chill until firm. To serve, garnish with whipped cream and sprinkle with cinnamon.

Makes 4 servings.

CHOCOLATE STRAWBERRY PARFAITS

4	egg yolks	4
½ cup	sugar	125 mL
⅓ cup	flour	75 mL
2 cups	milk, hot	500 mL
1 teaspoon	vanilla	5 mL
½ cup	semi-sweet chocolate chips	125 mL
1 cup	whipping cream	250 mL
2 cups	fresh strawberries	500 mL

Beat egg yolks and sugar until light. Add flour and beat until smooth. Gradually beat hot milk into egg mixture. Return to heat, stirring constantly, while mixture comes to a boil; simmer 5 minutes. Off heat add vanilla and chocolate chips. Stir until chocolate is melted. Cool. Beat cream until stiff and fold half into the cooled chocolate custard. Slice strawberries and sprinkle with a little sugar. Arrange layers of chocolate mixture and strawberries in tall parfait glasses, topping off each glass with whipped cream and a perfect strawberry.

Makes 6 servings.

Summertime Strawberry Parfaits

1	package (4 serving size) vanilla pudding and pie filling	1
1¼ cups	milk	300 mL
1 cup	sour cream	250 mL
2 teaspoons	vanilla	10 mL
¼ cup	fruit sugar	50 mL
4 cups	fresh, hulled, halved strawberries	1 L
	whole strawberries	
	mint leaves	

Combine pudding mix and milk in a saucepan. Cook according to package directions. Stir in sour cream and vanilla. Cover surface with plastic wrap; cool. Chill. Sprinkle sugar over berries; toss lightly to combine. Alternate layers of chilled pudding mixture and strawberries in parfait glasses, ending with a pudding layer. Garnish top of each parfait with a whole strawberry and mint leaves.

Makes 4 to 6 servings.

CLASSIC CARAMEL CUSTARD

1 cup	sugar	250 mL
2 tablespoons	water	30 mL
3	egg yolks	3
3	eggs	3
2 ½ cups	milk, hot	625 mL
1 teaspoon	vanilla	5 mL

Preheat oven to 325°F (160°C).

Heat ½ cup (125 mL) sugar with 2 tablespoons (30 mL) water in a heavy skillet. Swirl gently while sugar dissolves and syrup turns a light caramel colour. Pour at once into a warm 4 cup (1 L) baking dish or metal ring mold, tilting to film bottom and sides. Turn upside down over a plate and set aside to harden. Beat remaining sugar into eggs and egg yolks until mixture is light and frothy. Slowly beat in hot milk. Add vanilla. Strain custard into caramel-lined mold and set mold into a pan containing hot water to come about half way up the sides. Bake 40-60 minutes or until a tester inserted in the centre of the custard comes out clean.

Makes 4-6 servings.

COCONUT CRUNCH CUSTARDS

2 cups	milk	500 mL
⅓ cup	granulated sugar	75 mL
1½ teaspoons	vanilla	7 mL
5	eggs	5
½ cup	dessicated coconut	125 mL
2 tablespoons	brown sugar	30 mL
1 tablespoon	butter	15 mL
¼ teaspoon	ground cinnamon	1 mL

Scald milk. Add granulated sugar and vanilla. Beat eggs until light. Gradually stir in milk mixture. Pour into six (6-ounce/150 mL) custard cups. Set in a pan of hot water and bake in preheated 300°F (150°C) oven 50 to 55 minutes or until a knife inserted near centre comes out clean. Cool. Mix together coconut, brown sugar, butter and cinnamon. Sprinkle evenly over top of each custard. Broil until coconut is golden brown, about 1 minute. Chill custards before serving.

Makes 6 servings.

Baked Lemon Custard

2 teaspoons	grated lemon rind	10 mL
3 cups	milk	750 mL
6	eggs	6
½ cup	sugar	125 mL
1 teaspoon	vanilla extract	5 mL

Heat lemon rind with milk until milk is hot. Beat eggs until foamy. Add sugar and beat to blend. Whisk a little hot milk into beaten eggs. Return to pan; cook and stir until custard coats a spoon. Stir in vanilla. Strain into a buttered 1 quart (1 L) mold. Place mold in a pan of hot water. Bake in a preheated 325°F (160°C) oven 1 hour or until knife inserted in the centre comes out clean.

Makes 6 servings.

Mocha custard

½ cup	sugar	125 mL
2 tablespoons	cocoa	30 mL
1 ½ teaspoons	instant coffee	7 mL
2 cups	milk	500 mL
	pinch salt	
4	eggs	4
½ teaspoon	vanilla	2 mL
1 tablespoon	orange-flavoured liqueur (optional)	15 mL

Preheat oven to 350°F (180°C).

Blend sugar with cocoa and coffee in a saucepan; gradually stir in milk and salt. Bring to boil, stirring until cocoa and coffee are dissolved. Remove from heat. Beat eggs in a medium-sized bowl until light; gradually add hot milk mixture, beating constantly. Flavour with vanilla and liqueur. Strain into a 5 cup (1.25 L) baking dish. Set dish in a pan containing hot water to come about half way up sides. Bake 45 minutes, or until a toothpick inserted in centre comes out clean. Cool and refrigerate. A chocolate layer forms on top of the chilled custard.

Makes 4-6 servings.

Pumpkin Custard

4	eggs, beaten	4
1½ cups	mashed cooked pumpkin	375 mL
½ cup	sugar	125 mL
½ teaspoon	cinnamon	2 mL
½ teaspoon	ginger	2 mL
¼ teaspoon	salt	1 mL
2 cups	milk	500 mL
	juice of 1 small orange	
	whipped cream	
	maple syrup	
	grated orange rind	

Preheat oven to 350°F (180°C).

Beat eggs with pumpkin, sugar, cinnamon, ginger, salt and orange juice. Slowly beat in hot milk. Pour mixture into 6 cup (1.5 L) baking dish and cover with foil. Set dish in large pan containing hot water to reach halfway up the sides. Bake about 1 hour or until custard is firm. Remove dish from water bath and chill. Serve with whipped cream, a sprinkle of maple syrup and grated orange rind.

Makes 8 servings.

ORANGE DREAM CUSTARD

4		eggs	4
½	cup	sugar	125 mL
2¼	cups	milk	550 mL
1		grated orange rind	1
½	teaspoon	vanilla	2 mL

FRESH FRUIT SAUCE

2	tablespoons	sugar	30 mL
¼	cup	water	50 mL
		juice of lemon and two oranges	
2	teaspoons	cornstarch	10 mL
		orange segments	
		fresh mint sprigs (optional)	

Pre-heat oven to 300°F (150°C).

Lightly oil a shallow baking mold (4 cup/1 L) or individual baking cups (4 oz./125 mL). Beat eggs lightly; gradually whisk in sugar. Heat milk just below boiling point; slowly stir into egg mixture; add orange rind and vanilla. Pour mixture into prepared pan and set in a large shallow pan in the oven.

Pour hot water into the larger pan to reach halfway up the filled mold or cups. Bake about 50 minutes, or until a testing toothpick comes out clean. Chill thoroughly. Meanwhile, make sauce by combining sugar and water in a small saucepan. Bring to a boil; simmer for 2 minutes. Blend cornstarch with fruit juices; stir into syrup. Bring to a boil, stirring constantly; cook until lightly thickened. Cool and chill for 1 hour.

To serve, unmold dessert onto a cold plate; serve with fruit sauce, fresh orange segments and fresh mint sprigs.

Makes 6-8 servings.

Topsy-Turvy Custards

1 cup	sugar	250 mL
2 cups	milk	500 mL
1 teaspoon	vanilla	5 mL
5	eggs	5

Place ½ cup (125 mL) of the sugar in a small frypan; cook over medium heat, stirring constantly, until sugar melts and turns golden. Divide this syrup among six 6-ounce (150 mL) custard cups. Scald milk; add remaining ½ cup (125 mL) sugar and vanilla. Beat eggs until light. Gradually stir in milk mixture. Pour mixture through a fine sieve and divide among prepared custard cups. Set in a pan of hot water and bake in preheated 300°F (150°C) oven 40 to 45 minutes or until a knife inserted near centre comes out clean. Cool then chill well. To serve, loosen edges with a spatula and turn out of cup onto serving plate.

Makes 6 servings.

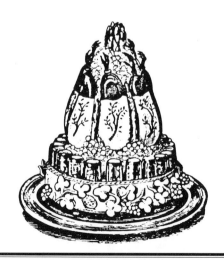

Plum Custard Delight

⅓ cup	cornstarch	75 mL
⅓ cup	sugar	75 mL
	pinch salt	
3 cups	milk	750 mL
1	egg	1
1 teaspoon	almond extract	5 mL

Combine cornstarch, sugar and salt in saucepan. Stir in milk. Cook and stir over medium heat, until mixture just boils and thickens. Remove from heat. Whisk ¼ cup (50 mL) hot milk mixture into beaten egg. Return to saucepan. Cook, stirring for 2 minutes. Blend in almond extract. Pour into individual serving dishes. Chill until set. Serve with Plum Sauce.

Makes 6 servings.

PLUM SAUCE

1 (14 ounce)	can of plums	1 (398 mL)
⅓ cup	honey	75 mL
1 tablespoon	lemon juice	15 mL
2 teaspoons	cornstarch	10 mL

Drain plums reserving ½ cup (125 mL) juice. Slice plums. Combine juice, honey, lemon juice and cornstarch in a small saucepan. Cook and stir until thick and clear. Add plums. Heat through. Serve either hot or cold.

*If using fresh plums, pit and slice; combine with honey. Bring to a boil and simmer 5 minutes. Add combined cornstarch and lemon juice. Continue cooking until sauce is thick and clear.

CRÉME PRALINE

½ cup	sugar	125 mL
1 tablespoon	water	15 mL
2 ounces	flaked almonds, toasted	50 g
1	envelope unflavoured gelatin	1
½ teaspoon	cornstarch	2 mL
4	egg yolks	4
1½ cups	milk, hot	375 mL
2 teaspoons	vanilla	10 mL
2 tablespoons	sherry (optional)	30 mL
½ cup	whipping cream	125 mL
8	lady fingers	8
	whipped cream	
	fresh berries	

Combine ¼ cup (50 mL) sugar and 1 tablespoon (15 mL) of water in a heavy skillet. Swirl over moderate heat until sugar dissolves and becomes caramel coloured. Stir in toasted almonds. Bring to boil then turn out onto a baking sheet. Cool until hard and crush with a rolling pin or in a blender. Set aside. Sprinkle gelatin over ¼ cup (50 mL) water. Set aside to soften. Beat remaining sugar, egg yolks and cornstarch until light, gradually beat in hot milk. Set in pan over hot water and cook, stirring constantly until custard coats a metal spoon. Stir in softened gelatin, vanilla, praline and sherry if desired. Chill until mixture reaches consistency of unbeaten egg whites. Beat cream until stiff and fold into custard. Dip lady fingers into a little sherry diluted with a splash of water and arrange in the base of an 8″ (1.2 L) cake pan. Cover with praline custard. Chill until set. Turn out onto a platter and decorate with whipped cream and fresh berries.

Makes 6 servings.

Orange-Lemon Cream

2	envelopes unflavoured gelatin	2
½ cup	orange juice	125 mL
¼ cup	lemon juice	50 mL
3	eggs, separated	3
⅔ cup	sugar	150 mL
	pinch salt	
2 cups	milk, hot	500 mL
1 tablespoon	grated orange rind	15 mL

Sprinkle gelatin over juices to soften; set aside. In the top of a double boiler beat egg yolks with sugar and salt until thick. Gradually blend in hot milk and orange rind. Set over hot water and cook, stirring, until custard coats a metal spoon. Remove from heat and stir in softened gelatin. Chill until mixture thickens to consistency of unbeaten egg whites. Whisk egg whites until stiff but not dry and fold into cool custard. Pour into a 4 cup (1 L) mold and chill until set.

Makes 4-6 servings.

Raspberry cream

1	large orange	1
2 cups	milk	500 mL
½ cup plus	sugar	125 mL
2 tablespoons	sugar	30 mL
4 tablespoons	cornstarch	60 mL
2	envelopes unflavoured gelatin	2
1 cup	puréed raspberries*	250 mL
1 tablespoon	orange liqueur	15 mL
1 teaspoon	vanilla extract	5 mL
½ cup	whipping cream	125 mL
	raspberries	

Remove thin layers of orange peel. Squeeze orange reserving juice. Pour 1 cup (250 mL) milk into a saucepan. Add ½ cup (125 mL) sugar and orange peel. Blend cornstarch with half the reserved milk. Sprinkle gelatin into remaining milk. Stir and let stand to soften. Combine raspberry purée with orange juice, liqueur and vanilla. Heat milk in saucepan to boiling point. Stir in cornstarch mixture and whisk to prevent lumps from forming. Bring to a boil. Add gelatin and stir to dissolve. Mix in raspberry purée. Cool. Whip cream until stiff. Fold into raspberry cream. Fill individual dishes or one 4 cup (1L) mold, with cream. Chill until set. Garnish with additional whipped cream and raspberries.

Makes 8 servings.

*May be made with fresh or frozen raspberries. Strain raspberry purée if you wish to remove seeds.

STRAWBERRY CREAM

1 pound	cream cheese	500 g
1½ cups	milk	375 mL
1 teaspoon	vanilla	5 mL
1	envelope unflavoured gelatin	1
¼ cup	sugar	50 mL
½ cup	boiling water	125 mL
2 cups	strawberries, halved	500 mL
3 tablespoons	orange liqueur	45 mL
3 tablespoons	sugar	45 mL

In a large bowl blend cream cheese until smooth. Gradually add milk and vanilla; continue blending until smooth and creamy. In a small bowl, combine gelatin and sugar. Add boiling water, stir until dissolved. In food processor or blender blend strawberries, liqueur and sugar. Combine gelatin mixture with cream cheese mixture. Add pureed strawberries. Pour into parfait glasses. Chill until set. Garnish with whole strawberries.

Makes 8 servings, 5 cups (1.2 L).

Spanish Cream

1½	envelopes unflavoured gelatin	1½
½ cup	sugar	125 mL
¼ teaspoon	salt	1 mL
¼ teaspoon	cinnamon	1 mL
¼ teaspoon	grated nutmeg	1 mL
4	eggs, separated	4
2 cups	milk	500 mL
2	large seedless oranges	2
¼ cup	dry sherry	50 mL
1 tablespoon	lemon juice	15 mL
½ cup	whipping cream	125 mL
	chopped nuts	

Combine gelatin, sugar and salt in top of a double boiler. Add cinnamon and nutmeg. Beat egg yolks lightly and blend. Stir into sugar mixture. Add milk. Cook and stir over simmering water until mixture thickens. Chill. Peel, section and drain oranges. Cut in half and mix with sherry. Chill. When cream mixture has cooled and thickened fold in oranges and lemon juice. Beat egg whites until stiff. Beat whipping cream until stiff. Fold both into orange mixture. Spoon into serving dishes. Chill until set. Sprinkle with chopped nuts to serve.

Makes 8 servings.

ALMOND CREAM DESSERT

1	envelope unflavoured gelatin	1
½ cup	water	125 mL
1 cup	milk	250 mL
¾ cup	sugar	175 mL
½ teaspoon	almond extract	2 mL
1½ cups	sour cream	375 mL

Sprinkle gelatin over cold water; leave 5 minutes to soften. Heat milk. Add sugar and gelatin, stir until dissolved. Chill until softly set. Stir almond flavouring into sour cream. Beat gelatin mixture until foamy and blend in sour cream. Pour into 4 cup (1L) mold. Chill until firm. Serve with raspberry or apricot sauce.

Makes 6 servings.

MOCHA CREAM

¼	cup	sugar	50 mL
2	teaspoons	cornstarch	10 mL
1		envelope unflavoured gelatin	1
3		eggs, separated	3
1½	cups	milk	375 mL
¾	cup	semi-sweet chocolate chips	175 mL
2	teaspoons	instant coffee granules	10 mL
1	teaspoon	vanilla	5 mL
1	tablespoon	sugar	15 mL
½	cup	whipping cream	125 mL
1	tablespoon	coffee liqueur or rum (optional)	15 mL

In a heatproof bowl, combine sugar, cornstarch and gelatin; add egg yolks; beat until light. Reserve. Heat milk; add chocolate chips and instant coffee; blend until smooth.

Whisk flavoured milk into egg mixture; set bowl over a pan of hot water over medium heat for about 10 minutes, stirring until mixture is lightly thickened. Do not boil. Remove from heat; stir in vanilla.

Set in refrigerator for an hour until softly jelled, but not firm. Beat egg whites until stiff, whisking in remaining sugar; fold into chilled mocha cream. Fold in stiffly beaten whipping cream, then optional liqueur. Pour into goblets; refrigerate overnight until set. Garnish with additional lightly sweetened whipped cream and shaved chocolate.

Variation: May be poured into a 6 cup (1.5 L) mold. Chill; unmold onto a chilled platter to serve.

Makes 6 servings.

Rum Bavarian

1 tablespoon	rum	15 mL
½ cup	chopped raisins	125 mL
1	envelope unflavoured gelatin	1
1¼ cups	milk	300 mL
3	eggs, separated	3
¾ cup	sugar	175 mL
½ cup	whipping cream	125 mL
	pinch salt	

Pour rum over raisins and leave to marinate. Sprinkle gelatin over ¼ cup (50 mL) cold milk; set aside. Heat remaining milk. In the top of a double boiler beat egg yolks, ½ cup (125 mL) sugar and salt until thick. Gradually blend in hot milk. Set over hot water and cook, stirring constantly until custard coats a metal spoon. Stir in gelatin. Leave custard to cool until it reaches the consistency of unbeaten egg whites. Beat egg whites until stiff with remaining sugar. Fold into custard. Beat cream until stiff and fold into custard with the marinated raisins. Turn into a 4 cup (1L) mold and chill.

Makes 4-6 servings.

STRAWBERRY BAVARIAN

2	envelopes unflavoured gelatin	2
½ cup	cold water	125 mL
6	egg yolks	6
¾ cup	sugar	175 mL
	pinch salt	
2¼ cups	milk, hot	550 mL
2 teaspoons	vanilla	10 mL
1 cup	whipping cream	250 mL
2 cups	fresh strawberries	500 mL
1 tablespoon	grated orange rind	15 mL

Sprinkle gelatin over water to soften; set aside. In the top of a double boiler beat egg yolks with sugar and salt until thickened. Gradually blend in hot milk. Set over hot water and cook, stirring until custard coats a metal spoon. Remove from heat and stir in vanilla and softened gelatin. Chill until mixture thickens to consistency of unbeaten egg whites. Beat cream until stiff and fold into custard. Remove 2 cups (500 mL) custard to another bowl. Reserve 4-5 strawberries and crush the rest, to make about 1 cup (250 mL) purée. Fold into 2 cups (500 mL) custard and flavour with orange rind. Pour half the vanilla custard into a 6 cup (1.5 L) mold. Chill until set. Pour half the strawberry custard on top; chill. Repeat layers and chill several hours or overnight. Unmold onto a serving platter. Surround with reserved berries and garnish with additional whipped cream, if desired.

Makes 8 servings.

APRICOT BAVARIAN

1	cup	dried apricots	250 mL
1½	cups	water	375 mL
½	cup	sugar	125 mL
1		grated lemon rind	1
6	ounces	vanilla pudding mix (not instant)	185 g
2	cups	milk	500 mL
2		egg yolks	2
1		envelope unflavoured gelatin	1
¼	cup	cold water	50 mL
2		egg whites	2
1	cup	whipping cream	250 mL

In a small saucepan combine apricots, water, sugar and lemon rind. Bring to boil; cover and simmer for 15 minutes. Drain syrup and add enough water to make 2 cups (500 mL). Chop apricots coarsely and set aside.

In a medium saucepan blend pudding mix, milk, apricot syrup and egg yolks. Cook over medium heat, stirring constantly, until pudding thickens and comes to a full boil.

Sprinkle gelatin over cold water and let stand until softened; add to hot pudding and stir until dissolved. Add chopped apricots and chill until pudding starts to set.

Beat egg whites until stiff but not dry. Whip cream. Fold egg whites and cream into pudding mixture. Pour into serving dish and chill until firm.

Makes 8 servings.

Moo

1 (3-ounce)	package fruit flavoured jelly powder	1 (85 g)
1 cup	boiling water	225 mL
2 cups	milk	450 mL

Dissolve jelly powder in boiling water. Cool to room temperature. Very gradually stir in milk. Pour into dessert dishes; chill until firm.

Makes about 3 cups/700 mL.

CRUNCHY CARAMEL MOUSSE

1½ cups	granulated sugar	375 mL
⅔ cup	water – divided	150 mL
3 tablespoons	rum or water	45 mL
1	envelope unflavoured gelatin	1
3	egg yolks	3
1½ cups	milk, hot	375 mL
1 teaspoon	pure vanilla extract	5 mL
2 cups	whipping cream	500 mL
2 ounces	semi-sweet or bittersweet chocolate, melted	56 g
2	Crispy Crunch bars (45 g each), crushed	2

Combine sugar with ⅓ cup (75 mL) water in a large, heavy saucepan. Heat gently, stirring, until sugar dissolves. Boil mixture 5 to 10 minutes, watching carefully, without stirring, until sugar turns a deep caramel colour. Do not burn. Remove from heat and standing back, add remaining ⅓ cup (75 mL) water. Mixture will bubble up furiously. Stir to dissolve caramel; return to heat for a minute if necessary. Reserve. Sprinkle gelatin over rum in a small saucepan. Rest 5 minutes to soften gelatin; heat gently to dissolve. Reserve. In the top of a double boiler combine egg yolks, milk, caramel and gelatin. Cook over simmering water until mixture thickens slightly – about 7 minutes. Add vanilla; cool over a larger bowl filled with ice and water. Mixture should be cool but not yet set. Prepare serving bowl by drizzling half of chocolate around the bottom and up the sides of a 3 quart (3 L) glass bowl or individual wine glasses. Sprinkle with half of crushed chocolate bars. Reserve. Whip cream and fold ⅔ of it into caramel base. Pour into bowl. Refrigerate mousse and extra cream one hour. Decorate top of mousse with reserved cream, drizzle with remaining chocolate; sprinkle with remaining crushed chocolate bars. Chill two hours.

Makes 6 servings.

Delicious chocolate mousse

1	envelope unflavoured gelatin	1
½ cup	sugar	125 mL
	pinch salt	
1¼ cups	milk	300 mL
6 ounces	semi-sweet chocolate chips	175 g
1 teaspoon	vanilla	5 mL
1 cup	whipping cream	250 mL

Combine gelatin, sugar and salt in a saucepan. Stir in milk and chocolate chips. Stir constantly over medium heat until gelatin dissolves and chocolate melts. Remove from heat and beat until chocolate is thoroughly blended. Add vanilla. Chill until mixture has consistency of unbeaten egg whites. Beat cream until stiff; fold into chocolate mixture. Turn into a 5 cup (1.25 L) serving bowl. Chill until firm. Decorate with additional whipped cream and shaved semi-sweet chocolate if desired.

Makes 6 servings.

CRANBERRY MOUSSE

CRANBERRY SAUCE

1 pound	fresh or frozen cranberries	500 g
2 cups	sugar	500 mL
2 cups	water	500 mL
1	orange	1

In a saucepan, combine cranberries, sugar, water and juice and grated rind of orange. Bring to a boil, stirring occasionally. Cover, reduce heat and simmer 2-5 minutes until cranberries are tender and just beginning to pop. Remove 1½ cups (375 mL) cranberries with a slotted spoon. Set aside. Blend remainder of cranberries and liquid to make a sauce.

CRANBERRY MOUSSE

2	envelopes unflavoured gelatin	2
2 cups	milk	500 mL
¼ cup	sugar	50 mL
1½ tablespoons	cornstarch	22 mL
1 teaspoon	grated orange rind	5 mL
1 cup	whipping cream	250 mL
1½ cups	cooked cranberries (above)	375 mL

Sprinkle gelatin over ½ cup (125 mL) cold milk. Set aside. Blend sugar and cornstarch with a little cold milk; heat remaining milk. Beat hot milk into cornstarch mixture. Return to heat, stirring constantly until mixture comes to a boil and thickens. Simmer 5 minutes. Add softened gelatin and orange rind. Chill until mixture is consistency of unbeaten egg whites. Beat cream until stiff. Fold into chilled mixture. Gently, but thoroughly fold in cooked cranberries. Pour into a 6 cup (1.5 L) mold. Cover and chill until set. Serve with Cranberry Sauce.
Makes 8 servings.

BLENDER CHOCOLATE ALMOND MOUSSE

2	eggs	2
¼ cup	soft butter	50 mL
1 (6 ounce)	package semi-sweet chocolate chips	1(170g)
1 tablespoon	almond liqueur	15 mL
1 teaspoon	vanilla	5 mL
1 teaspoon	unflavoured gelatin	5 mL
1 cup	milk	250 mL
	whipped cream	
	chopped toasted almonds	

Place eggs, butter, chocolate chips, almond liqueur and vanilla in blender container. Cover and blend at medium speed until smooth and creamy. Sprinkle gelatin over milk. Let stand 5 minutes to soften. Heat milk mixture to just below boiling; stirring constantly. Turn blender speed to low and slowly pour hot milk mixture through top opening. Blend 1 or 2 minutes or until smoothly combined. Stop blender and scrape down sides if necessary. Pour into 4 or 5 dessert dishes and chill until set. Garnish with whipped cream and toasted almonds if desired.

Makes 4 or 5 servings.

Buttery Orange Crepes

1 cup	butter	250 mL
3⅓ cups	marmalade	900 mL
¼ cup	orange liqueur	50 mL
24	crêpes	24

Melt butter in a large frypan. Add marmalade and orange liqueur. Cook over medium heat, stirring constantly, until smoothly combined and heated through. Fold each crêpe in half, then in half again to form triangles. Arrange folded crêpes in sauce in frypan. Continue cooking and spooning sauce over crêpes until sauce thickens slightly and crêpes are hot.

Makes 8 servings.

CREPES

1¼ cups	milk	300 mL
3	eggs	3
2 tablespoons	soft butter	30 mL
¾ cup	all-purpose flour	175 mL
2 tablespoons	sugar	30 mL
¼ teaspoon	salt	1 mL

Place milk, eggs and butter in blender container. Add flour, sugar and salt. Cover and blend at high speed 30 seconds. Scrape down sides of container and blend an additional 30 seconds or until batter is smooth. Refrigerate, covered, 2 hours or overnight. Heat a 6-inch (15 cm) frypan or crêpe pan until a drop of water sizzles and bounces. Brush melted butter over bottom of pan. Add 2 tablespoons (30 mL) batter and rotate quickly to spread batter completely over bottom of pan. Cook about 1 minute or until lightly browned. Turn and cook 30 seconds. Remove from pan. Repeat with remaining batter. To store, separate each crêpe with waxed paper. Wrap in foil. Refrigerate or freeze until needed.

Makes about 24 crêpes.

Pineapple carrot dessert crepes

3	eggs	3
1¼ cups	all purpose flour	300 mL
2 tablespoons	icing sugar	30 mL
	pinch salt	
1 cup	milk	250 mL
⅓ cup	pineapple juice	75 mL
1 tablespoon	lemon juice	15 mL
2 tablespoons	melted butter	30 mL
½ cup	grated carrot	125 mL
	butter for cooking crepes	
	icing sugar	
	toasted coconut	

Combine all ingredients, except grated carrot, in container of a blender or food processor. Cover and blend mixture until smooth. Stir in carrots. Let rest one hour. Set a small crêpe pan over moderately high heat; brush with butter. Pour in enough batter to cover bottom of pan with a thin even layer. Hold over heat for 2-3 minutes, until underside is lightly browned. Flip crêpe over and briefly cook other side. Repeat with remaining batter. Stack cooked crêpes one on the other, separated with small squares of waxed paper. Enclose a spoonful of Pineapple Crêpe Filling in each crêpe. Place seam side down in a buttered baking dish. Sprinkle with icing sugar and toasted coconut. Set briefly under a preheated broiler before serving.

Makes 6 servings.

PINEAPPLE CREPE FILLING

1 cup	whipping cream	250 mL
1 tablespoon	sugar	15 mL
2 tablespoons	rum, optional or	30 mL
	1 teaspoon (5 mL) vanilla	
1 (14 ounce)	can pineapple tidbits, drained	1 (398 mL)

Beat cream until stiff; add sugar and flavour with rum or vanilla. Fold in drained pineapple.

Makes about 2 cups (500 mL).

SWEET CHEESE CREPES

CREPES

$\frac{1}{2}$ cup	water	125 mL
$\frac{1}{2}$ cup	milk	125 mL
1 cup	flour	250 mL
2	eggs	2
$\frac{1}{4}$ teaspoon	salt	1 mL
2 tablespoons	butter, melted	30 mL

Combine all ingredients and blend until smooth. Let stand at least 2 hours or longer. Pour a scant $\frac{1}{4}$ cup (50 mL) batter into a greased 6" (15 cm) frypan. Swirl pan to distribute batter evenly. Cook until golden on bottom and dry on top. Remove from pan. Repeat until all crepes are cooked. Makes 16 crepes.

FILLING

2 tablespoons	sugar	30 mL
3	eggs, beaten	3
1 tablespoon	lemon juice	15 mL
$\frac{1}{2}$ cup	raisins	125 mL
2 cups	dry cottage cheese	500 mL
	(parchment wrap)	
$\frac{1}{4}$ cup	butter	50 mL

Combine sugar and eggs. Beat well. Add lemon juice, raisins and cottage cheese and mix well. Distribute evenly onto crepes. Fold in sides and roll up. Melt half of butter in a 10" (25 cm) skillet. Place 8 crepes, seam side downward, in pan and sauté until golden and firm. Turn to brown other side. Remove from pan and keep warm. Add remaining butter, and repeat with next batch of crepes. Top with custard sauce.

continued on next page

SWEET CHEESE CREPES (continued)

CUSTARD SAUCE

2 cups	milk	500 mL
2	eggs, well-beaten	2
¼ cup	sugar	50 mL
1 teaspoon	vanilla extract	5 mL

Combine all ingredients. Heat slowly in the top of a double boiler, or in a heavy saucepan until custard thickens and just coats a spoon.

Arrange crepes on a warm serving platter. Pour some sauce down the centre of the crepes. Serve remaining sauce separately.

Makes 8 servings.

WHOLE WHEAT CREPES WITH RHUBARB SAUCE

CREPES

4	eggs	4
¼ teaspoon	salt	1 mL
1 cup	whole wheat flour	250 mL
1 cup	all-purpose flour	250 mL
2¼ cups	milk	550 mL
¼ cup	butter, melted	50 mL
1 tablespoon	sugar	15 mL

Whirl all ingredients together in blender for about 1 minute. Scrape down sides; blend 15 seconds more. Refrigerate at least 1 hour. Bake in traditional crêpe pan or in electric crêpe-maker. Crêpes may be frozen for future use. Makes about 30 crêpes.

RHUBARB SAUCE

2 cups	sliced rhubarb	500 mL
1 cup	sugar	250 mL
1 tablespoon	water	15 mL
1 teaspoon	grated orange rind	5 mL
1 cup	whipping cream (for filling)	250 mL

Mix ingredients in saucepan; simmer 10 minutes. Cool. Serve crêpes filled with whipped cream. Spoon rhubarb sauce over top.
Makes 1½ cups (375 mL), enough for 6 servings.

FLOATING FRUIT MERINGUE

2	seedless peeled and thinly sliced oranges	2
2	bananas, peeled and sliced	2
2 cups	milk	500 mL
4	eggs, separated	4
6 tablespoons	sugar	90 mL
1 teaspoon	vanilla extract	5 mL
	pinch nutmeg	

Lightly grease 4-2 cup (500 mL) baking dishes. Line bottom with sliced fruit. Heat milk until hot. Whisk egg yolks with ¼ cup (50 mL) sugar, until smooth and creamy. Add a little hot milk to egg yolks, return to saucepan. Cook and stir until custard coats the back of spoon. Stir in vanilla and nutmeg. Divide custard evenly between dishes. Whip egg whites until frothy. Gradually beat in remaining sugar until meringue is stiff and shiny. Spoon meringue on top of custard. Bake in 350°F (180°C) oven until lightly browned, approximately 10 minutes.

Makes 4 servings.

SNOW SALAD DESSERT

2 tablespoons	sugar	30 mL
1 teaspoon	cornstarch	5 mL
1 cup	milk	250 mL
4	egg yolks, beaten	4
1 (14 ounce)	can crushed pineapple, drained	1 (398 mL)
1 (11 ounce)	can mandarin oranges, drained	1 (312 mL)
1	envelope unflavoured gelatin	1
1 cup	whipping cream	250 mL
1 cup	miniature marshmallows	250 mL
1 teaspoon	grated lemon rind	5 mL

Combine sugar, cornstarch and cold milk in a small saucepan. Stir over medium heat until mixture comes to a boil. Gradually beat hot mixture into egg yolks. Return to heat, stirring constantly, until sauce is heated through – do not boil. Cool. Drain fruit, reserving ¼ cup (50 mL) fruit juice. Sprinkle gelatin over fruit juice, heat gently to dissolve. Set aside. Beat cream until stiff. Gradually beat in gelatin. Gently combine cool custard with whipped cream, fruit, marshmallows, and lemon rind. Turn into a serving bowl and chill until set.

Makes 6 servings.

CREAMED PEACH RICE MOLD

1½ cups	cooked long grain rice	375 mL
1⅔ cups	milk	400 mL
1 tablespoon	butter	15 mL
½ cup	sugar	125 mL
¼ teaspoon	salt	1 mL
1	envelope unflavoured gelatin	1
¼ cup	cold water	50 mL
½ teaspoon	vanilla extract	2 mL
1 (14 ounce)	can sliced peaches	1 (398 mL)
½ teaspoon	ground cinnamon	2 mL
1 cup	whipping cream	250 mL

Combine rice, milk, butter, sugar, salt, in saucepan. Cook over medium heat stirring often about 20 minutes or until thickened. Sprinkle gelatin over cold water. Let stand 5 minutes. Stir into hot rice mixture about 1 minute to dissolve gelatin. Stir in vanilla. Chill 1 hour or until mixture starts to thicken. Drain peaches, reserving juice. Chop and fold into rice mixture. Whip cream and fold into rice. Turn into a 5 cup (1.2 L) ring mold. Chill until set. Unmold and serve with Orange Peach Sauce.

Makes 8 servings.

ORANGE PEACH SAUCE

	reserved peach syrup	
	orange juice	
⅓ cup	sugar	75 mL
2 teaspoons	cornstarch	10 mL

Combine peach syrup with enough orange juice to make 1 cup (250 mL). Pour into saucepan. Bring to a boil. Blend sugar and cornstarch. Stir into syrup. Cook over medium heat until clear and thickened. Serve warm or chilled.

Streusel Topped Baked Apples

3	large cooking apples	3
⅓ cup	raisins	75 mL
½ cup	firmly-packed brown sugar	125 mL
½ cup	all-purpose flour	125 mL
3 tablespoons	chilled butter	45 mL

Halve and core apples. Place an equal amount of raisins in the hollow of each apple half. Combine sugar and flour. Cut in butter until mixture resembles coarse crumbs. Pack about 3 tablespoons (45 mL) crumb mixture in a mound over top of each apple half. Place in 9-inch (1.5 L) cake pan. Bake in preheated 350°F (180°C) oven 50 to 60 minutes or until apples are tender. Serve with Creamy Vanilla Sauce.

Makes 6 servings.

CREAMY VANILLA SAUCE

1	(4 serving size) package vanilla pudding and pie filling	1
3 cups	milk	750 mL
1½ teaspoons	vanilla	7 mL
¼ teaspoon	ground cinnamon	1 mL
¼ teaspoon	ground nutmeg	1 mL

Combine pudding mix and milk in a saucepan. Cook according to package directions. Stir in vanilla, cinnamon and nutmeg. Serve warm.

Makes about 3 cups/750 mL.

DESSERT FONDUES

MILK CHOCOLATE

1 cup	hot milk	250 mL
2 (6-ounce)	packages semi-sweet chocolate chips	2 (170 g)
2 tablespoons	coffee liqueur	30 mL

Place hot milk, chocolate chips and liqueur in blender container. Cover and blend at medium speed about 1 minute or until smooth. Pour sauce into small pot and keep warm over a candle flame.

Makes about 2 cups/500 mL sauce.

CREAMY VANILLA

½ cup	butter	125 mL
1 cup	sugar	250 mL
½ cup	light cream	125 mL
1 teaspoon	vanilla	5 mL

Melt butter in a saucepan. Remove from heat, add sugar, cream and vanilla. Simmer, stirring constantly, about 5 minutes or until sugar is dissolved. Pour sauce into a small pot and keep warm over a candle flame.

Makes about 1½ cups/375 mL sauce.

Serve with pieces of fresh fruit (strawberries, melon balls, mandarin orange sections, nectarines, etc.), cubes of pound cake and plain cookies for dipping.

CHOCOLATE FROSTED PEARS

1 cup	water	250 mL
1 cup	sugar	250 mL
1 tablespoon	lemon juice	15 mL
1½ teaspoons	vanilla	7 mL
6	ripe pears, peeled	6
1	package (4 serving size) vanilla pudding and pie filling	1
2 cups	milk	500 mL
½ cup	whipping cream	125 mL
½ cup	chocolate sauce	125 mL

Combine water and sugar in a saucepan. Bring to a boil. Reduce heat; add lemon juice, vanilla and pears. Cover and cook over medium heat until pears are just tender, 5 to 10 minutes. Cool then chill. Combine pudding mix and milk in a saucepan. Cook according to package directions. Cover surface with plastic wrap; cool. Whip cream until softly stiff. Fold in cooked pudding. Chill. To serve, drain pears; stand on end in individual dishes. Pour equal amounts of pudding mixture around base. Drizzle chocolate sauce over top.

Makes 6 servings.

Profiterole Pyramid

PROFITEROLE PASTRY

1	cup	milk	250 mL
½	cup	butter	125 mL
1	cup	all-purpose flour	250 mL
4		large eggs	4

In a heavy saucepan, bring milk and butter to a boil; while boiling, add flour all at once. Stir rapidly until mixture forms a ball. Remove from heat. Beat in eggs thoroughly, one at a time. Place heaping teaspoons of dough on greased cookie sheet, 2 inches (5 cm) apart.

Bake in a 425°F (220°C) oven for 20 minutes. Reduce heat to 350°F (180°C); continue baking until puffs are well risen and dry, about 10-15 minutes. Cool.

FRENCH VANILLA CREAM

⅔	cup	sugar	150 mL
½	cup	all-purpose flour	125 mL
½	teaspoon	salt	2 mL
3	cups	milk, scalded	750 mL
6		egg yolks, lightly beaten	6
2	teaspoons	vanilla	10 mL
2	tablespoons	butter	30 mL

Mix dry ingredients in a large saucepan; add scalded milk gradually, stirring constantly. Cook together over medium heat until thickened. Remove from heat. Stir a little hot mixture into beaten egg yolks; blend this into remaining hot mixture and cook over low heat until thickened. Remove from heat; blend in vanilla and butter. Cool.

To assemble: When all ingredients are cool, slit puffs and fill with French Vanilla Cream, using a small spoon or pastry bag. To serve, arrange puffs in a pyramid on a serving dish and garnish with chocolate sauce and whipped cream (optional).

Makes 18-20 small profiteroles.

COUNTRY CRUNCH APPLE DESSERT

1 cup	quick-cooking rolled oats	250 mL
½ cup	whole bran cereal	125 mL
⅓ cup	whole wheat flour	75 mL
⅓ cup	firmly-packed brown sugar	75 mL
¼ teaspoon	baking soda	1 mL
¾ teaspoon	ground cinnamon	3 mL
½ cup	butter	125 mL
1	package (4 serving size) butterscotch pudding and pie filling	1
1½ cups	milk	375 mL
1 (19-ounce)	can apple pie filling	1 (540 mL)

Combine oats, cereal, flour, sugar, baking soda and cinnamon. Cut in butter until mixture resembles coarse crumbs. Press half the mixture evenly over bottom of an 8 x 8 x 2-inch (2 L) square baking dish. Bake in preheated 350°F (180°C) oven 10 minutes. Combine pudding mix and milk in a saucepan. Cook according to package directions. Pour over crust. Spoon apple pie filling over pudding. Top with remaining crumbs. Return to oven and bake 30 to 35 minutes. Serve warm.

Makes 8 servings.

Plum delicious cobbler

3 (14-ounce)	cans prune plums	3 (398 mL)
¼ cup	sugar	50 mL
2 tablespoons	corn starch	30 mL
¼ teaspoon	ground cinnamon	1 mL
1 tablespoon	grated orange rind	15 mL
⅓ cup	orange juice	75 mL
1⅓ cups	all-purpose flour	325 mL
3 tablespoons	sugar	45 mL
2½ teaspoons	baking powder	12 mL
½ teaspoon	salt	2 mL
⅓ cup	chilled butter	75 mL
¾ cup	milk	175 mL
½ teaspoon	vanilla	2 mL

Drain plums, reserving syrup. Pit plums; place in a 2-quart (2 L) shallow rectangular baking dish. Combine ¼ cup (50 mL) sugar, corn starch, cinnamon and orange rind in a saucepan. Stir in orange juice and plum syrup. Cook over medium heat, stirring constantly, until mixture comes to a boil and thickens. Reduce heat and cook 2 minutes longer, stirring occasionally. Set aside. Combine flour, 3 tablespoons (45 mL) sugar, baking powder and salt. Cut in butter until mixture resembles coarse crumbs. Add milk and vanilla. Stir just until moistened. Bring plum syrup to a boil. Pour over pitted plums. Drop batter by spoonfuls onto hot fruit, spacing evenly. Bake in preheated 400°F (200°C) oven 25 to 30 minutes. Serve warm.
Makes 6 to 8 servings.

Fresh Peach Cobbler

1 cup	firmly-packed brown sugar	250 mL
4 teaspoons	corn starch	20 mL
6 cups	sliced, peeled fresh peaches	1.5 L
2 cups	variety baking/biscuit mix	500 mL
2 tablespoons	granulated sugar	30 mL
1 teaspoon	grated lemon rind	5 mL
½ cup	milk	125 mL

Combine brown sugar and corn starch in a saucepan. Add peaches and toss to coat. Cook over low heat, stirring constantly, until mixture comes to a boil and thickens. Reduce heat and cook 2 minutes longer, stirring occasionally; set aside. Combine baking mix, sugar and lemon rind. Add milk. Stir with a fork just until moistened. Return peach mixture to a boil; pour into a 2-quart (2 L) shallow rectangular baking dish. Drop batter by spoonfuls onto hot fruit, spacing evenly. Bake in preheated 400°F (200°C) oven 20 to 25 minutes. Serve warm with CREAMY LEMON SAUCE.
Makes 6 to 8 servings.

CREAMY LEMON SAUCE

½ cup	sugar	125 mL
2 tablespoons	corn starch	30 mL
1¼ cups	milk	300 mL
2 tablespoons	butter	30 mL
1 teaspoon	grated lemon rind	5 mL
3 tablespoons	lemon juice	45 mL

Combine sugar, corn starch and milk in a small saucepan. Cook over medium heat, stirring constantly, until mixture comes to a boil and thickens. Reduce heat and cook 2 minutes longer, stirring occasionally. Remove from heat. Stir in butter, lemon rind and juice. Serve warm.
Makes about 1¾ cups/425 mL.

Strawberry Rhubarb Cobbler

1 pint	strawberries, sliced	500 mL
1 pound	rhubarb, cut into 1"/2.5 cm chunks (approx. 3 ½ cups)	500 g
2 tablespoons	lemon juice	30 mL
⅔ cup	sugar	150 mL
1 tablespoon	butter, cut into bits	15 mL

TOPPING

1 ⅓ cups	all purpose flour	325 mL
2 teaspoons	baking powder	10 mL
3 tablespoons	brown sugar	45 mL
1 teaspoon	grated lemon rind	5 mL
3 tablespoons	uncooked oatmeal	45 mL
3 tablespoons	chopped walnuts (preferably toasted)	45 mL
1 teaspoon	cinnamon	5 mL
⅓ cup	butter, cut into bits	75 mL
1 cup	milk	250 mL
2 tablespoons	icing sugar, sifted	30 mL

Preheat oven to 400°F (200°C). Butter a 9"x 9" (2.5 L) baking dish. Combine strawberries, rhubarb, lemon juice, sugar and butter. Place in bottom of pan. For the topping combine flour with baking powder, sugar, lemon rind, oatmeal, nuts and cinnamon. Cut in butter until it is in tiny bits. Sprinkle mixture with milk. Stir together just until a heavy batter is formed. Drop batter by spoonfuls over top of the rhubarb. Bake 35-40 minutes. Allow to cool before serving. Sprinkle with icing sugar. (Rhubarb mixture may be quite runny at first but will firm up slightly when cool.)
Makes 6 servings.

CHOCOLATE MILK CREME CARAMEL

1 cup	sugar	250 mL
3	eggs	3
5	egg yolks	5
2 cups	partially skimmed chocolate milk	500 mL
2 cups	whipping cream	500 mL
	pinch salt	
2 tablespoons	instant coffee	30 mL
¾ cup	ground or crushed hazelnuts	175 mL

Preheat oven to 325°F (160°C).

Heat ½ cup (125 mL) sugar with 2 tablespoons (30 mL) water in a heavy skillet. Swirl gently while sugar dissolves and syrup turns a light caramel colour. Pour at once into a warm 8 cup (2 L) mold or individual custard cups. Tilt to coat bottom and sides of mold. Set aside to cool. Beat eggs and egg yolks with remaining sugar until light. Combine chocolate milk and cream and heat just to boiling. Slowly beat into egg mixture and add salt and coffee. Strain chocolate custard into caramel-lined mold and set into a pan containing hot water to come about half way up the sides. Bake 1¼ - 1½ hours, or until a toothpick inserted in centre comes out clean. (Individual custard cups about 1 hour.) Chill. Sprinkle hazelnuts evenly on top.

Makes 8 servings.

IBERIAN CREME CARAMEL

¾	cup	sugar	175 mL
2	cups	milk	500 mL
5		eggs	5
1	teaspoon	vanilla	5 mL

Heat ½ cup (125 mL) of the sugar in a small frypan; cook and stir over medium heat until it forms a light brown syrup. Divide syrup among six 6-ounce (150 mL) custard cups. Scald milk. Beat eggs with remaining ¼ cup (50 mL) sugar and vanilla. Gradually stir in milk. Pour mixture through a fine sieve and divide among prepared custard cups. Place cups in a shallow pan; pour hot water to within ½″ (12 mm) of the tops of the cups. Bake in preheated 300°F (150°C) oven 40 to 45 minutes or until a knife inserted in the centre comes out clean. Cool then chill several hours or overnight. To serve, loosen edges with a spatula and turn out onto serving dishes.

Makes 6 servings.

GRILLED CINNAMON PEARS AND FRESH STRAWBERRIES

3	large ripe pears	3
2 tablespoons	melted butter	30 mL
6 teaspoons	brown sugar	30 mL
6 teaspoons	brandy (optional)	30 mL
1 tablespoon	sugar	15 mL
1½ teaspoons	cinnamon	7 mL
	fresh strawberries	

Peel and halve pears, scoop out cores leaving a well in the centre of each pear half. Brush flat side of cut pears with butter; place on broiler rack flat side down. Brush hump with butter. Place under preheated broiler 2-3 minutes. Turn pear halves flat side up. Fill each cavity with a teaspoon of brown sugar and a teaspoon of brandy. Combine white sugar and cinnamon and sprinkle over pears. Return under broiler for 2 minutes. Arrange pear halves, flat side down, in a shallow dish. Spoon Custard Sauce over and around pears and garnish with fresh strawberries.
Makes 6 servings.

CUSTARD SAUCE

2 cups	milk	500 mL
2 tablespoons	cornstarch	30 mL
1 tablespoon	sugar	15 mL
1	egg, beaten	1
1 tablespoon	butter	15 mL
1 teaspoon	vanilla	5 mL

Mix ½ cup (125 mL) milk with cornstarch and sugar. Heat remaining milk. Gradually whisk hot milk into cornstarch mixture. Set over medium heat, stirring constantly, until sauce comes to a boil. Lower heat and simmer for 5 minutes. Beat half of sauce into beaten egg, then return to saucepan and cook for another minute. Remove from heat and stir in butter and vanilla.
Makes 2 cups/500 mL.

PARTY SPECIAL FRUIT SALAD

1	package (4 serving size) vanilla pudding and pie filling	1
2 cups	milk	500 mL
1 teaspoon	grated orange rind	5 mL
½ cup	whipping cream	125 mL
2 tablespoons	sugar	30 mL
½ teaspoon	vanilla	2 mL
	Fresh Fruit Salad	

Combine pudding mix, milk and orange rind in a saucepan. Cook according to package directions. Cover surface of pudding with plastic wrap. Cool then chill. Whip cream until softly stiff. Beat in sugar and vanilla. Stir pudding until smooth. Fold in whipped cream. Spoon over servings of Fresh Fruit Salad.

Makes 3 cups/750 mL sauce.

FRESH FRUIT SALAD

½ cup	sugar	125 mL
1 tablespoon	corn starch	15 mL
1 cup	orange juice	250 mL
2 tablespoons	orange liqueur	30 mL
10 cups	mixed cut-up fresh fruit eg. apples, pears, oranges, strawberries, etc.	2.5 L

Combine sugar and corn starch. Stir in orange juice. Cook over medium heat, stirring constantly, until mixture comes to a boil and thickens. Reduce heat, cover and cook an additional 2 minutes. Cool. Stir in liqueur. Chill. Before serving pour syrup over prepared fruit.

Makes 10 servings.

CHOCOLATE RUM & RAISIN CHEESECAKE

¼ cup	raisins	50 mL
¼ cup	rum or 1½ teaspoons (7 mL) rum extract in ¼ cup (50 mL) water	50 mL
1¼ cups	graham cracker crumbs	300 mL
¼ cup	sugar	50 mL
⅓ cup	butter, melted	75 mL
3	eggs, separated	3
2 cups	milk	500 mL
2	envelopes unflavoured gelatin	2
1 tablespoon	instant coffee granules	15 mL
3 ounces	semi-sweet chocolate chips	100 mL
2 (8-ounce)	packages cream cheese, softened	2 (250 g)
½ cup	sugar	125 mL

Soak raisins in rum or rum extract; set aside. Combine graham cracker crumbs, sugar and butter; press firmly into base of a 9" (2.5 L) springform pan; refrigerate. Beat together egg yolks and milk, sprinkle in gelatin and whisk to combine. Stir constantly over low heat until gelatin dissolves and mixture thickens, about 5-10 minutes. Add coffee and chocolate chips; continue stirring over low heat until melted and smooth; remove from heat. Beat cream cheese and sugar until light and fluffy; gradually beat in gelatin mixture, raisins and rum. Chill until mixture mounds slightly when dropped from a spoon. Beat egg whites until stiff but not dry; fold into gelatin mixture. Pour into springform pan; chill until set, about 2 to 3 hours. When set, decorate with chocolate curls and sprinkle with icing sugar, if desired.

Makes 6 to 8 servings.

CITRUS CHEESECAKE

CRUST

1½ cups	chocolate wafer crumbs	375 mL
2 tablespoons	sugar	30 mL
¼ cup	melted butter	50 mL

FILLING

1 cup	orange juice	250 mL
1 tablespoon	orange rind	15 mL
2 tablespoons	lemon juice	30 mL
2 tablespoons	gelatin	30 mL
2	eggs, separated	2
1 cup	milk	250 mL
¾ cup	sugar	175 mL
2 pounds	cottage cheese	1 kg
½ cup	whipping cream	125 mL
	fresh fruit	
	apricot glaze	

To prepare crust combine chocolate crumbs, sugar and butter. Press into 9 ½ inch (3 L) springform pan. Bake in 350°F (180°C) oven for 8 minutes. Cool.

For filling combine juices and rind. Soften gelatin in ¼ cup juice. Whisk together egg yolks, ½ cup (125 mL) sugar and milk, and heat in a double boiler. Add gelatin and continue cooking until dissolved. Cool. Stir in remaining juices. Beat cottage cheese until smooth. Mix with gelatin mixture. Beat egg whites with remaining sugar until stiff. Beat cream. Fold in egg whites, then whipped cream into orange mixture. Spoon into prepared base. Chill well. Top with fruit and brush on Apricot Glaze.

Makes 10 servings.

APRICOT GLAZE

½ cup	apricot jam	125 mL
2 tablespoons	water	30 mL

Combine ingredients in small saucepan. Cook until jam has melted.

MOCHA SWIRL CHEESECAKE

CRUST

1 cup	graham cracker crumbs	250 mL
¼ cup	melted butter	50 mL
½ cup	chopped pecans	125 mL
½ teaspoon	cinnamon	2 mL

FILLING

2 cups	milk	500 mL
2	envelopes unflavoured gelatin	2
1 tablespoon	instant coffee	15 mL
4	eggs, separated	4
2 (8 ounce)	packages cream cheese	2 (250 g)
½ cup	sugar	125 mL
1 cup	chocolate chips	250 mL
¼ cup	coffee-flavoured liqueur	50 mL

To prepare crust combine graham cracker crumbs, melted butter, pecans and cinnamon. Press into the bottom of a 9½″ (2.5 L) spring form pan. Bake in a 350°F (180°C) oven for 8 minutes. Cool. For filling, mix gelatin and coffee with 1 cup (250 mL) milk. Beat egg yolks until light. Add remaining milk. Stir this into gelatin mixture. Cook over low heat in a saucepan until gelatin is dissolved. Remove from heat. Beat cream cheese with ¼ cup (50 mL) sugar until smooth. (This is easily done in a food processor.) Gradually whisk in gelatin mixture. Beat until smooth. Chill until mixture mounds slightly on a spoon. Melt chocolate with liqueur. Cool slightly. Beat egg whites until soft peaks form. Gradually add remaining sugar and beat until stiff and shiny. Fold whites into gelatin mixture. Add chocolate to two cups (500 mL) gelatin mixture. Spoon both mixtures into prepared crust alternating spoonfuls. Swirl with a knife to give a marbled effect. Chill until firm.

Makes 8-10 servings.

Pina Colada Cheese Cake

1	envelope unflavoured gelatin	1
2 tablespoons	rum	30 mL
¾ cup	sugar	175 mL
3	eggs, separated	3
1 cup	coconut milk, hot (see below)	250 mL
1 (14 ounce)	can pineapple tidbits	1 (398 mL)
1 pound	cream cheese	500 g
1	Coconut Graham Crust	1
1 (14 ounce)	can pineapple rings, drained	1 (398 mL)
	maraschino cherries	

Sprinkle gelatin over rum. Set aside. Beat egg yolks with ½ cup (125 mL) sugar until light. Slowly add hot coconut milk. Set in top of double boiler over simmering water; cook stirring constantly, until mixture thickens slightly. Stir in gelatin. Beat egg whites until stiff; gradually beat in remaining sugar and beat until stiff peaks form. Thoroughly drain pineapple tidbits and purée in blender until as smooth as possible. Beat cream cheese until light; gradually beat in gelatin mixture. Add puréed pineapple. Gently but thoroughly blend in beaten egg whites. Pour mixture into prepared Coconut Graham Crust and refrigerate until set. Before serving carefully remove springform ring and decorate cheese cake with pineapple rings and cherries. Chill.

COCONUT GRAHAM CRUST

1 cup	graham cracker crumbs	250 mL
¼ cup	unsweetened flaked coconut, toasted	50 mL
2 tablespoons	sugar	30 mL
¼ cup	melted butter	50 mL

Preheat oven to 350°F (180°C).
Combine ingredients and press in an even layer onto the greased bottom of 9″ (23 cm) springform pan. Do not put on ring. Bake 8-10 minutes. Cool. Lightly oil ring of pan and attach to base.

continued on next page

PINA COLADA CHEESE CAKE (continued)

COCONUT MILK

Combine 1½ cups (375 mL) milk with 1 cup (250 mL) unsweetened flaked coconut in blender container. Cover and blend 1 minute. Press through a fine sieve to obtain approximately 1 cup (250 mL) of coconut milk.

Makes one cheesecake.

BUTTERMILK CHEESE DESSERT RING

1		envelope gelatin	1
2	tablespoons	lemon juice	30 mL
¼	cup	orange juice	50 mL
1½	cups	creamed cottage cheese	375 mL
1	teaspoon	grated orange rind	5 mL
1	cup	buttermilk	250 mL
¼	cup	brown sugar	50 mL
¼	teaspoon	nutmeg	1 mL
2	cups	sliced fresh strawberries, or other fresh fruit	500 mL
		fresh mint (optional)	

Soften gelatin in lemon and orange juices; dissolve over very low heat (or boiling water). In a blender or food processor, blend cottage cheese, orange rind, buttermilk, sugar and nutmeg together. Add dissolved gelatin, stirring to blend well.

Pour into a rinsed 4 cup (1 L) mold; refrigerate until firm. Unmold onto serving plate. Fill centre with strawberries. Garnish with fresh mint.

Makes 6 servings.

BANANA CHIFFON PIE

16	cream filled chocolate cookies	16
1	package (6 servings) banana pudding and pie filling mix	1
½ cup	sugar	125 mL
2 ⅓ cups	milk	575 mL
2	envelopes unflavoured gelatin	2
1 tablespoon	butter	15 mL
	pinch nutmeg	
2	egg whites	2
½ cup	whipping cream	125 mL
2	bananas	2
	whipping cream	

Crush cookies to form crumbs. Press into 9-inch (23 cm) pie plate; reserving ⅓ cup (75 mL) for top. Combine pudding mix, sugar, milk and gelatin, in a saucepan. Whisk and cook over medium heat until mixture comes to a boil and thickens. Remove from heat. Stir in butter and nutmeg. Cover surface with plastic wrap and chill until mixture mounds slightly. Beat egg whites until stiff and whip cream until stiff. Fold into pudding. Slice bananas and spread on bottom of crust. Pour filling over top. Chill a few hours or overnight. Garnish with reserved crumbs and whipped cream.

Makes 6 servings.

BANANA SPLIT PIE

1½ cups	Graham cracker crumbs	375 mL
¼ cup	sugar	50 mL
⅓ cup	butter, melted	75 mL
1	package (6 serving size) Banana Pudding & Pie Filling	1
2 cups	milk	500 mL
2	bananas	2
1 cup	whipping cream	250 mL
	cherries	
	walnuts	
	Chocolate Syrup	

Preheat oven to 350°F (180°C).

Combine Graham cracker crumbs, sugar and butter. Press into a buttered 9″ (1 L) pie plate and bake 10 minutes. Set aside to cool. Prepare banana filling according to package instructions, but using 2 cups (500 mL) milk. Cool slightly and spoon into pie shell. Chill. Just before serving slice bananas and arrange on top of filling. Whip cream until stiff (sweetening to taste if desired). Pile on top of bananas and decorate with cherries, walnuts and chocolate syrup.

Makes one 9″/1 L pie.

CHOCOLATE COCONUT PIE

1	envelope unflavoured gelatin	1
2 cups	milk	500 mL
1	egg, separated	1
1 tablespoon	cornstarch	15 mL
⅓ cup	sugar	75 mL
⅓ cup	semi-sweet chocolate chips	75 mL
1 teaspoon	vanilla	5 mL
¼ cup	instant skim milk powder	50 mL
¼ cup	ice water	50 mL
1	9 inch (23 cm) Coconut Crust (see below)	1
	whipped cream	
	chocolate shavings	

Sprinkle gelatin over ½ cup (125 mL) milk; set aside. Heat remaining milk. In a bowl, combine egg yolk, cornstarch and sugar. Beat in hot milk. Return to saucepan and cook over medium heat, stirring constantly while mixture thickens and comes to a boil. Add chocolate chips, gelatin and vanilla and blend thoroughly. Chill until slightly thickened. Beat egg white with dry milk and ice water until stiff peaks form. Fold into chocolate custard and pour into prepared coconut crust. Chill.

Makes one 9 inch/1 L pie.

COCONUT CRUST

3 tablespoons	butter	45 mL
1 cup	unsweetened flaked coconut	250 mL
¼ cup	flour	50 mL
2 tablespoons	sugar	30 mL
2 tablespoons	milk	30 mL

Preheat oven to 350°F (180°C).

Melt butter in a skillet. Add coconut, and flour and toss over medium heat until lightly golden. Stir in sugar and milk. Press mixture over bottom and sides of a 9 inch (23 cm) pie plate. Bake 8 minutes. Set aside.

Makes one 9 inch/1 L crust.

COCONUT ALMOND PIE

1	9-inch (23 cm) baked graham cracker crust	1
1	envelope unflavoured gelatin	1
½ cup	sugar	125 mL
¼ teaspoon	salt	1 mL
	pinch nutmeg	
3	eggs, separated	3
1 cup	milk	250 mL
1 teaspoon	vanilla extract	5 mL
½ teaspoon	almond extract	2 mL
¼ cup	toasted sweetened shredded coconut	50 mL
¼ cup	toasted slivered almonds	50 mL

Combine gelatin, sugar, salt and nutmeg in top of a double boiler. Beat egg yolks. Add to pot. Gradually whisk in milk. Cook over simmering hot water stirring constantly until custard thickens (coats the back of a spoon). Stir in vanilla and almond extract. Chill until partially set. Whisk whites until stiff. Fold into almond custard. Chill until set. Sprinkle with almonds and toasted coconut before serving.

Makes 6 servings.

Deluxe Banana Cream Pie

1-9″	baked pie crust	1-23 cm
4	egg yolks	4
½ cup	sugar	125 mL
⅓ cup	flour	75 mL
2 cups	milk	500 mL
¼ teaspoon	nutmeg	1 mL
1 teaspoon	vanilla extract	5 mL
2 tablespoons	butter	30 mL
2	bananas, peeled, and sliced	2
	banana slices	
	maraschino cherries	
	whipped cream	

Beat egg yolks and sugar until lightly thickened and creamy. Fold in flour. Heat milk until just hot. Add a little of hot milk to egg yolks. Mix well. Return to pot and cook and stir over medium heat until nicely thickened. Remove from heat. Stir in nutmeg, vanilla extract and butter. Cool slightly. Arrange bananas on bottom of prepared pie crust. Pour filling on top. Refrigerate until set. Garnish with additional sliced banana, maraschino cherries and whipped cream.

Makes 6 servings.

EGGNOG PIE

1-9″	crumb or pastry pie crust (baked)	1-1.2 L
1½ tablespoons	unflavoured gelatin	25 mL
¼ cup	cold water	50 mL
2 cups	milk	500 mL
⅓ cup	sugar	75 mL
¼ teaspoon	salt	2 mL
3	eggs, separated	3
6 tablespoons	sugar	100 mL
2 tablespoons	rum, or ½ teaspoon (2 mL) artificial rum flavouring	30 mL
1 teaspoon	vanilla	5 mL
	nutmeg	

Sprinkle gelatin over cold water and let stand until softened. Heat milk, ⅓ cup (75 mL) sugar and salt in saucepan; add gelatin and stir until dissolved. Add a little of the hot mixture to the beaten egg yolks, then return to saucepan and cook for 2 minutes. Chill until slightly thickened. Beat egg whites until frothy and gradually beat in 6 tablespoons (100 mL) sugar until stiff and shiny. Fold into chilled custard, along with flavourings. Pour into prepared crust, sprinkle with nutmeg and chill until set. For a festive touch, garnish with whipped cream, shaved chocolate and sliced red and green cherries, if desired.

Makes 6-8 servings.

Grapefruit Custard Pie

1-9"	unbaked pastry shell	1-23 cm
2	eggs	2
½ cup	sugar	125 mL
½ cup	flour	125 mL
	pinch salt	
1¼ cups	milk, hot	300 mL
½ cup	fresh grapefruit juice	125 mL
½ teaspoon	vanilla	2 mL
	pinch grated nutmeg	
2	large grapefruit	2
¼ cup	brown sugar	50 mL
¼ teaspoon	cinnamon	1 mL

Preheat oven to 425°F (220°C).

Bake pastry shell 12 minutes. Remove, prick bottom of shell and bake 2-3 minutes more. Set aside to cool. In a bowl combine eggs, sugar, flour and salt. Slowly beat in hot milk. Set in top of double boiler over hot water, stirring constantly until mixture comes to a boil. Add grapefruit juice and cook, stirring for 10 minutes. Stir in vanilla and nutmeg. Chill. Spoon chilled mixture into pie shell. With a sharp knife, remove peel and pits from grapefruit and cut out sections of fruit slicing between the membranes. Arrange segments on top of filling in a decorative pattern. Combine brown sugar and cinnamon and sprinkle over top. Cover exposed pastry rim with strips of foil and set pie under preheated broiler 6" (15 cm) from heat until sugar melts. Remove foil and let pie cool.

Makes one pie.

GRASSHOPPER PIE

¾ cup	icing sugar	175 mL
⅓ cup	unsweetened cocoa	75 mL
½ teaspoon	salt	2 mL
⅓ cup	butter	75 mL
1 cup	finely-chopped toasted nuts	250 mL
1	envelope unflavoured gelatin	1
¾ cup	granulated sugar	175 mL
2	eggs, separated	2
¾ cup	milk	175 mL
¼ cup	mint liqueur	50 mL
1 cup	whipping cream green food colouring (optional)	250 mL

Sift together icing sugar, cocoa and ¼ teaspoon (1 mL) of the salt. Melt butter in a saucepan. Continue cooking until butter is hot and bubbly. Remove from heat. Blend cocoa mixture into hot butter. Stir in nuts. Press onto bottom and sides of a 9-inch (1 L) pie plate. Chill. Combine gelatin, ½ cup (125 mL) of the granulated sugar and remaining ¼ teaspoon (1 mL) salt in a saucepan. Beat egg yolks and milk together; gradually add to gelatin mixture. Cook over low heat, stirring constantly, until gelatin and sugar are dissolved. Stir in liqueur. Chill until slightly thickened. Beat egg whites until frothy. Gradually beat in remaining ¼ cup (50 mL) granulated sugar. Continue to beat until stiff peaks form. Whip cream until softly stiff. Fold both into gelatin mixture. Tint a pale green if desired. Chill until mixture mounds from a spoon. Spoon into chilled pie shell. Chill until firm.

Makes one 9-inch/1 L pie.

HARVEST APPLE CRISP PIE

3 cups	peeled, sliced apples	750 mL
½ cup	water	125 mL
1	package (4 serving size) vanilla pudding and pie filling	1
2 cups	milk	450 mL
⅓ cup	firmly-packed brown sugar	75 mL
¼ cup	all-purpose flour	50 mL
¼ cup	quick-cooking rolled oats	50 mL
½ teaspoon	ground cinnamon	2 mL
½ teaspoon	ground nutmeg	2 mL
3 tablespoons	butter	45 mL
1	baked and cooled 9-inch (1 L) pie shell	1

Combine apples and water in a saucepan. Bring to a boil. Cover and simmer 10 minutes; drain. Combine pudding mix and milk in a saucepan. Cook according to package directions; cover surface with plastic wrap; set-aside. Combine sugar, flour, oats, cinnamon and nutmeg in a small bowl. Cut in butter until mixture resembles coarse crumbs. Place apple slices in bottom of pie shell. Pour pudding over apples. Sprinkle sugar mixture over top. Bake in preheated 375°F (190°C) oven 15 minutes or until topping is golden. Cool; chill well before serving.

Makes one 9-inch/1 L pie.

Impossible Coconut Pie

1 cup	unsweetened shredded coconut	250 mL
½ cup	flour	125 mL
1 cup	sugar	250 mL
½ cup	soft butter	125 mL
½ teaspoon	baking powder	2 mL
4	eggs, beaten	4
2 cups	milk	500 mL
2 teaspoons	vanilla extract	10 mL

Preheat oven to 350°F (180°C).

Combine all ingredients in a food processor bowl and process until smooth, or combine in a large bowl and beat well. Pour into a buttered 9″ (1 L) pie dish. Bake for about 1 hour or until toothpick inserted in centre comes out clean. Serve with whipped cream and fresh berries.

Makes one cake.

LEMON BLUEBERRY PIE

CRUST

1½ cups	all purpose flour	375 mL
½ teaspoon	salt	2 mL
1 tablespoon	sugar	15 mL
1 tablespoon	grated orange rind	15 mL
½ cup	shortening	125 mL
¼ cup	butter	50 mL
1 tablespoon	lemon juice	15 mL
2 tablespoons	ice water	30 mL

FILLING

2 cups	blueberries	500 mL
1½ cups	sugar	375 mL
5 tablespoons	flour	75 mL
1 tablespoon	lemon juice	15 mL
⅓ cup	lemon juice	75 mL
3 tablespoons	butter	45 mL
4	eggs, separated	4
1 cup	milk	250 mL
	pinch salt	
	grated rind of two lemons	

To prepare pastry: blend flour, salt, sugar and orange rind together. Cut in shortening and butter until mixture resembles coarse crumbs. Sprinkle with lemon juice and water. Stir and form into a ball. Roll and line a 10″ (25 cm) lightly greased deep dish pie plate. Chill.

To prepare filling: combine blueberries, ½ cup (125 mL) sugar, 2 tablespoons (30 mL) flour and 1 tablespoon (15 mL) lemon juice. Let stand until sugar dissolves. Spread into chilled pastry shell. Bake in 450°F (230°C) oven for 15 minutes. While crust is baking cream butter and remaining sugar together. Add egg yolks one at a time. Stir in remaining flour, milk, salt, remaining lemon juice and rind. Beat egg whites until stiff. Fold into cake mixture. Spoon on top of blueberry mixture. Reduce heat to 350°F (180°C) and bake approximately 45 minutes or until knife, inserted in side of pie comes out clean. Chill.

Makes 10 servings.

MANDARIN PIE

CRUST

½ cup	butter	125 mL
1 cup	flour	250 mL
3 tablespoons	brown sugar	45 mL
¼ teaspoon	ground ginger	1 mL
½ cup	finely chopped nuts	125 mL

FILLING

¾ cup	milk	175 mL
1 cup	sour cream	250 mL
1 tablespoon	grated orange rind	15 mL
1	package instant vanilla pudding mix (4 servings)	1
1 (10 ounce)	can mandarin oranges, well drained	1 (284 mL)
½ cup	orange marmalade	125 mL
2 tablespoons	water	30 mL

For crust: melt butter in a 9-inch (23 cm) pie plate. Combine flour, sugar, ginger and nuts. Mix well with butter. Bake in a 350°F (180°C) oven for approximately 25 minutes, stirring occasionally until crumbs are dried and beginning to brown. Cool slightly to handle. Press into bottom and sides of pie plate to make crust. Reserve ¼ cup (50 mL) for top. Beat milk, sour cream, orange rind and pudding mix in a bowl until thick. Pour into crust. Chill until firm. To serve arrange orange segments on top of pie. Melt marmalade with water. Brush over orange segments. Sprinkle with reserved crumbs. Chill 30 minutes to one hour before serving.

Makes 6 servings.

MAPLE SUGAR CHIFFON PIE

1-9″	baked pie shell, cooled	1-23 cm
2 cups	milk	500 mL
2 tablespoons	cornstarch	30 mL
½ cup	maple syrup	125 mL
2 tablespoons	maple syrup	30 mL
4	eggs, separated	4
1	envelope unflavoured gelatin	1
¼ cup	water	50 mL
1 teaspoon	vanilla	5 mL
	whipped cream	

Blend milk, cornstarch and ½ cup (125 mL) maple syrup together in a saucepan. Cook and stir over medium heat until thickened. Beat egg yolks until lightly thickened and lemony. Beat some of the cornstarch mixture with the eggs. Return to saucepan and continue cooking for 2 minutes. Soften gelatin in water. Add to custard and stir until dissolved. Stir in vanilla. Chill until custard begins to set. Beat egg whites until almost stiff. Continue beating, adding remaining maple syrup. Beat until stiff. Fold into custard and spoon into pie crust. Chill well. Serve sprinkled with Maple Walnut Crunch and whipped cream.
Makes 6 servings.

MAPLE WALNUT CRUNCH

¼ cup	butter	50 mL
½ cup	chopped walnuts	125 mL
⅓ cup	maple syrup	75 mL

Cook walnuts in butter until lightly browned. Add maple syrup and continue cooking over medium heat (just boiling) until syrup reaches hard crack stage 300°F (150°C) on candy thermometer. Remove from heat. Pour onto cookie sheet. Let set. Chop or break to serve.

MILE-HIGH BLACK BOTTOM PIE

CRUST

1¾ cups	graham cracker crumbs	425 mL
¼ cup	sugar	50 mL
½ cup	melted butter	125 mL

To make crust combine all ingredients. Press firmly into bottom and sides of a deep 10-inch (25 cm) pie plate. Bake in 350°F (180°C) oven for 15 minutes. Cool.

FILLING

4 teaspoons	unflavoured gelatin	20 mL
¼ cup	cold water	50 mL
2 ounces	unsweetened chocolate	56 g
2¼ cups	milk	550 mL
4	eggs, separated	4
¾ cup	sugar	175 mL
1 tablespoon	cornstarch	15 mL
¼ teaspoon	salt	1 mL
½ teaspoon	cream of tartar	2 mL
1 tablespoon	vanilla extract	15 mL
2 cups	whipping cream	500 mL
	semi-sweet chocolate	

For filling soften gelatin in cold water and set aside. Melt chocolate in a pot over hot water. Heat milk until hot. Beat egg yolks until smooth and creamy. Combine ½ cup (125 mL) sugar and cornstarch. Whisk into egg yolks. Add a little hot milk to egg yolks. Return to pan and cook over hot water until thickened. Remove 1⅓ cups (325 mL) of custard and blend with melted chocolate. Pour into prepared crust. Refrigerate. Add gelatin to remaining custard. Stir to dissolve. Add vanilla. Refrigerate until cool. Whisk egg whites with salt and cream of tartar until stiff. Whip cream until stiff. Fold both into vanilla custard. Spoon over chocolate layer in pie crust. Chill well. Garnish with chocolate shavings.
Makes 8-10 servings.

APPLE BACON PANCAKE PIE

1	tablespoon	butter	15 mL
1		onion, sliced	1
½	pound	bacon, sliced	225 g
2		pears, peeled and sliced	2
2		apples, peeled and sliced	2
½	teaspoon	thyme	2 mL
¾	cup	flour	175 mL
1	cup	milk	250 mL
2		eggs, beaten	2
½	teaspoon	salt	2 mL

Heat butter in a large frying pan. Add onion and bacon and cook until softened. Stir in pears, apples and thyme, and cook until heated through. Place flour in a small bowl. Add ½ the milk and beat well. Add eggs and remaining milk and beat until smooth. Whisk in salt. Spread fruit mixture over bottom of a 9″ x 13″ (3 L) baking pan. Pour batter on top. Bake in 350°F (180°C) oven for 30 minutes until puffed and lightly browned. Cut in squares and serve with warmed maple syrup.

MINCEMEAT CUSTARD PIE

⅓ cup	sugar	75 mL
¼ teaspoon	salt	1 mL
¼ teaspoon	ground nutmeg	1 mL
¼ teaspoon	ground cinnamon	1 mL
2	eggs, beaten	2
1¼ cups	milk	300 mL
1 teaspoon	vanilla	5 mL
1 cup	prepared mincemeat	250 mL
1	9-inch (1 L) unbaked pie shell	1

Combine sugar, salt, nutmeg and cinnamon in mixer bowl. Add eggs, milk and vanilla; mix thoroughly. Press mincemeat evenly over bottom of unbaked pie shell. Slowly pour custard mixture over mincemeat. Bake in preheated 425°F (220°C) oven 10 minutes. Reduce heat to 350°F (180°C) and continue baking 30 to 35 minutes or until knife inserted near centre comes out clean. Cool before serving.

Makes 6 servings.

MOCHA CHIFFON PIE

1⅓ cups	chocolate wafer crumbs	325 mL
3 tablespoons	butter, melted	45 mL
1	envelope unflavoured gelatin	1
1 tablespoon	instant coffee	15 mL
¾ cup	sugar	175 mL
¼ teaspoon	salt	1 mL
3	eggs, separated	3
1 cup	milk	250 mL
2 tablespoons	chocolate or coffee liqueur	30 mL
1 cup	whipping cream	250 mL
	chocolate curls	

Combine wafer crumbs and butter. Press onto bottom and sides of a 9-inch (1 L) pie plate. Chill. Combine gelatin, coffee, ½ cup (125 mL) of the sugar and salt in a saucepan. Beat egg yolks and milk together; gradually add to gelatin mixture. Cook over low heat, stirring constantly, until gelatin, coffee and sugar are dissolved. Stir in liqueur. Chill until slightly thickened. Beat egg whites until frothy. Gradually beat in remaining sugar. Continue to beat until stiff peaks form. Whip cream until softly stiff. Fold both into gelatin mixture. Chill until mixture mounds from a spoon. Pile in prepared pie shell. Chill until firm. Garnish with chocolate curls if desired.

Makes one 9-inch/1 L pie.

Orange Chocolate Pie

1½ cups	chocolate wafer crumbs	375 mL
½ cup	melted butter	125 mL
4	eggs, separated	4
½ cup	sugar	125 mL
2 teaspoons	cornstarch	10 mL
1½ cups	hot milk	375 mL
1	envelope unflavoured gelatin	1
2 ounces	unsweetened chocolate, melted	56 g
2 tablespoons	rum or brandy	30 mL
2 teaspoons	grated orange rind	10 mL

Combine chocolate wafer crumbs and melted butter and press evenly into a 9″ (23 cm) pie plate. Bake in 350°F (180°C) oven for 10 minutes. Cool. Reserve one tablespoon (15 mL) sugar and beat remaining sugar with egg yolks until smooth and creamy. Beat in cornstarch. Gradually whisk in hot milk. Sprinkle gelatin on hot mixture and whisk in thoroughly. Pour into saucepan and cook over medium heat, stirring constantly until sauce is lightly thickened. Do not boil. Remove from the heat, stir in melted chocolate and rum and orange rind. Beat egg whites with one tablespoon (15 mL) sugar until stiff and fold into chocolate mixture. Spoon into chocolate crust and chill several hours or overnight until set. Decorate with whipped cream and chocolate shavings.

Makes 6 servings.

Peanut Butter Pie

1	9-inch (23 cm) baked pie crust	1
¾ cup	crunchy peanut butter	175 mL
¼ cup	honey	50 mL
¼ teaspoon	cinnamon	1 mL
2 cups	milk	500 mL
¼ cup	sugar	50 mL
2 tablespoons	sugar	30 mL
¼ cup	cornstarch	50 mL
4	eggs, separated	4
1 teaspoon	vanilla	5 mL
¼ teaspoon	nutmeg	1 mL

Combine peanut butter and honey. Blend well and spread on bottom of prepared pie crust. Sprinkle with cinnamon. In a small saucepan, combine milk, cornstarch, sugar. Bring just to a boil stirring constantly. Beat egg yolks until light and creamy. Add a little of the hot milk to eggs. Return to saucepan. Cook and stir until well thickened. Flavour with vanilla and nutmeg. Cool slightly, stirring. Spoon on top of peanut butter. Beat egg whites until frothy. Gradually add sugar and beat until stiff and shiny. Spread meringue on top of peanut butter making sure to cover inner edge of pie crust. Bake in 350°F (180°C) oven until meringue is lightly browned, approximately 10 minutes. Chill well.

Makes 6 servings.

PEPPERMINT PIE

1 cup	peppermint candy pieces	250 mL
1	envelope unflavoured gelatin	1
¼ cup	water	50 mL
2	eggs, separated	2
1¼ cups	milk, hot	300 mL
½ cup	whipping cream	125 mL
	drop red food colouring	
	pinch salt	
¼ cup	sugar	50 mL
1-9"	baked pie shell	1-23 cm
1½ ounces	semi-sweet chocolate	42 g
1 tablespoon	butter	15 mL
6 tablespoons	icing sugar	90 mL
1	egg yolk	1

Whirl peppermint candy pieces in a blender or food processor in two batches. Process enough to measure ½ cup (125 mL) finely crushed candy. Sprinkle gelatin over water; set aside. Beat milk into lightly beaten egg yolks. Cook, stirring constantly, until mixture thickens slightly and coats a metal spoon. Add softened gelatin and crushed candy and stir until smooth. Chill until mixture reaches consistency of unbeaten egg whites. Whip cream until stiff and fold into mixture. Add a few drops of red food colouring. Beat egg whites and salt until soft peaks form; gradually add sugar and beat until stiff. Fold into peppermint mixture. Spoon into baked pie shell and chill until firm. Just before serving melt chocolate with butter. When cooled slightly beat in egg yolk and sifted icing sugar. Spread glaze over chilled pie.
Makes one 9 inch/1L pie.

PUMPKIN PIE DELITE

1 (250 g)	package soft cream cheese	1 (250 g)
3	eggs	3
1 cup	sugar	250 mL
1 teaspoon	vanilla	5 mL
1	9-inch (1 L) unbaked pie shell	1
1 (14-ounce)	can pumpkin	1 (398 mL)
1 teaspoon	ground cinnamon	5 mL
½ teaspoon	ground ginger	2 mL
1 cup	milk	250 mL
	whipped cream	
	crushed peanut brittle	

Beat cheese until light and fluffy; blend in 1 of the eggs, ½ cup (125 mL) of the sugar and vanilla. Spread cheese mixture in bottom of unbaked pie shell. Beat remaining 2 eggs; blend in pumpkin, remaining ½ cup (125 mL) sugar, cinnamon, ginger and milk. Pour mixture carefully over cream cheese to form a second layer. Bake in preheated 425°F (220°C) oven 10 minutes. Reduce temperature to 350°F (180°C). Bake 30 to 35 minutes longer or until set. Cool. Garnish with whipped cream and peanut brittle.
    Makes 6 servings.

Pumpkin Chiffon Pie

1-9″	baked pie crust	1-23 cm
1⅓ cups	milk	325 mL
¾ cup	brown sugar	175 mL
1 (14 ounce)	can pumpkin purée	1 (398 mL)
	pinch salt	
	pinch cloves	
1 teaspoon	ginger	5 mL
½ teaspoon	cinnamon	2 mL
½ teaspoon	nutmeg	2 mL
3	eggs, separated	3
2	envelopes unflavoured gelatin	2
⅓ cup	cold water	75 mL
1 teaspoon	vanilla extract	5 mL

Combine milk, sugar, pumpkin, salt, cloves, ginger, cinnamon and nutmeg in a saucepan. Cook over medium heat until mixture has just reached boiling point. Beat egg yolks lightly. Add some of the hot mixture to eggs then return to pot. Heat through. Soften gelatin in water and add to pumpkin mixture. Stir until dissolved. Chill until pumpkin mounds slightly. Beat egg whites until stiff. Fold into pumpkin gently. Pour into prepared pie shell. Chill several hours before cutting.
     Makes 6 servings.

RHUBARB CHEESECAKE PIE

1 (250 g)	package soft cream cheese	1 (250 g)
2 tablespoons	butter	30 mL
1¼ cups	sugar	300 mL
2 tablespoons	flour	30 mL
1	egg	1
⅔ cup	milk	150 mL
1 tablespoon	grated lemon rind	15 mL
¼ cup	lemon juice	50 mL
1	9-inch (1 L) graham wafer crust	1
2 cups	fresh sliced rhubarb	500 mL

Beat together cream cheese and butter. Beat in ½ cup (125 mL) of the sugar, flour and egg. Gradually beat in milk. Add lemon rind and juice. Pour into wafer pie crust. Bake in preheated 350°F (180°C) oven 30 to 35 minutes. Cool completely. Combine rhubarb and remaining ¾ cup (175 mL) sugar in a saucepan. Cook over low heat, stirring constantly, until sugar dissolves. Bring to a boil. Reduce heat and simmer until rhubarb is tender. Cool. Spread over top of pie and chill well before serving.

Makes one 9-inch/1 L pie.

Snow Pie

1	envelope unflavoured gelatin	1
²/₃ cup	sugar	150 mL
¹/₃ cup	cornstarch	75 mL
	pinch salt	
3	eggs, beaten	3
2 ½ cups	milk	625 mL
12	large marshmallows, quartered	12
2 teaspoons	vanilla	10 mL
1 cup	whipping cream	250 mL
1	Shortbread Cookie Crust pie shell (see next page) sweetened whipped cream chocolate shavings maraschino cherries	1

Sprinkle gelatin over ¼ cup (50 mL) water to soften; set aside. Combine sugar, cornstarch and salt in a bowl. Mix to a paste with ½ cup (125 mL) cold milk. Heat remaining milk. Whisk into cornstarch mixture then return to pan and stir over medium heat while mixture comes to a boil and thickens. Add a little hot mixture to beaten eggs then return to pan and stir over medium heat for 2-3 minutes. Add softened gelatin and marshmallows; stir until melted. Add vanilla and blend in thoroughly. Cover custard with plastic wrap and set aside to cool. Whip cream until stiff and fold into chilled mixture. Pour into prepared shortbread crust, sprinkle with reserved crumbs and chill until serving. Decorate pie with additional lightly sweetened whipped cream, chocolate shavings and cherries if desired.

Makes 10-12 servings.

continued on next page

Snow Pie (continued)

SHORTBREAD COOKIE CRUST

1½ cups	shortbread cookie crumbs	375 mL
½ cup	icing sugar	125 mL
½ cup	unsweetened shredded coconut	125 mL
¼ cup	melted butter	50 mL
½ teaspoon	vanilla	2 mL

Preheat oven to 350°F (180°C).

Toss all ingredients together. Press ⅔ of mixture on bottom and sides of a greased 10-inch (26 cm) pie plate; reserve remaining crumbs for top of pie. Bake 10 minutes. Set aside.

Makes one 10-inch (26 cm) pie shell.

Strawberry Ginger Cream Pie

CRUST

| 8 ounces | gingersnap cookies, crushed into crumbs (approx. 2 cups/500 mL) | 250 g |
| ⅓ cup | butter (preferably unsalted) | 75 mL |

FILLING

1 quart	fresh strawberries, hulled	1 L
1	envelope unflavoured gelatin	1
⅓ cup	rum or water	75 mL
2	eggs	2
½ cup	sugar	125 mL
2 tablespoons	all purpose flour	30 mL
1½ cups	milk, hot	375 mL
1 teaspoon	pure vanilla extract	5 mL
1 cup	whipping cream	250 mL
⅓ cup	finely chopped candied ginger	75 mL

GARNISH

| ½ cup | strawberry or raspberry jam or jelly, strained sprigs of fresh mint – optional | 125 mL |

Combine cookie crumbs with melted butter; pat into bottom and sides of a 9″ (23 cm) pie dish. Bake at 350°F (180°C) for 10 minutes. Cool. Sprinkle gelatin over rum in a saucepan. Soften 5 minutes; heat gently to dissolve. Beat eggs with sugar in the top of a double boiler. Beat in flour; slowly whisk in hot milk. Cook over simmering water until custard thickens. Beat in gelatin and vanilla; cool over a larger bowl filled with ice and water. Mixture should be cool but not yet set. Slice ⅓ of the berries; place in cooled pie shell. Halve remaining berries and reserve for the top. Whip cream until light and fold into gelatin base along with ginger. Spoon into pie shell and refrigerate one hour. Arrange reserved berries over top of pie. Brush berries with jam. Decorate with sprigs of mint. Chill 2 hours longer.

Makes one 9″/23 cm pie.

SUNSHINE CITRUS PIE

9"		pie shell, baked	1 L
1	cup	sugar	250 mL
2	tablespoons	flour	30 mL
¼	cup	cornstarch	50 mL
½	teaspoon	salt	2 mL
1¼	cups	milk	300 mL
½	cup	orange juice	125 mL
2		eggs, lightly beaten	2
½	teaspoon	vanilla	2 mL
⅛	teaspoon	nutmeg	0.5 mL
2		grapefruit	2
1		orange	1
2	tablespoons	brown sugar	30 mL

In a medium saucepan mix sugar, flour, cornstarch and salt. Blend in milk and orange juice. Cook together over medium heat, stirring until mixture comes to a full boil. Remove from heat. Stir a little hot mixture into beaten eggs, blend this into remaining hot mixture and cook over low heat for 2 minutes. Remove from heat; blend in vanilla and nutmeg. Cool for 10 minutes; pour into baked pie shell.

Peel and section grapefruit and orange. Arrange in pinwheel or other decorative pattern on top of pie filling. Sprinkle with brown sugar. Cover pie crust edge with foil. Place under preheated broiler until sugar melts; remove foil. Cool at least 4 hours before serving.

Makes one 9" (1 L) pie.

SWEET POTATO AND APPLE PIE

1	9-inch (23 cm) unbaked pie shell	1
2 cups	mashed sweet potatoes	500 mL
1 cup	brown sugar	250 mL
½ teaspoon	grated nutmeg	2 mL
1½ teaspoons	cinnamon	7 mL
2 tablespoons	rum or brandy	30 mL
3 tablespoons	butter	45 mL
2	eggs, separated	2
¾ cup	milk	75 mL
½ cup	whipping cream	125 mL
1 teaspoon	grated lemon rind	5 mL
1½ cups	peeled, sliced apples	375 mL
½ cup	chopped nuts	125 mL

Combine sweet potatoes, ½ cup (125 mL) sugar, spices and rum. Melt 1 tablespoon (15 mL) of butter. Beat egg yolks. Add butter and egg yolks to sweet potatoes. Mix well. Blend in milk and cream. Stir in lemon rind. Whip egg whites until stiff and fold in. Pour into prepared pie shell. Spread apple slices on top. Sprinkle with remaining brown sugar and cinnamon, and nuts. Drizzle with melted butter. Bake in 375°F (190°C) oven for 20 minutes. Reduce heat to 350°F (180°C) and continue to bake pie for 45 minutes or until set.

Makes 6-8 servings.

ALMOND SOUFFLE

1 tablespoon	butter	15 mL
1 tablespoon	sugar	15 mL
3 tablespoons	flour	45 mL
¾ cup	milk	175 mL
⅓ cup	sugar	75 mL
4	eggs, separated	4
2 tablespoons	butter	30 mL
	pinch salt	
¼ teaspoon	cream of tartar	1 mL
1 tablespoon	sugar	15 mL
1 tablespoon	vanilla extract	15 mL
½ teaspoon	almond extract	2 mL
½ cup	finely ground toasted almonds	125 mL

Preheat oven to 400°F (200°C).

Butter the inside of a 5 cup (1.2 L) souffle mold. Dust with 1 tablespoon (15 mL) of sugar. Whisk flour with milk and ⅓ cup (75 mL) sugar in a small saucepan. Stir over moderate heat until the mixture comes to a boil. Boil 30 seconds. Remove from heat and beat for 2 minutes. Beat egg yolks. Gradually add to milk mixture. Beat in 1 tablespoon (15 mL) of butter and dot surface of sauce with remaining butter. Beat egg whites, salt and cream of tartar in a separate bowl until soft peaks form; add sugar and beat to form stiff peaks. Stir in flavourings and ground almonds. Gently fold egg white mixture into custard base. Do not over mix. Turn into prepared souffle dish. Reduce heat to 375°F (190°C) and bake 35-40 minutes or until knife inserted in centre comes out clean.

Makes 6-8 servings.

Note: This is very good served with Orange Honey Syrup, Page 562.

CHILLY LIME SOUFFLE

2	envelopes unflavoured gelatin	2
1½ cups	milk	375 mL
4	eggs, separated	4
¼ cup	sugar	50 mL
1 (6¼-ounce)	can frozen limeade concentrate, thawed	1 (177 mL)
¼ cup	lime juice	50 mL
	green food colouring (optional)	
1 teaspoon	cream of tartar	5 mL
¾ cup	whipping cream	175 mL
	lime slices	

Fasten a 2-inch (5 cm) foil collar around a 5-cup (1.25 L) souffle dish. Sprinkle gelatin over milk in a saucepan. Let stand 10 minutes to soften. Cook over low heat, stirring constantly, until gelatin is dissolved; cool. Beat together egg yolks and sugar. Add limeade concentrate, lime juice and few drops food colouring if desired. Stir in milk mixture. Chill until mixture mounds from a spoon. Beat egg whites with cream of tartar until stiff but not dry. Whip cream until softly stiff. Fold both into gelatin mixture. Turn into prepared dish. Chill until set. Remove collar to serve and garnish with fresh lime slices if desired.
    Makes 8 servings.

HOT CHOCOLATE SOUFFLE

3 tablespoons	corn starch	45 mL
½ cup	sugar	125 mL
1 cup	milk	250 mL
2	squares unsweetened chocolate, melted	2
2 tablespoons	butter	30 mL
4	eggs, separated	4
1½ teaspoons	vanilla	7 mL
¼ teaspoon	salt	1 mL
	icing sugar	
	sweetened whipped cream	

Combine corn starch and sugar in a saucepan. Gradually stir in milk. Cook over medium heat, stirring constantly, until mixture comes to a boil and thickens. Reduce heat and cook an additional 2 minutes, stirring occasionally. Remove from heat. Add chocolate and butter. Beat yolks well. Gradually stir a small amount of hot sauce into beaten egg yolks. Return all to saucepan and blend thoroughly. Add vanilla. Cool slightly. Add salt to egg whites; beat until stiff but not dry. Stir a large spoonful of beaten egg whites into chocolate sauce. Fold in remaining egg whites. Turn into a greased 6-cup (1.5 L) souffle dish that has been dusted with icing sugar. Bake in preheated 375°F (190°C) oven 35 to 40 minutes. Sprinkle surface of souffle with icing sugar and serve immediately with whipped cream if desired.

Makes 6 servings.

MAPLE FUDGE TARTS

¼ cup	butter	50 mL
⅓ cup	cornstarch	75 mL
1 cup	maple syrup	250 mL
½ cup	milk	125 mL
12 medium or 18 small tart shells, baked		

Melt butter; blend in cornstarch. Stir in maple syrup and milk. Bring to boil; stir and cook until thick. Continue cooking for 2 or 3 minutes more, stirring occasionally. Remove from heat; cool 15 minutes. Spoon into baked shells. Garnish with toasted almond slices.

Makes 12 medium or 18 small tarts.

COLONIAL MAPLE TARTS

2 tablespoons	butter	30 mL
1 cup	firmly packed brown sugar	250 mL
3	egg yolks	3
½ cup	milk	125 mL
1 cup	maple syrup	250 mL
¼ teaspoon	nutmeg	1 mL
3	egg whites	3
12 large or 18 medium unbaked tart shells		

Cream together the butter and sugar. Add egg yolks and beat well. Stir in milk, maple syrup and nutmeg. Beat the egg whites until they are stiff but not dry. Fold the egg whites into syrup mixture. Pour into tart shells. Bake at 450°F (230°C) for 10 minutes. Reduce heat to 350°F (180°C) and bake 30-35 minutes or more until the crust is golden brown and the filling is set. Serve with whipped cream.

Makes 12 large or 18 medium tarts.

MUD PIE TARTS

3 tablespoons	butter	45 mL
3 cups	miniature marshmallows	750 mL
3¾ cups	crisp rice cereal	925 mL
½ cup	sugar	125 mL
3 tablespoons	unsweetened cocoa	45 mL
3 tablespoons	corn starch	45 mL
1½ cups	milk	375 mL
1½ tablespoons	butter	25 mL
¾ teaspoon	vanilla	3 mL
	whipped cream	
	chocolate curls	

Melt butter in a large saucepan; add marshmallows. Cook over low heat, stirring constantly, until marshmallows are melted and mixture is thoroughly combined. Remove from heat. Add rice cereal and stir until coated. Press equal amounts of marshmallow cereal mixture into 12 large buttered muffin cups to form tart shells*. Let stand until set; remove tart shells from cups. Combine sugar, cocoa and corn starch in a saucepan. Gradually stir in milk. Cook over medium heat, stirring constantly, until mixture comes to a boil and thickens. Reduce heat and cook 2 minutes longer, stirring occasionally. Stir in butter and vanilla. Cover surface with plastic wrap; cool. Chill well. Spoon equal amounts into prepared tart shells. Garnish with whipped cream and chocolate curls.

Makes 1 dozen tarts.

* 12 large baked and cooled tart shells may be substituted for crisp rice cereal cups.

Black Forest Torte

CAKE

1¾ cups	all purpose flour	425 mL
½ cup	cocoa	125 mL
¼ teaspoon	baking soda	1 mL
1 teaspoon	salt	5 mL
1 tablespoon	baking powder	15 mL
⅔ cup	vegetable oil	150 mL
1 cup	sugar	250 mL
1 cup	buttermilk	250 mL
6	eggs	6

FILLING

½ cup	sherry	125 mL
2 cups	cherry pie filling	500 mL
3 cups	whipping cream	750 mL
½ cup	sugar	125 mL
	maraschino cherries	
	grated semi-sweet chocolate	

To prepare cakes, grease two 9″ (1.5 L) cake pans. Combine flour, cocoa, baking soda, salt and baking powder in a large bowl. Whisk in oil and mix until just moistened. Blend sugar and buttermilk together. Add to flour mixture. Stir well. Beat in eggs one at a time using medium speed or mixer until batter is smooth. Divide batter evenly between cake pans. Bake in 350°F (180°C) oven for 35-40 minutes until cake is firm and cooked through. Remove from pans. Cool on cake rack. To assemble cut cakes in half. Drizzle each section with sherry. Whip cream until stiff adding sugar a little at a time. Place one layer on plate spread with ⅓ cherry pie filling and 1 cup (250 mL) of whipped cream. Repeat with 2nd and 3rd layers. Top final layer with whipped cream. Cover sides with whipped cream. Garnish with cherries and grated chocolate. Refrigerate overnight, before serving.

Makes 10 servings.

Mocha Meringue Torte

MERINGUE

6	egg whites	6
1/4 teaspoon	cream of tartar	1 mL
2 teaspoons	vanilla	10 mL
	pinch salt	
1 1/4 cups	fruit sugar	300 mL

PASTRY CREAM

2 cups	hot milk	500 mL
6	egg yolks	6
1/2 cup	sugar	125 mL
1/3 cup	flour	75 mL
1 teaspoon	grated lemon rind	5 mL
1 teaspoon	vanilla extract	5 mL
2 ounces	melted semi-sweet chocolate	56 g
1 tablespoon	instant coffee	15 mL
1 tablespoon	hot water	15 mL
2 cups	whipping cream	500 mL
	fresh strawberries	

To make meringue cover ungreased cookie sheets with parchment or waxed paper. Draw 2-9" (23 cm) circles. Preheat oven to 275°F (140°C). In a large bowl beat egg whites until foamy. Add cream of tartar and salt and vanilla. Continue beating adding sugar a little at a time until stiff peaks form, and egg whites are shiny. Spoon onto circles dividing amount evenly and spreading to 1/2" (1 cm) thickness. Bake 1 hour. Turn off heat and let cool in oven.

To prepare pastry cream, beat egg yolks with sugar until light and creamy. Fold in flour. Add a little hot milk to egg. Return to pot. Cook and stir until well thickened. Dissolve coffee in water. Add to custard, with chocolate, vanilla and lemon rind. Cover with plastic wrap and chill well. To assemble torte place one meringue on plate. Spread with 1/2 of pastry cream. Repeat with next layer. Whip cream until stiff. Use to ice torte. Garnish top with fresh strawberries. Let set overnight before serving.

Makes 8 servings.

Note: This works particularly well as individual meringues. Makes 8 small meringues. Fill with pastry cream, fresh berries and top with whipped cream.

Orange Walnut Torte

CAKE

⅔ cup	all purpose flour	150 mL
2 ½ teaspoons	baking powder	12 mL
¼ teaspoon	salt	1 mL
2 cups	graham wafer crumbs	500 mL
½ cup	butter, softened	125 mL
¾ cup	sugar	175 mL
3	eggs, separated	3
1 cup	milk	250 mL
¾ cup	finely chopped walnuts	175 mL
1 teaspoon	vanilla extract	5 mL

FILLING

4 cups	miniature marshmallows	1 L
1 cup	orange juice	250 mL
2 teaspoons	lemon juice	10 mL
1 tablespoon	grated orange rind	15 mL
1 ¾ cups	whipping cream	425 mL
	chopped nuts	

To prepare cake combine flour, baking powder, salt and wafer crumbs in a large bowl. Mix well. Cream butter with sugar until light and fluffy. Add egg yolks, one at a time beating after each addition. Add dry ingredients alternately with milk mixing lightly. Beat egg whites until stiff. Fold in vanilla, nuts and egg whites. Spoon into 2-9″ (1.5 L) greased cake pans, lined with waxed paper. Bake in 375°F (190°C) oven for 30-35 minutes until firm and lightly browned. Remove from pans and cool.

Prepare filling: heat marshmallows and orange juice in a double boiler stirring occasionally until marshmallows have melted and mixture is smooth. Add lemon juice and orange rind. Chill. Beat cream until stiff. Fold into chilled marshmallow mixture. To assemble: spread filling between cake layers and on top. Garnish with chopped nuts. Refrigerate until ready to serve.

Makes 8-10 servings.

DEVIL'S FOOD TORTE

CAKE

1 teaspoon	lemon juice	5 mL
1¾ cups	milk	425 mL
½ cup	shortening	125 mL
1¼ cups	sugar	300 mL
1 teaspoon	salt	5 mL
2 teaspoons	vanilla extract	10 mL
3	egg yolks	3
4 ounces	melted unsweetened chocolate	112 g
3 cups	all purpose flour	750 mL
1½ teaspoons	baking soda	7 mL
1 teaspoon	baking powder	5 mL

ICING

3 cups	sugar	750 mL
½ cup	corn syrup	125 mL
½ cup	water	125 mL
3	egg whites	3
1 teaspoon	vanilla extract	5 mL
1 teaspoon	grated orange rind	5 mL
	fresh fruit	
	nuts or grated chocolate	

To make cake lightly grease three 9" (1.5 L) cake pans. Combine lemon juice and milk. Blend shortening with salt, sugar, vanilla and egg yolks. Beat until smooth. Stir in chocolate. Combine dry ingredients and add to shortening alternating with milk. Spoon into cake pans and bake, for 30 minutes in a 350°F (180°C) oven until cooked through. Remove from oven. Unmold and set on rack to cool. To prepare icing boil sugar, corn syrup and water until syrup makes a long fine thread when poured from a spoon. Beat egg whites until just stiff. Gradually add corn syrup mixture, beating continuously. Add vanilla and orange rind. Beat until good spreading consistency. To assemble torte spread icing between cake layers, on top and sides. Garnish with fruit, nuts or chocolate. Let icing set before slicing.
Makes 10 servings.

Fresh Peach Torte

CAKE

1 cup	all purpose flour	250 mL
1 teaspoon	baking powder	5 mL
¼ teaspoon	salt	1 mL
1 tablespoon	melted butter	15 mL
½ cup	milk	125 mL
2	eggs	2
¾ cup	sugar	175 mL
1 teaspoon	grated lemon rind	5 mL
1 tablespoon	lemon juice	15 mL

Grease one 9 inch (2 L) round cake pan. Line bottom with waxed paper. Mix together flour, baking powder and salt. Combine milk and butter. Beat eggs with sugar and lemon rind and juice. Fold dry ingredients into eggs alternating with milk. Pour into cake pan. Bake in 350°F (180°C) oven for 25-30 minutes until cooked through. Cool 10 minutes. Remove from pans and peel off paper.

FILLING

½ cup	sugar	125 mL
3 tablespoons	cornstarch	45 mL
½ teaspoon	salt	2 mL
1½ cups	milk	625 mL
2	egg yolks, lightly beaten	2
2 tablespoons	butter	30 mL
½ teaspoon	almond extract	2 mL

Combine sugar, cornstarch and salt in a small saucepan. Stir in milk gradually and cook over medium heat until nicely thickened. Add a little of the mixture to beaten egg yolks. Return to saucepan and cook 2 more minutes. Whisk in butter and almond extract. Place in bowl, cover and chill until set.

continued on next page

FRESH PEACH TORTE (continued)

TOPPING

3 cups	sliced fresh peaches	750 mL
3 tablespoons	sugar	45 mL
1 cup	whipping cream	250 mL

Toss peaches with sugar. Whip cream until stiff. Cut cake in two. Place filling on top of bottom layer and half of the peaches. Top with second layer and cover with whipped cream and remaining peaches. Let rest 15-20 minutes before serving.

Makes 8 servings.

FRESH PEACH FLAN

1 cup	flour	250 mL
2 tablespoons	icing sugar	30 mL
½ cup	butter	125 mL
1	package (6 serving size) banana pudding & pie filling mix	1
1	envelope unflavoured gelatin	1
2½ cups	milk	625 mL
4	ripe peaches, peeled and pitted	4
⅓ cup	apricot jam	75 mL

Sift flour and icing sugar in a bowl. Cut in butter until mixture resembles coarse meal. Form into a ball and press into bottom and sides of 9 inch (23 cm) flan pan. Bake in 425°F (240°C) oven for about 10 minutes or until golden brown. Cool. Combine pudding and pie filling mix, gelatin and milk in a saucepan. Cook and stir over medium heat until mixture comes to a full boil. Remove from heat and cover surface of pudding with wax paper. Chill. Beat chilled pudding until smooth. Pour into flan shell. Slice peaches ¼ inch (.6 cm) thick over bowl to catch the juice. Reserve 1 tablespoon (15 mL). Drain slices on paper towel. Arrange in attractive design on surface of pudding. Heat apricot jam with peach juice until melted. Brush over peaches. Chill two hours.

Makes 8 servings.

LEMON CHERRY FLAN

2	envelopes unflavoured gelatin	2
½ cup	sugar	125 mL
2 tablespoons	sugar	30 mL
2	eggs separated	2
1 cup	milk	250 mL
1 cup	creamed cottage cheese	250 mL
½ cup	yogurt	125 mL
⅓ cup	lemon juice	75 mL
	grated rind of one lemon	
1 teaspoon	vanilla	5 mL
	sliced fresh fruit	

Lightly oil a round 9 inch (1.5 L) cake pan. Line with a circle of waxed paper. Mix together gelatin and ½ cup (125 mL) sugar in a saucepan. Whisk egg yolks with milk and add to gelatin mixture. Cook and stir over medium heat until lightly thickened. DO NOT BOIL. Whisk cottage cheese, yogurt, lemon juice, rind and vanilla in blender or processor until smooth. Add to milk mixture. Let cool until lightly thickened. Beat egg whites with remaining sugar until stiff. Fold into milk/cheese mixture. Spoon into prepared pan and chill until set. Invert onto a platter and serve with Cherry Sauce. Makes 6 servings.

CHERRY SAUCE:

1-14 ounce	can red sour or sweet pitted cherries	1 (398 mL)
1 tablespoon	cornstarch	15 mL
2 tablespoons	sugar	30 mL
1 teaspoon	rum (optional)	5 mL

Drain cherries, reserving juice. Combine cornstarch and sugar in a saucepan. Whisk in cherry juice. Cook and stir over medium heat until thick. Stir in cherries. Heat through. Stir in rum.

Raspberry Flan

SHORTBREAD BASE

¾ cup	butter	175 mL
⅓ cup	icing sugar	100 mL
1½ cups	flour	375 mL

Blend ingredients to make a soft dough. Pat into a 10" (25 cm) quiche or fluted flan pan. Prick well. Bake in a 350°F (180°C) oven for 15-20 minutes. Cool.

CUSTARD LAYER

˙4	egg yolks	4
½ cup	sugar	125 mL
¼ cup	flour	50 mL
1½ cups	milk	375 mL
1 teaspoon	grated lemon rind	5 mL
1 teaspoon	vanilla	5 mL

Mix egg yolks, sugar and flour in a heavy saucepan. Blend in milk. Cook over medium-low heat, stirring constantly until mixture thickens and boils. Remove from heat; add lemon rind and vanilla. Cool slightly and spread filling in flan shell.

RASPBERRIES AND GLAZE

2 cups	raspberries	500 mL
1 tablespoon	cornstarch	15 mL
⅓ cup	orange juice	75 mL
½ cup	red currant jelly	125 mL

Spread raspberries evenly over custard. Cook glaze ingredients over medium heat until thick and clear. Spoon over raspberries. Chill and serve.

Makes about 10 servings.

Blender Melon Dessert

1	cantaloupe	1
2 teaspoons	sugar	10 mL
	grated rind of 1 lemon	
1/8 teaspoon	cinnamon	0.5 mL
1/8 teaspoon	salt	0.5 mL
6 tablespoons	butter, melted	90 mL
2 cups	milk	500 mL
2-4 tablespoons	white rum	30-60 mL
3 tablespoons	lemon juice	45 mL

Remove seeds and rind from cantaloupe and dice. Add sugar, lemon rind, cinnamon, and salt. In a large saucepan, sauté mixture in butter for 2 minutes. Add milk and simmer, stirring occasionally, for 10 minutes. Let cool. Pour mixture into blender, and blend until smooth. Stir in rum and lemon juice. Pour into an ice cube tray, and leave until partly frozen. Remove from tray and rewhip in blender. Repeat process once more. Store in freezer. Remove 10-15 minutes before serving, and allow to soften to a slush. This makes an ideal summer dessert.

Makes 4 servings.

BUTTERMILK STRAWBERRY ICE

3 cups	buttermilk	750 mL
⅓ cup	honey	75 mL
3 tablespoons	lemon juice	45 mL
1 teaspoon	grated lemon rind	5 mL
1½ teaspoons	vanilla extract	7 mL
4 cups	sliced or mashed (fresh or frozen) strawberries	1 L
4	egg whites	4

Thoroughly combine the first six ingredients. Pour into a 9" x 13" (3.5 L) metal pan, cover with foil and freeze overnight .or until firm. Spoon frozen mixture into blender or food processor; whirl until smooth. Beat egg whites until stiff but not dry, fold into strawberry-buttermilk mixture. Refreeze for an additional 2 to 3 hours or until firm. Spoon into large goblets; garnish with additional strawberries.
Makes 6 servings.

LEMON ICE

2 cups	milk	500 mL
⅓ cup	sugar	75 mL
2-3 tablespoons	grated lemon rind	30-45 mL
½ cup	lemon juice	125 mL

Thoroughly combine all ingredients; pour into a shallow metal pan. (The mixture will appear curdled.) Freeze until firm, about 2 to 3 hours. Whirl frozen lemon mixture in a blender or food processor until smooth. Scoop into parfait glasses or hollowed out lemon shells and refreeze for 2 to 3 hours or until ready to serve.
Makes 4 servings.

FRENCH VANILLA ICE CREAM

4	egg yolks	4
¾ cup	sugar	175 mL
¼ teaspoon	salt	1 mL
1½ cups	milk	375 mL
2½ cups	whipping cream	625 mL
1 tablespoon	vanilla	15 mL
	ice cream maker	

Beat egg yolks slightly; beat in sugar and salt. Gradually stir in milk. Cook in a medium saucepan over low heat, stirring constantly, until mixture will coat a metal spoon. Remove from heat and cool. Stir in cream and vanilla and chill completely (overnight if possible). Freeze mixture in ice cream maker according to manufacturers directions.

Makes 1½ quarts/1.5 L.

FROZEN CARAMEL CREAM

1½ cups	sugar	375 mL
½ cup	boiling water	125 mL
2 tablespoons	corn starch	30 mL
1½ cups	milk	375 mL
2	eggs, separated	2
1 cup	whipping cream	250 mL
¾ cup	chopped, toasted pecans	175 mL

Measure 1¼ cups (300 mL) of the sugar into a heavy saucepan. Cook over medium heat, stirring constantly, until sugar melts and turns golden. Remove from heat; carefully stir in boiling water. Return to heat; bring to a boil. Combine corn starch and milk; add to caramel syrup. Cook over medium heat, stirring constantly, until mixture comes to a boil and thickens. Beat egg yolks slightly; stir a little of the hot mixture into the yolks. Return all to saucepan; cook and stir 2 minutes longer. Cool. Beat egg whites until frothy. Gradually beat in remaining ¼ cup (75 mL) sugar and continue beating until stiff peaks form. Whip cream until softly stiff. Fold meringue, whipped cream and nuts into sauce. Turn into 9-inch (2.5 L) square cake pan. Cover and freeze until firm. Remove to refrigerator 15 minutes before serving.
Makes 8 to 10 servings.

FROZEN MOCHA MARVEL

17	crushed cream filled chocolate cookies	17
3 tablespoons	melted butter	45 mL
1	package (4 servings) instant vanilla pudding mix	1
1 teaspoon	vanilla	5 mL
1¾ cups	milk	425 mL
1 cup	whipping cream	250 mL
½ teaspoon	unflavoured gelatin	2 mL
2 tablespoons	hot water	30 mL
2 teaspoons	instant coffee	10 mL
1 teaspoon	brown sugar	5 mL
1 ounce	unsweetened chocolate, melted	28 g

Combine crushed cookies and butter. Press into an 8″ square (2 L) baking dish reserving ¼ cup (50 mL) for top. Mix pudding mix, milk and vanilla in a bowl. Beat slowly until blended. Gradually increase beating speed and beat on high until thickened. Dissolve gelatin in 1 tablespoon (15 mL) water. Beat cream until almost stiff. Add gelatin and continue beating until stiff. Fold into milk mixture. Dissolve coffee in remaining water. Add brown sugar. Stir into half the whipped dessert. Spread this mixture over crumb crust. Stir chocolate into remaining cream. Spread this over coffee layer. Top with reserved crumbs. Cover and freeze. Remove from freezer 15 minutes before serving. Cut in squares.

Makes one 8-inch/2 L square pan.

FROZEN NEOPOLITAN TORTE

1	package (4 serving size) chocolate pudding and pie filling	1
4 cups	milk	1 L
6	egg yolks	6
1 ounce	unsweetened chocolate	28 g
½ teaspoon	vanilla	2 mL
1 (18 ounce)	package frozen strawberries	1 (510 g)
1	package (4 serving size) vanilla pudding and pie filling	1
1	package (4 serving size) butterscotch pudding and pie filling	1
1 teaspoon	instant coffee	5 mL
3 cups	whipping cream	750 mL
½ cup	sugar	125 mL
1 tablespoon	sugar	15 mL
¼ cup	walnut pieces	50 mL

To make chocolate layer, prepare chocolate pudding according to package directions except use only 1½ cups (375 mL) milk. Beat two egg yolks lightly. Stir in half of the hot chocolate pudding. Return to pan and stir while mixture returns to the boil. Remove from heat and cool with surface covered in plastic wrap to prevent a skin from forming. Grate chocolate and stir into pudding. Add vanilla. Set aside. To make strawberry layer, heat strawberries and force them through a sieve to form a purée. Add enough milk to make 1¾ cups (425 mL) liquid. Prepare vanilla pudding according to package directions using strawberry milk as liquid. Beat two egg yolks lightly and add to pudding as above. Set aside to cool covered in plastic wrap. To make mocha layer, stir instant coffee into butterscotch pudding mix and prepare according to package directions using only 1½ cups (375 mL) milk. Beat two egg yolks lightly and add to pudding as above. Set aside to cool covered in plastic wrap. Beat 2 cups (500 mL) cream until stiff gradually adding ½ cup (125 mL) sugar. Fold ⅓ cream into each of the prepared puddings. Spoon butterscotch pudding into a 10" (3 L) tube pan to make an even layer. Spoon strawberry pudding on top and finish with chocolate layer. Cover pan with foil and set in freezer for at least 3 hours. Unmold about 20 minutes before serving. Beat remaining cream until stiff with a tablespoon (15 mL) sugar. Decorate torte with whipped cream and walnut pieces.

Makes 12-16 servings.

MINTED MERINGUE ICE

3 ½ cups	milk	875 mL
1 cup	sugar	250 mL
1 teaspoon	peppermint extract	5 mL
4	egg whites	4
2 ounces	semi sweet chocolate	56 g
	fresh mint leaves	

Heat milk with sugar, stirring until sugar is dissolved. Bring to a boil. Add peppermint extract. Cool. Pour mixture into freezer container; cover and set in freezer until mixture reaches semi-frozen state. Beat egg whites until very stiff. Break up semi-frozen ice, then fold in beaten egg whites. Return to freezer until firm. Serve ice in goblets, topped with grated chocolate and sprigs of mint.

Makes 8 servings.

MOCHA TREAT

2¼ cups	crushed chocolate wafers	550 mL
¼ cup	melted butter	50 mL
2 teaspoons	instant coffee	10 mL
1 cup	milk	250 mL
1 quart	vanilla ice cream	1 L
1	package (4 servings) instant chocolate pudding mix	1
1 teaspoon	vanilla extract	5 mL

Combine crushed chocolate wafers with melted butter. Press into 10 inch square (3 L) baking pan, reserving ½ cup (125 mL) for top. Mix instant coffee with milk and stir until dissolved. Place ice cream in a chilled bowl and whisk until just softened. Add milk, pudding mix and vanilla. Beat just to blend. Pour into prepared crust. Sprinkle with reserved crumbs. Cover with foil and freeze. Cut in squares to serve.

Makes 1-10 inch/3 L square pan.

STRAWBERRY BANANA SHERBET

1	teaspoon	unflavoured gelatin	5 mL
1	tablespoon	water	15 mL
1	cup	puréed strawberries	250 mL
½	cup	sugar	125 mL
¼	cup	lemon juice	50 mL
1	teaspoon	grated lemon rind	5 mL
1		banana, mashed	1
2	cups	cold milk	500 mL

Soften gelatin in water; combine with strawberries and sugar in a small pan and stir over low heat until gelatin and sugar are dissolved. Add lemon juice and banana. Stir mixture into cold milk. Pour into a metal tray, cover with foil and freeze until firm (2-3 hours). Whirl frozen mixture in a blender or food processor until smooth. Spoon into parfait glasses and serve at once or pack into a container, cover and re-freeze. Soften slightly before scooping.

Makes 6 servings.

STRAWBERRY YOGURT POPS

1 (15-ounce)	package frozen sliced strawberries, thawed	1 (425 g)
2 cups	strawberry yogurt	500 mL
1½ cups	milk	375 mL

Place undrained strawberries in blender container. Cover and blend at high speed until smooth. Combine strawberry purée, yogurt and milk. Divide mixture evenly among sixteen 3-ounce (100 mL) paper cups. Freeze until partially frozen. Insert a wooden stick into the centre of each. Freeze until firm. To serve, peel off paper cup.

Makes 16 pops.

*Raspberry Flan,
page 600*

SURPRISE SHERBET

2 cups	buttermilk	500 mL
1 cup	sugar	250 mL
1 (19 ounce)	can applesauce	1 (540 mL)
1	lemon	1
	fresh mint leaves	

Mix together buttermilk, sugar, applesauce and juice of lemon. Pour into two refrigerator trays. Cover and freeze until firm. Spoon into sherbet glasses, sprinkle with grated lemon rind and decorate with mint.

Makes 6 servings.

NOTES

INDEX

A

A #1 Seafood Chowder, 45
Acorn Squash, Beef Stuffed, 107
ALMOND
  Blender Chocolate Almond
    Mousse, 532
  Cream Dessert, 523
  Pie, Coconut, 562
  Soufflé, 587
  Syrup Cake, 441
  Turkey Leftovers with, 345
Amandine Quiche, Chicken, 165
American Style Enchiladas, 103
Appetizer, Avocado, 2
Appetizer Coquilles, 1
APPLES
  Bacon Pancake Pie, 573
  Bake, Turnip, 394
  Crisp Pie, Harvest, 567
  Curry, Chicken, 239
  Dessert, Country Crunch, 545
  Filling, Spicy, 474
  'n' Oats Muffins, 464
  Pancake, 474
  Pie, Sweet Potato and, 586
  Raisin Cottage Pudding with
    Brown Sugar Sauce, 493
  Streusel Coffeecake, 426
  with Creamy Vanilla Sauce,
    Streusel Topped, Baked, 541
APRICOTS
  Bavarian, 527
  Bread Pudding, 488
  Dessert, Rice and, 496
  Glaze, 554
  Milkshake, Creamy, 18
  Pudding, Creamy Rhubarb, 494
ASPARAGUS
  Cashew Casserole, 354
  Cheddar Quiche, 144
  Cheese Soufflé Roll, 145
  Chicken and Pasta Supper, 213
  Continental, 353

Asparagus cont'd
  Cream of Asparagus
    Soup (Diet), 65
  Fresh Asparagus Mimosa, 375
  Rarebit, 143
  Springtime Crab and
    Asparagus Tart, 206
Autumn Vegetable Soup, 46
Avocado Appetizer, 2
Avocado Soup, Curried, 74

B

BACON
  and Rice Casserole, Peameal, 328
  and Tomato Stack-Ups, 147
  Bread, Swiss
    Cheese (Appetizer), 421
  Cheese and Onion Tart, 146
  Cheese Roulade with
    Mushrooms and, 157
  Clam Chowder with Bacon
    and Croutons, 58
  Harvest Soup, Potato, 98
  Liver with Bacon 'n' Onion
    Sauce, 126
  Pancake Pie, Apple, 573
  Potage, Bean and, 48
  Sauce, Smoke-House, 405
  Soufflé, Swiss Cheese and, 208
  Tomato Bacon Wraps, 393
Baked Brunch Eggs, 149
Baked Cheddar Strata, 148
Baked Lemon Custard, 512
Baked Lemon Sponge Pudding, 486
Baked Salmon à la Russe, 259
Baked Scallops, Bubbly, 260
Baked Sausage Pizza, 313
Baked Winter Squash Ring, 355
BANANA
  Bananaberry Blender
    Milkshake, 18
  Buttermilk (Blender Breakfast), 17
  Chiffon Pie, 559
  Cocoa, Hot, 33
  Deluxe Banana Cream Pie, 563

INDEX

Banana cont'd
Flip, Peachy, 40
Frappé, 15
French Toast, 483
Orange (Blender Breakfast), 16
Scotch-er-oo Pudding, 497
Shake, Strawberry, 24
Sherbet, Strawberry, 609
Split Pie, 560
Barbecued Beef Loaf, 104
Barbecued Pork with Pineapple
Peanut Sauce, Oriental, 327
Basic Pork Mixture, 339

BAVARIAN
Apricot, 527
Rum, 525
Strawberry, 526
Bean and Bacon Potage, 48
Bean Beef Casserole, Mexican, 127
Beans, Green, with
Water Chestnuts, 377

BEEF
Barbecued Beef Loaf, 104
Beef and Rice Crisp, 108
Beef Casserole with Puffed
Topping, 105
Beef Liver with Toast Points, 106
Beef Stuffed Acorn Squash, 107
Cabbage Caraway Beef Bake, 110
Cheeseburger Casserole, 115
Cheeseburger Meat Balls, 114
Cheeseburger Pie, 113
Corn 'n' Beef Soup, 62
Corned Beef and
Cabbage Bake, 117
Corned Beef Casserole, 116
Corned Beef Oven Omelette, 166
Curried Beef Bake, 120
Curried Beef Loaf, 119
Greek Beef and
Macaroni Casserole, 122
Hamburger Hot Cakes, 123
Hearty Beef Goulash, 124

Beef cont'd
Honey Garlic Appetizer
Meat Balls, 9
Mexican Bean Beef Casserole, 127
Mexican Casserole, 128
My Own Mini Meat Loaf, 129
Peppery Beef Stroganoff, 130
Saucy Short Ribs, 131
Sauerbraten Sauced
Beef Loaf, 132
Savoury Franks and Cabbage, 336
Scalloped Hamburger
Casserole, 134
Shepherd's Pie, 135
Spaghetti Meat Bake, 136
Spicy Beef Casserole, 137
Swedish Meat Balls, 138
Sweet and Sour Pineapple
Meat Balls, 139
Wiener and Spaghetti
Casserole, 142

BEETS
Borscht, 47
Cold Beet Borscht
with Cucumbers, 61
Cold Pink Beet Soup, 60
Berry Puff Pancakes, 475

BISCUITS
Buttermilk, 410
Mini Cheese, 412
Puffy Cheese, 413
Savoury, 234
Whole Wheat Sesame, 411

BISQUE
Cauliflower with
Cheesy Croutons, 51
Creole, 71
Easy Salmon, 75
Herbed Fresh Tomato, 84
Lobster, 90
Zucchini Buttermilk, 102
Black Forest Torte, 592

INDEX

BLENDER BREAKFASTS
Banana Buttermilk, 17
Banana Orange, 16
Breakfast-in-a-Glass, 17
Pineapple Grapefruit Whiz, 16
Pineapple Refresher, 17
Strawberry Frost, 16
Blender Chocolate Almond
Mousse, 532
Blender Melon Dessert, 601
Blender Milkshakes (see Milkshakes)
Blintzes, Cheese, 158
Blue Cheese Dip (Appetizer), 3
Blueberry Muffins, Lemon, 467
Blueberry Pie, Lemon, 569
Blueberry Streusel Coffeecake, 432
Bobotie, 307
Borscht, 47
Braised Pork Loin, 314
Branberry Muffins, 461

BREAD
Carrot Tea Loaf, 424
Cranberry Nut, 422
Crisp-Crusted Soda, 423
Egg Twist, 417
Herbed Parmesan Batter, 415
Kulich Easter Bread, 416
Lemon Nut, 425
Nuts and Seeds, 418
Savoury Cheddar, 419
Simply Seasoned Casserole, 420
Swedish Tea Ring, 430
Swiss Cheese Bacon, 421
Bread Pudding, Apricot, 488
Bread Pudding, Tipsy, 491
Breaded Cauliflower with Sour
Cream Sauce, 356
Breaded Chicken Breasts with
Spinach and Dill Sauce, 214
Breakfast Egg Casserole, 150
Breakfast-in-a-Glass (Blender), 17
Breakfasts, Blender,
(See Blender Breakfasts), 16
Breast of Chicken Pronto, 215

BROCCOLI
à la Suisse, 357
Cheese Delight, 358
Chicken and
Broccoli Casserole, 236
Company Broccoli Soufflé, 366
Creamy Broccoli
and Zucchini, 371
Creamy Broccoli Potage, 68
Ham and Broccoli Royale, 321
Ham and Broccoli Spuds, 325
Parmesan, 382
Soup, 49
Soup, Cheesy, 54
Tomato and Broccoli Quiche, 209
Tuna Broccoli Casserole, 303
Brown Sugar Sauce, Apple Raisin
Cottage Pudding with, 493
Brown Sugar Nuggets, 458
Brunch Eggs, Baked, 149
Brunch, Cheddar Pinwheel, 153
Brunch, Scrambled Easter Egg, 199
Bubbly Baked Scallops, 260
Buffet Chicken and Shrimp à la
King, 216
Buffet Potato Casserole, 359

BUTTERMILK
Banana (Blender Breakfast), 17
Biscuits, 410
Cheese Dessert Ring, 558
Cinnamon Coffeecake, 427
Cooler, Orange, 18
Liver, 109
Low-Cal Zesty Dip (Appetizer), 3
Strawberry Ice, 602
Zucchini Buttermilk Bisque, 102
Butterscotch Sundae
Cake, 442
Buttery Orange Crêpes, 533

INDEX

C

CABBAGE
and Liver Fingers, 111
Caraway Beef Bake, 110
Cheesy Cabbage Bake, 365
Corned Beef and
Cabbage Bake, 117
Fruit Salad, 360
Rice Duo, 361
Rolls, 112
Savoury Franks and, 336
Café au Lait, 34

CAKES
Almond Syrup, 441
Butterscotch
Sundae, 442
Chocolate Cream
Cheese Cupcakes, 457
Coconut Meringue, 440
Country Kitchen
Chocolate, 449
Double Fudge Chocolate, 448
Grandma's Chocolate, 434
Happy Face Cupcakes with
Cheesy Frosting, 456
Lemon Ice-Box, 445
Milk and Honey Cake With Honey
Coconut Topping, 433
Orange Date, 439
Orange Raisin, 447
Orange Rum, 453
Peach Upside Down, 446
Peaches 'n' Cream Shortcake, 452
Pineapple Tote, 438
Poppy Seed, 435
Raspberry Delight, 454
Rhubarb, 436
Strawberries 'n' Cream, 451
Triple Treat Lemon, 437
Canadiana Cheddar Cheese Soup, 50
Canadiana Shake, 19
Cannelloni, 151
Cappuccino Coffee, Italian, 32
Cappuccino Parfaits, 507

Caramel Cream, Frozen, 604
Caramel Custard, Classic, 510
Caramel Mousse, Crunchy, 529
Caraway Beef Bake, Cabbage, 110
Caribbean Chicken, 217

CARROTS
Cheese Scalloped, 364
Dessert Crêpes, Pineapple, 534
Ring, Elegant, 374
Soup, Golden, 79
Tea Loaf, 424
Wheat Muffins, 462
Cashew Casserole, Asparagus, 354
Cashew Curry, Quick, 312

CASSEROLES
Asparagus Cashew, 354
Beef Casserole with
Puffed Topping, 105
Bobotie, 307
Breakfast Egg, 150
Buffet Potato, 359
Cauliflower, 362
Cheese 'n' Chili, 160
Cheese Vegetable, 159
Cheeseburger, 115
Cheesy Chicken, 219
Cheesy Seafood, 261
Chicken and Broccoli, 236
Chicken Casserole with
Savoury Biscuits, 234
Chicken Noodle Doodle, 224
Chicken, Vegetable and Rice, 235
Chicken Wild Rice, 233
Chowder, 262
Cordon Bleu Chicken, 242
Corned Beef, 116
Country Ham, 315
Creamy Vegetable, 369
Egg, Cheese and Olive, 170
Greek Beef and Macaroni, 122
Ham and Egg, 326
Italian Egg Noodle, 176
Likeable Liver and Rice, 125

INDEX

Casseroles Cont'd
Macaroni Casserole with
Double Cheese, 183
Macaroni Tuna, 274
Mexican, 128
Mexican Bean Beef, 127
Peameal Bacon and Rice, 328
Pork Chop, 331
Savoury Veal, 133
Scalloped Fish, 287
Scalloped Hamburger, 134
Sea Captain's, 289
Seafood, 296
Shrimp Fromage, 297
Spicy Beef, 137
Thermidor en Casserole, 300
Tomato, Tuna,
Egg, Mozzarella, 301
Tote Along, 210
Tuna Broccoli, 303
Tuna Casserole Deluxe, 305
Turkey, 342
Turkey Buffet, 341
Turkey Noodle, 346
Wiener and Spaghetti, 142
Zucchini, 395

CAULIFLOWER
and Ham Chowder, Fresh, 77
Bisque with Cheesy Croutons, 51
Breaded with Sour
Cream Sauce, 356
Casserole, 362
Parmesan Sauced, 383
Soup, Cheese, 52

CELERY
à la Suisse, 363
Creamed, with Pecans, 368
Crunch, Tuna, 302

CHEDDAR CHEESE
Asparagus Cheddar Quiche, 144
Baked Cheddar Strata, 148
Canadiana Cheddar
Cheese Soup, 50

Cheddar Cheese cont'd
Cheddar Cheese Soufflé, 152
Cheddar Chicken Divan, 218
Cheddar Pinwheel Brunch, 153
Cheddar Sauce Supreme, 396
Cheddar Streusel Coffeecake, 428
Eggs Benedict au Gratin, 171
Fran's Frugal Cheddar
and Ham Flan, 319
Protein Power Cheddar Soup, 99
Salmon Cheddar Macaroni, 284
Savoury Cheddar Bread, 419

CHEESE
Asparagus Cheese
Soufflé Roll, 145
Asparagus Continental, 353
Asparagus Rarebit, 143
Bacon, Cheese
and Onion Tart, 146
Blue Cheese Dip, 3
Broccoli à la Suisse, 357
Broccoli Cheese Delight, 358
Buttermilk Cheese
Dessert Ring, 558
Cauliflower Bisque
and Cheesy Croutons, 51
Celery à la Suisse, 363
Cheese and Vegetable Strata, 155
Cheese Bake, 154
Cheese Biscuits, Puffy, 413
Cheese Blintzes, 158
Cheese Cauliflower Soup, 52
Cheese 'n' Chili Casserole, 160
Cheese Roulade with Mushrooms
and Bacon, 156
Cheese Sauce, Pork Pastitsio
with, 333
Cheese Scalloped Carrots, 364
Cheese Soufflé in
Pepper Cups, 161
Cheese Spinach Roll, 162
Cheese Vegetable Casserole, 159
Cheese Zucchini Soup, 53
Cheeseburger Casserole, 115

INDEX

Cheese cont'd
Cheeseburger Meat Balls, 114
Cheeseburger Pie, 113
Cheesy Biscuit Bake, 163
Cheesy Broccoli Soup, 54
Cheesy Cabbage Bake, 365
Cheesy Chicken Casserole, 219
Cheesy Croutons, 51
Cheesy Frosting, 456
Cheesy Seafood Casserole, 261
Cheesy Shrimp Toasts, 4
Cheesy Tomato Pie, 164
Cheesy Vegetable Soup, 55
Chocolate Cream
   Cheese Cupcakes, 457
Cod au Gratin, 264
Dairy Delicious Cottage Mold, 168
Egg, Cheese
   and Olive Casserole, 170
Golden Cheese
   and Rice Bake, 172
Gouda Dunk (Appetizer), 8
Happy Face Cupcakes with
   Cheesy Frosting, 456
Herbed Parmesan
   Batter Bread, 415
Hot Cheese Cocktail
   Dip (Appetizer), 10
Low-Cal Zesty Dip (Appetizer), 3
Macaroni and Cheese
   Parmigiano, 180
Macaroni and Cheese Soufflé, 182
Macaroni Casserole
   with Double Cheese, 183
Macaroni Ring, 184
Mexican Cheese Dip
   (Appetizer), 11
Mini Cheese Biscuits, 412
Noodles au Gratin, 190
Pasta Primavera au Gratin, 193
Poached Fish Parmesan, 277
Puffy Cheese Biscuits, 413
Savoury Cheese Muffins, 471
Scalloped Potatoes with
   Cheese and Herbs, 386

Cheese cont'd
Sea Shells and Cheese, 200
Shrimp Fromage Casserole, 297
Spaghettini with Herbed
   Cheese Sauce, 202
Stuffed Eggs in Cheese Sauce, 207
Sweet Cheese Crêpes, 535
Swiss Cheese
   and Bacon Soufflé, 208
Swiss Cheese Bacon Bread, 421
Tomato, Tuna, Egg, Mozzarella
   Casserole, 301

CHEESECAKE
Chocolate Rum and Raisin, 553
Citrus, 554
Mocha Swirl, 555
Pie, Rhubarb, 581
Pina Colada, 556
Cherry Flan, Lemon, 599
Chestnuts, Green Beans
   with Water, 377

CHICKEN
à la King, 226
and Broccoli Casserole, 236
and Pasta Supper, Asparagus, 213
Apple Curry, 239
Breaded Chicken Breasts with
   Spinach and Dill Sauce, 214
Breast of Chicken Pronto, 215
Breasts Supreme, 232
Buffet Chicken and
   Shrimp à la King, 216
Caribbean, 217
Casserole with Savoury
   Biscuits, 234
Cheddar Chicken Divan, 218
Cheesy Chicken Casserole, 219
Chunky Chicken Chowder, 57
Coconut Curried, 240
Cold Day Chicken Curry, 241
Cordon Bleu Chicken
   Casserole, 242
Creamed Chicken in
   Noodle Ring, 243

INDEX

Chicken cont'd
Creamy Chicken Chow Mein, 244
Crêpes, 225
Crunchy, 245
Crunchy Chicken Salad Mold, 246
Drumsticks Italiano, 227
East Indian, 247
Easy Chicken Divan, 248
Espanol, 228
Fricassée, 229
Hasty Tasty Chicken
   Chop Suey, 249
Hawaiian, 250
Home-Style Chicken Stew, 251
in Wine Sauce, 223
Jellied Chicken Loaf, 252
Lemon Chicken Pilaf, 253
Livers on Toast Points, 237
'n' Spoon Bake, 231
Noodle Doodle Casserole, 224
Paprika, 221
Picnic Oven Fried, 254
Poached Chicken
   with Mushrooms, 255
Puff, 220
Quiche Amandine, 165
Rice Bake, 230
Ring with Cranberry Sauce, 238
Savoury Coated Chicken
   and Sauce, 256
Sherried Chicken and Mushrooms
   in Toast Cups, 257
Vegetable and Rice Casserole, 235
Versatile Chicken
   and Ham Pâté, 14
Vol-au-Vent, 222
Wild Rice Casserole, 233
Chili Casserole, Cheese 'n', 160
Chilled Vichyssoise Supreme, 56
Chilly Lime Soufflé, 588

CHOCOLATE
Almond Mousse, Blender, 532
Cake, Country Kitchen, 449
Cake, Double Fudge, 448

Chocolate cont'd
Cake, Grandma's, 434
Chip Muffins, 463
Coconut Pie, 561
Continental Hot Chocolate, 32
Cream Cheese Cupcakes, 457
Delicious Chocolate Mousse, 530
Easy Icing, 434
Frosted Pears, 543
Frosting, 449
Hot Chocolate Soufflé, 589
Meringue, 502
Mile-High Black Bottom Pie, 572
Milk Chocolate Fondue, 542
Milk Crème Caramel, 549
Orange Blossom (Beverage), 25
Orange Chocolate Pie, 576
Peanutty Shake, 21
Regal Chocolate Sauce, 406
Rice Pudding, 487
Rum and Raisin Cheesecake, 553
Strawberry Parfaits, 508
Chop Suey, Hasty Tasty
   Chicken, 249
Chow Mein, Creamy Chicken, 244
Chowder Casserole, 262

CHOWDERS
(See Soups)
Christmas Puddings with Custard
   Sauce, Easy Steamed, 492
Chunky Chicken Chowder, 57
Cinnamon Buttermilk
   Coffeecake, 427
Cinnamon-Nutmeg Butter, 482
Cinnamon Popovers With Honey
   Butter, 480
Citrus Cheesecake, 554
Citrus Pie, Sunshine, 585
Clam Bakes, Individual, 273
Clam Chowder with Bacon and
   Croutons, 58
Clam Sauce, Linguine with, 179
Clam Sauce, Spaghetti with, 203
Clam Soup, 59

INDEX

Clam Soup, Hearty, 83
Classic Caramel Custard, 510
Classic White Sauce, 397
COBBLER
Fresh Peach with
Creamy Lemon Sauce, 547
Plum Delicious, 546
Strawberry Rhubarb, 548
Cocktail Dip, Hot Cheese
(Appetizer), 10
Cocoa, Hot Banana, 33
COCONUT
Almond Pie, 562
Chocolate Coconut Pie, 561
Crunch Custards, 511
Curried Chicken, 240
Impossible Coconut Pie, 568
Meringue Cake, 440
Cod au Gratin, 264
Coffee, Italian Cappuccino, 32
Coffee Nog, Open House, 28
COFFEECAKE
Apple Streusel, 426
Honey Pecan, 429
Blueberry Streusel, 432
Cheddar Streusel, 428
Cinnamon Buttermilk, 427
Cold Beet Borscht
with Cucumbers, 61
Cold Day Chicken Curry, 241
Cold Pink Beet Soup, 60
Colonial Maple Tarts, 590
Company Broccoli Soufflé, 366
Continental Hot Chocolate, 32
Cookies, Whirly-Twirly Pinwheel, 458
Coquilles Appetizer, 1
Coquilles St Jacques, 5
Cordon Bleu Chicken Casserole, 242
Corn Chowder, 63
Corn 'n' Beef Soup, 62
Corn Pancakes, 479
Corned Beef and Cabbage Bake, 117
Corned Beef Casserole, 116

Corned Beef Oven Omelette, 166
Cottage Curried Shrimp, 265
Cottage Pudding with Brown Sugar
Sauce, Apple Raisin, 493
Country Crunch Apple Dessert, 545
Country Ham Casserole, 315
Country Kitchen Chocolate
Cake, 449
Country Kitchen Pancake Sauce, 398
Country Style Pork Chops, 316
CRAB
and Asparagus Tart,
Springtime, 206
Crabmeat Luncheon Dish, 268
Custard Baltimore, 267
Imperial, 266
on Melon Appetizer, 6
Quiche, 167
Soup, Delicate Curried, 73
Soup, Martha Washington's
Cream of, 92
CRANBERRY
Mousse, 531
Nut Bread, 422
Nut Stuffing, 367
Sauce, 531
Sauce, Chicken Ring with, 238
Crazy Raisin Rice Pudding, 495
Cream Cheese Cupcakes,
Chocolate, 457
CREAM
Cake, Strawberries 'n' Cream, 451
Dessert, Almond, 523
Frozen Caramel, 604
Mocha, 524
Orange-Lemon, 519
Raspberry, 520
Shortcake, Peaches 'n', 452
Spanish, 522
Strawberry, 521
Cream of Asparagus Soup (Diet), 65
Cream of Rutabaga Soup, 66
Cream of Tomato and Leek Soup, 64

INDEX

Cream of Zucchini Soup, Lo-Cal, 91
Creamed Celery with Pecans, 368
Creamed Chicken in Noodle
 Ring, 243
Creamed Peach Rice Mold, 540
Creamed Peas, Mushroom, 380
Creamed Pork Tenderloin, 317
Creamed Pumpkin Soup, 67
Creamy Broccoli and Zucchini, 371
Creamy Broccoli Potage, 68
Creamy Chicken Chow Mein, 244
Creamy Herbed Pork Chops, 318
Creamy Italian Minestrone, 70
Creamy Lemon Sauce, 547
Creamy Mushroom Soup, 72
Creamy Onion Sauce, 399
Creamy Onion Soup, 69
Creamy Onions and Potatoes, 370
Creamy Rhubarb-Apricot
 Pudding, 494
Creamy Rice Pudding, 501
Creamy Rum Sauce, Pina Colada
 Pancakes with, 476
Creamy Vanilla Fondue, 542
Creamy Vanilla Sauce, 541
Creamy Veal Ragout, 118
Creamy Vegetable Casserole, 369
Crème Caramel, Chocolate Milk, 549
Crème Caramel, Iberian, 550
Crème Praline, 518
Creole Bisque, 71

CREPES
 Buttery Orange, 533
 Chicken, 225
 Pineapple Carrot Dessert, 534
 Pineapple Filling, 534
 Seafood, 291
 Sweet Cheese, 535
 Vegetable, 211
 Whole Wheat Crêpes
  with Rhubarb Sauce, 537
Cricket Dessert Sipper, 26
Crisp-Crusted Soda Bread, 423
Crispy-Topped Fish, 269

Croquettes, Piquant Ham, 329
Croutons, Cauliflower Bisque with
 Cheesy, 51
Croutons, Clam Chowder
 with Bacon and, 58
Crunchy Caramel Mousse, 529
Crunchy Chicken, 245
Crunchy Chicken Salad Mold, 246
Cucumber Sauce, Dilly, Fresh
 Salmon with, 272
Cucumber Soup, Iced, 86
Cucumbers, Cold Beet
 Borscht with, 61
Cupcakes, Happy Face
 with Cheesy Frosting, 456
Cupcakes, Chocolate
 Cream Cheese, 457

CURRY
 Avocado Soup, 74
 Beef Bake, 120
 Beef Loaf, 119
 Chicken-Apple, 239
 Chicken, Coconut, 240
 Cold Day Chicken, 241
 Crab Soup, Delicate, 73
 Liver Strips, 121
 Pork, 339
 Pronto Shrimp, 278
 Quick Cashew, 312
 Sauce, 390, 400
 Shrimp, Cottage, 265
 Tuna, 270

CUSTARDS
 Baked Lemon, 512
 Classic Caramel, 510
 Coconut Crunch, 511
 Crab Custard Baltimore, 267
 Mocha, 513
 Orange Dream, 515
 Pie, Grapefruit, 565
 Pie, Mincemeat, 574
 Plum Custard Delight, 517
 Pumpkin, 514

Custards Cont'd
Sauce, Easy Steamed Christmas
Puddings with, 492
Topsy-Turvey, 516

D
Dairy Delicious Cottage Mold, 168
Danish Rum Pudding
with Raspberry Sauce, 498
Date Cake, Orange, 439
Date 'n' Nut Muffins, 472
Deep Dish Pork Pie, 340
Deep Dish Tuna Quiche, 169
Deep Dish Vegetable Pie, 373
Deep Fried Onion Rings, 372
Delicate Curried Crab Soup, 73
Delicious Chocolate Mousse, 530
Delmonico Potatoes and Ham, 320
Deluxe Banana Cream Pie, 563
Dessert Sippers, 26
Devilled Eggs, Hot, 175
Devil's Food Torte, 595
Dill Sauce, Breaded Chicken Breasts
with Spinach and, 214
Dilly Cucumber Sauce,
Fresh Salmon with, 272
Dip, Blue Cheese, 3
Dip, Hot Cheese Cocktail, 10
Dip, Low-Cal Zesty (Appetizer), 3
Dip, Mexican Cheese, 11
Dippity-Do Cheese Fondue, 7
Double Fudge Chocolate Cake, 448
Dressing, Fruit Salad, 407
Dressing, Tangy Tomato, 408
Dumplings, Peach, 460

E
East Indian Chicken, 247
Easy Chicken Divan, 248
Easy Salmon Bisque, 75
Easy Steamed Christmas Puddings
with Custard Sauce, 492
EGGS
Baked Brunch, 149
Benedict au Gratin, 171

Eggs Cont'd
Breakfast Egg Casserole, 150
Cheese and Olive Casserole, 170
Ham and Egg Casserole, 326
Hot Devilled, 175
Night-Before Scrambled, 189
Pizza Oven Omelette, 196
Poached Eggs in Spinach
Yogurt Sauce, 197
Sauce, Salmon Loaf with, 281
Scrambled Easter Egg Brunch, 199
Stuffed Eggs in Cheese Sauce, 207
Tomato, Tuna, Egg,
Mozzarella Casserole, 301
Twist Bread, 417
Eggnog Alexander, Fireside, 28
Eggnog Pie, 564
Eggnog Royale, 27
Elegant Carrot Ring, 374
Enchiladas, American Style, 103
English Muffins, Ham on, 323

F
Festive Raspberry Trifle, 503
Fireside Eggnog Alexander, 28
FISH
Baked Salmon à la Russe, 259
Cheesy Seafood Casserole, 261
Chowder Casserole, 262
Cod au Gratin, 264
Crispy-Topped, 269
Curried Tuna, 270
Deep Dish Tuna Quiche, 169
Easy Salmon Bisque, 75
Fish in a Pouch
with Mushroom Sauce, 271
Fresh Salmon with
Dilly Cucumber Sauce, 272
Kettle of Fish Chowder, 88
Macaroni Tuna Casserole, 274
Neptune, 275
Poached Fish Parmesan, 277
Poached Fish
with Vegetable Sauce, 276

INDEX

Fish cont'd
Quick Fish Bake with
Hot Tartar Sauce, 279
Salmon Cheddar Macaroni, 284
Salmon Florentine, 283
Salmon Loaf with Egg Sauce, 281
Salmon Mini Quiche
(Appetizer), 13
Salmon Soufflé, 280
Salmon Supreme, 282
Saucy Salmon in Toast Cups, 285
Scalloped Fish Casserole, 287
Sole Supreme, 298
Summer Tuna Salad Mold, 299
Thermidor en Casserole, 300
Tomato,Tuna,Egg,Mozzarella
Casserole, 301
Tuna à la King, 304
Tuna Broccoli Casserole, 303
Tuna Casserole Deluxe, 305
Tuna Celery Crunch, 302
Tuna in a Tub, 306

FLANS
Fran's Frugal Cheddar
and Ham, 319
Fresh Peach, 598
Lemon Cherry, 599
Raspberry, 600
Spinach, 204
Float, Mocha, 38
Floating Fruit Meringue, 538

FONDUES
Creamy Vanilla, 542
Milk Chocolate, 542
Franks and Cabbage, Savoury, 336
Fran's Frugal Cheddar and Ham
Flan, 319
Frappé, Raspberry, 42
French Onion Soup au Lait, 76
French Toast, Banana, 483
French Toast, Orange, 481
French Toast, Spice Islands, 482
French Toast, Turkey Sandwich, 344
French Vanilla Cream, 544

French Vanilla Ice Cream, 603
Fresh Asparagus Mimosa, 375
Fresh Cauliflower
and Ham Chowder, 77
Fresh Fruit Salad, 552
Fresh Fruit Sauce, 515
Fresh Fruit Trifle, 504
Fresh Lemon Cooler, 29
Fresh Peach Cobbler with Creamy
Lemon Sauce, 547
Fresh Peach Flan, 598
Fresh Peach Torte, 596,
Fresh Salmon With
Dilly Cucumber Sauce, 272
Fricassée, Chicken, 229
Frittata, Mushroom and Red
Pepper, 187
Frosting, Cheesy, Happy Face
Cupcakes with, 456
Frozen Caramel Cream, 604
Frozen Mocha Marvel, 605
Frozen Neapolitan Torte, 606

FRUIT
and Chops, Saucy, 334
Flip (Beverage), 30
Meringue, Floating, 538
Salad, Cabbage, 360
Salad Dressing, 360
Salad, Party Special, 552
Sauce,Fresh, 158
Trifle, Fresh, 504

G

Garden Loaf, 376
Garden Turkey Soup, 78
Garlic Appetizer Meat Balls, Honey, 9
Golden Carrot Soup, 79
Golden Cheese and Rice Bake, 172
Golden Penny Pancakes, 473
Gouda Dunk (Appetizer), 8
Goulash, Hearty Beef, 124
Goulash Soup, Hungarian, 85
Grandma's Chocolate Cake, 434
Grape Shake-up, Great, 23
Grapefruit Custard Pie, 565

INDEX

Grapefruit Whiz, Pineapple (Blender Breakfast), 16
Grasshopper Dessert Sipper, 26
Grasshopper Pie, 566
Great Grape Shake-up, 23
Greek Beef and Macaroni Casserole, 122
Green Beans with Water Chestnuts, 377
Green Split Pea Soup, 80
Grilled Cinnamon Pears and Fresh Strawberries, 551

H

Hallowe'en Pumpkin Muffins, 466

HAM
and Broccoli Royale, 321
and Broccoli Spuds, 325
and Egg Casserole, 326
and Mushrooms, Linguine with, 178
and Noodle Supper Supreme, 322
Bone Chowder, Hearty, 82
Casserole, Country, 315
Chowder, Fresh Cauliflower and, 77
Croquettes, Piquant, 329
Delmonico Potatoes and, 320
Flan, Fran's Frugal Cheddar and, 319
on English Muffins, 323
Pâté, Versatile Chicken and, 14
Peameal Bacon and Rice Casserole, 328
Piquant Ham Croquettes, 329
Tetrazzini, 324
Hamburger Casserole,Scalloped, 134
Hamburger Hot Cakes, 123
Happy Face Cupcakes with Cheesy Frosting, 456
Harvest Apple Crisp Pie, 567
Harvest Chowder, 81
Harvest Lasagne, 173
Hasty Tasty Chicken Chop Suey, 249

Hawaiian Chicken, 250
Hawaiian Punch, 31
Hearty Beef Goulash, 124
Hearty Clam Soup, 83
Hearty Ham Bone Chowder, 82
Hearty Spaghetti Sauce, 401
Herbed Dumpling Crust,Turkey Pot Pie with, 349
Herbed Fresh Tomato Bisque, 84
Herbed Mushroom Quiche, 174
Herbed Parmesan Batter Bread, 415
Holiday Dessert Pudding, 499
Home Style Chicken Stew, 251
Homemade Potato Salad, 378

HONEY
Butter, Cinnamon Popovers with, 480
Garlic Appetizer Meat Balls, 9
Milk and Honey Cake with Honey Coconut Topping, 433
Pecan Coffeecake, 429
Shake, Peanut Butter 'n', 23
Syrup, Orange, 481
Hot Banana Cocoa, 33
Hot Cheese Cocktail Dip (Appetizer), 10
Hot Chocolate, Continental, 32
Hot Chocolate Soufflé, 589
Hot Devilled Eggs, 175
How to Make The Perfect Milkshake, 22
Hungarian Goulash Soup, 85

I

Iberian Crème Caramel, 550

ICE
Buttermilk Strawberry, 602
Lemon, 602
Minted Meringue, 607
Ice-Box Cake, Lemon, 445
Ice Cream, French Vanilla, 603
Iced Cucumber Soup, 86
Iced Spinach Soup, 87
Icing,Boiled, 443

INDEX

Icing,Easy Chocolate, 434
Icing For Double Fudge Chocolate
Cake, 448
Impossible Coconut Pie, 568
Individual Clam Bakes, 273
Italian Cappuccino Coffee, 32
Italian Egg Noodle Casserole, 176

J
Jellied Chicken Loaf, 252
Jelly Muffins, Peanut Butter and, 469
Joggers' Nog, 35
Julep, Milk, 37

K
Kebabs, Seafood with Lemon
Sauce, 294
Kettle of Fish Chowder, 88
Kuchen, Peach, 459
Kulich Easter Bread, 416

L
LAMB
Bobotie, 307
Loaf, 308
Middle East Meat Pie, 310
Moussaka, 311
Quick Cashew Curry, 312
Stew, 309
Lasagne, Harvest, 173
Lasagne, Seafood Surprise, 201
Lasagne Special, 177
Layered Raspberry Squares, 455
Leek Soup, Cream
of Tomato and, 64
Leek Soup, Mushroom and, 94
Leftovers with Almonds, Turkey, 345

LEMON
Blueberry Muffins, 467
Blueberry Pie, 569
Cake, Triple Treat, 437
Cherry Flan, 599
Chicken Pilaf, 253
Cooler, Fresh, 29
Cream, Orange, 519

Lemon cont'd
Custard, Baked, 512
Ice, 602
Ice-Box Cake, 445
Nut Bread, 425
Pudding, 489
Sauce Creamy, Fresh Peach
Cobbler with, 547
Sauce, Seafood Kebabs with, 294
Sponge Pudding, Baked, 486
Lettuce Soup, 89
Likeable Liver and Rice
Casserole, 125
Lime Frost, Luscious, 29
Lime Refresher, 36
Lime Soufflé, Chilly, 588
Linguine with Clam Sauce, 179
Linguine with Ham and
Mushrooms, 178

LIVER
Beef Liver with Toast Points, 106
Buttermilk Liver, 109
Cabbage and Liver Fingers, 111
Chicken Livers
on Toast Points, 237
Curried Liver Strips, 121
Likeable Liver
and Rice Casserole, 125
Liver with Bacon 'n' Onion
Sauce, 126
Lobster and Shrimp Sauce,
Pasta with, 195
Lobster Bisque, 90
Lo-Cal Cream of Zucchini Soup, 91
Low-Cal Zesty Dip (Appetizer), 3
Luscious Lime Frost, 29

M
MACARONI
and Cheese Parmigiano, 180
and Cheese Soufflé, 182
Casserole, Greek Beef and, 122
Casserole with
Double Cheese, 183

INDEX

Macaroni cont'd
Mexican Casserole, 128
Pizza, 181
Ring, 184
Salmon Cheddar, 284
Supper, Skillet Sausage and, 337
Tuna Casserole, 274
Mandarin Pie, 570
Manicotti with Mushroom Sauce, 185
Maple Buttered Pancakes with Hot
  Maple Butter, 477
Maple Fudge Tarts, 590
Maple Tarts, Colonial, 590
Maple Nut Muffins, 465
Maple Sugar Chiffon Pie, 571
Martha Washington's
  Cream of Crab Soup, 92
Meat Balls,Cheeseburger, 114
Meat Balls, Honey Garlic Appetizer, 9
Meat Balls, Swedish, 138
Meat Balls, Sweet and Sour
  Pineapple, 139
Meat Loaf, My Own Mini, 129
Melon and Potato Soup, 93
Melon Appetizer, Crab on, 6
Melon Dessert, Blender, 601
Meringue Cake, Coconut, 440
Meringue, Chocolate, 502
Meringue, Floating Fruit, 538
Meringue Ice, Minted, 607
Meringue Torte, Mocha, 593
Mexican Bean Beef Casserole, 127
Mexican Casserole, 128
Mexican Cheese Dip (Appetizer), 11
Middle East Meat Pie, 310
Mile-High Black Bottom Pie, 572

MILK
  and Honey Cake with Honey
    Coconut Topping, 433
  Blender Breakfasts (See Blender)
  Chocolate Fondue, 542
  Chocolate Milk Crème
    Caramel, 549
  Chocolate Peanutty Shake, 21
  Julep, 37

MILKSHAKES
  Bananaberry, 18
  Canadiana Shake, 19
  Chocolate Peanutty, 21
  Creamy Apricot, 18
  Great Grape Shake-up, 23
  How to Make the Perfect, 22
  Orange Buttermilk Cooler, 18
  Peanut Butter 'n' Honey, 23
  Peaches 'n' Cream Shake, 19
  Strawberry Velvet Shake, 20
  Tin Roof, 18
Mincemeat Custard Pie, 574
Minestrone, Creamy Italian, 70
Mini Cheese Biscuits, 412
Minted Meringue Ice, 607
Mixed Vegetables Italian Style, 379

MOCHA
  Chiffon Pie, 575
  Cream, 524
  Custard, 513
  Float, 38
  Marvel, Frozen, 605
  Meringue Torte, 593
  Swirl Cheese Cake, 555
  Treat, 608

MOLDS
  Creamed Peach Rice, 540
  Crunchy Chicken Salad, 246
  Dairy Delicious Cottage, 168
  Summer Tuna Salad, 299
Monte Cristo Sandwiches, 186
Moo (Dessert), 528
Mornay Sauce, 402
Moussaka, 311

MOUSSE
  Blender Chocolate Almond, 532
  Cranberry, 531
  Crunchy Caramel, 529
  Delicious Chocolate, 530
Mozzarella Casserole,
  Tomato, Tuna, Egg, 301
Mud Pie Tarts, 591

INDEX

MUFFINS
Apple 'n' Oats, 464
Branberry, 461
Carrot Wheat, 462
Chocolate Chip, 463
Date 'n' Nut, 472
Hallowe'en Pumpkin, 466
Lemon Blueberry, 467
Maple Nut, 465
Peanut Butter and Jelly, 469
Peanut Butter Raisin, 468
Pumpkin Raisin, 470
Savoury Cheese, 471

MUSHROOMS
and Bacon,Cheese
Roulade with, 156
and Leek Soup, 94
and Red Pepper Frittata, 187
Creamed Peas, 380
Creamy Mushroom Soup, 72
Croustade Appetizers, 12
in Toast Cups, Sherried
Chicken and, 257
Linguine with Ham and, 178
Onion Quiche, 188
Poached Chicken with, 255
Quiche, Herbed, 174
Sauce, 123
Sauce, Easy, 129
Sauce, Fish in a Pouch with, 271
Sauce, Manicotti with, 185
Sauce, Savoury, 404
Sauce, Veal Burgers and, 140
Supreme, 381
Mustard Sauce, 403
My Own Mini Meat Loaf, 129

N
Night-Before Scrambled Eggs, 189
Nog, Joggers', 35
Nog, Open House Coffee, 28

NOODLES
au Gratin, 190
Chicken Noodle Doodle
Casserole, 224

Noodles cont'd
Italian Egg Noodle Casserole, 176
Neptune, 275
Pudding Deluxe, Sweet, 500
Ring, Creamed Chicken in, 243
Romanoff, 192
Squares Milano, 191
Supper Supreme, Ham and, 322
Turkey Noodle Casserole, 346
Nuggets, Brown Sugar, 458

NUTS
and Seeds Bread, 418
Cranberry Nut Bread, 422
Cranberry Nut Stuffing, 367
Date 'n' Nut Muffins, 472
Lemon Nut Bread, 425
Maple Nut Muffins, 465
Orange Walnut Torte, 594

O
Oatmeal Bars, 409
Oatmeal Pancakes, 478
Oats Muffins, Apple 'n', 464
Olive Casserole, Egg, Cheese, 170

OMELETTE
Corned Beef Oven, 166
Pizza Oven, 196

ONIONS
and Potatoes, Creamy, 370
Quiche, Mushroom, 188
Rings, Deep Fried, 372
Sauce, Creamy, 399
Sauce, Liver with Bacon 'n', 126
Soup au Lait, 95
Soup au Lait, French, 76
Soup,Creamy, 69
Tart, Bacon, Cheese and, 146
Open House Coffee Nog, 28

ORANGES
Banana (Blender Breakfast), 16
Blossom, Chocolate,
(Beverage), 25
Buttermilk Cooler, 18
Chocolate Pie, 576

INDEX

Oranges cont'd
Crêpes, Buttery, 533
Date Cake, 439
Dream Custard, 515
French Toast, 481
Glaze, 447
Honey Syrup, 481
Peach Sauce, 540
Raisin Cake, 447
Rum Cake, 453
Walnut Torte, 594
Oriental Barbecue Pork with
Pineapple Peanut Sauce, 327
Oyster Stew, 96

P
PANCAKES
Apple, 474
Berry Puff, 475
Corn, 479
Golden Penny, 473
Maple Buttered
with Hot Maple Butter, 477
Oatmeal, 478
Pie, Apple Bacon, 573
Pina Colada with
Creamy Rum Sauce, 476
Sauce, Country Kitchen, 398
Spicy Apple Filling
For Apple Pancake, 474
PARFAITS
Cappuccino, 507
Chocolate Strawberry, 508
Peach, 505
Strawberry Rice, 496
Summertime Strawberry, 509
Super Creamy Pecan, 506
Parmesan Batter Bread,Herbed, 415
Parmesan Broccoli, 382
Parmesan, Poached Fish, 277
Parmesan Sauced Cauliflower, 383
Parsley Rice, Seafood
Mornay with, 290
Party Special Fruit Salad, 552

PASTA
Asparagus, Chicken
and Pasta Supper, 213
Cannelloni, 151
Chicken Noodle Doodle
Casserole, 224
Creamed Chicken in
Noodle Ring, 243
Greek Beef
and Macaroni Casserole, 122
Ham and Noodle
Supper Supreme, 322
Ham Tetrazzini, 324
Harvest Lasagne, 173
Hearty Spaghetti Sauce, 401
Italian Egg Noodle Casserole, 176
Lasagne Special, 177
Linguine with Clam Sauce, 179
Linguine with Ham
and Mushrooms, 178
Macaroni and Cheese
Parmigiano, 180
Macaroni and Cheese Soufflé, 182
Macaroni Casserole
with Double Cheese, 183
Macaroni Pizza, 181
Macaroni Ring, 184
Macaroni Tuna Casserole, 274
Manicotti with
Mushroom Sauce, 185
Mexican Casserole, 128
Noodle Squares Milano, 191
Noodles au Gratin, 190
Noodles Neptune, 275
Noodles Romanoff, 192
Pasta and Sausage Fry Up, 194
Pasta Primavera au Gratin, 193
Pasta with Lobster
and Shrimp Sauce, 195
Pork Pastitsio, 333
Salmon Cheddar Macaroni, 284
Sea Shells and Cheese, 200
Seafood Surprise
and Macaroni Supper, 201

INDEX

Pasta Cont'd
Skillet Sausage
and Macaroni Supper, 337
Spaghetti Meat Bake, 136
Spaghetti with Clam Sauce, 203
Spaghettini with
Herbed Cheese Sauce, 202
Sweet Noodle Pudding
Deluxe, 500
Turkey Noodle Casserole, 346
Turkey Tetrazzini, 351
Wiener and Spaghetti
Casserole, 142
Pastitsio, Pork with Cheese
Sauce, 333
Pastry Rolls,Turkey, 347
Pâté, Versatile Chicken and Ham, 14
Pea Soup, Green Split, 80

PEACH
Cobbler with Creamy
Lemon Sauce, 547
Dumplings, 460
Flan, Fresh, 598
Kuchen, 459
Parfaits, 505
Peaches 'n' Cream Shake, 19
Peaches 'n' Cream Shortcake, 452
Peachy Banana Flip (Beverage), 40
Rice Mold, Creamed, 540
Torte, Fresh, 596
Upside Down Cake, 446
Peameal Bacon
and Rice Casserole, 328

PEANUT BUTTER
and Jelly Muffins, 469
'n' Honey Shake, 23
Pie, 577
Pudding, 485
Raisin Muffins, 468
Peanut Sauce, Oriental Barbecued
Pork with Pineapple, 327
Peanutty Shake, Chocolate, 21
P-Nutty Warm Up, (Beverage), 33

PEARS
Chocolate Frosted, 543
Condé Rice Pudding, 490
Grilled Cinnamon
and Fresh Strawberries, 551
Peas, Mushroom Creamed, 380
Pecan Coffeecake, Honey, 429
Pecan Parfaits, Super Creamy, 506
Pecans, Creamed Celery with, 368
Pepper Cups, Cheese Soufflé in, 161
Peppermint Pie, 578
Peppery Beef Stroganoff, 130
Picnic Oven Fried Chicken, 254

PIES
Apple Bacon Pancake, 573
Banana Chiffon, 559
Banana Split, 560
Cheesy Tomato, 164
Chocolate Coconut, 561
Coconut Almond, 562
Deep Dish Pork, 340
Deep Dish Vegetable, 373
Deluxe Banana Cream, 563
Eggnog, 564
Grapefruit Custard, 565
Grasshopper, 566
Harvest Apple Crisp, 567
Impossible Coconut, 568
Lemon Blueberry, 569
Mandarin, 570
Maple Sugar Chiffon, 571
Middle East Meat, 310
Mile-High Black Bottom, 572
Mincemeat Custard, 574
Mocha Chiffon, 575
Orange Chocolate, 576
Peanut Butter, 577
Peppermint, 578
Pumpkin Chiffon, 580
Pumpkin Pie Delite, 579
Rhubarb Cheesecake Pie, 581
Shepherd's, 135
Snow, 582
Strawberry Ginger Cream, 584

INDEX

Pies Çont'd
Sunshine Citrus, 585
Sweet Potato and Apple, 586
Turkey, 348
Pilaf, Lemon Chicken, 253
Pina Colada Cheese Cake, 556
Pina Colada Pancakes with Creamy
Rum Sauce, 476

PINEAPPLE
Carrot Dessert Crêpes, 534
Crêpe Filling, 534
Grapefruit Whiz
(Blender Breakfast), 16
Meat Balls, Sweet and Sour, 139
Peanut Sauce, Oriental Barbecue
Pork with, 327
Refresher (Blender Breakfast), 17
Stirfry, Pork and, 330
Tote Cake, 438
Piquant Ham Croquettes, 329

PIZZA
Baked Sausage, 313
Macaroni, 181
Oven Omelette, 196
Plum Custard Delight, 517
Plum Delicious Cobbler, 546
Plum Sauce, 517
P-Nutty Warm Up (Beverage), 33
Poached Chicken with
Mushrooms, 255
Poached Eggs in
Spinach Yogurt Sauce, 197
Poached Fish Parmesan, 277
Poached Fish with
Vegetable Sauce, 276
Popovers, Cinnamon
with Honey Butter, 480
Poppy Seed Cake, 435

PORK
Baked Sausage Pizza, 313
Braised Pork Loin, 314
Chop Casserole, 331
Cordon Bleu, 332
Country Style Pork Chops, 311

Pork Cont'd
Creamed Pork Tenderloin, 317
Creamy Herbed Pork Chops, 318
'n' Pineapple, Stirfry, 330
Oriental Barbecued Pork with
Pineapple Peanut Sauce, 327
Pastitsio, 333
Saucy Fruit and Chops, 334
Savoury Stuffed Sausage Loaf, 335
Skillet Sausage and
Macaroni Supper, 337
Tender Pork au Lait, 338
Versatile Freezer
Pork Dinners, 339
Basic Pork Mixture, 339
Curried Pork, 339
Deep Dish Pork Pie, 340
Pot Pie, Turkey with Herbed
Dumpling Crust, 349
Potage, Creamy Broccoli, 68
Potage Jardiniere, 97

POTATOES
Bacon Harvest Soup, 98
Buffet Casserole, 359
Creamy Onions and Potatoes, 370
Delmonico Potatoes
and Ham, 320
Ham and Broccoli Spuds, 325
Homemade Potato Salad, 378
Melon and Potato Soup, 93
Prestige Scalloped, 384
Ranch Style, 385
Scalloped Potatoes
with Cheese and Herbs, 386
Seaside Potatoes, 387
Sour Cream Baked Potatoes, 388
Sweet Potato and Apple Pie, 586
Turkey Sweet Potato Combo, 350
Praline, Crème, 518
Prestige Scalloped Potatoes, 384
Profiterole Pyramid with French
Vanilla Cream, 544
Pronto Shrimp Curry, 278
Protein Power Cheddar Soup, 99

INDEX

PUDDINGS
Apple Raisin Cottage Pudding
with Brown Sugar Sauce, 493
Apricot Bread, 488
Baked Lemon Sponge, 486
Banana Scotch-er-oo, 497
Chocolate Rice, 487
Crazy Raisin Rice, 495
Creamy Rhubarb-Apricot, 494
Creamy Rice, 501
Danish Rum Pudding with
Raspberry Sauce, 498
Easy Steamed Christmas Puddings
with Custard Sauce, 492
Holiday Dessert, 499
Lemon, 489
Peanut Butter, 485
Pears Condé Rice, 490
Pudding Pompadour, 502
Rice and Apricot Dessert, 496
Strawberry Rice Parfait, 496
Sweet Noodle
Pudding Deluxe, 500
Tipsy Bread, 491
Puffy Cheese Biscuits, 413
PUMPKIN
Chiffon Pie, 580
Custard, 514
Muffins, Hallowe'en, 466
Pie Delite, 579
Raisin Muffins, 470
Soup, Creamed, 67
Punch, Hawaiian, 31
Purple Cow (Beverage), 39

Q
QUICHE
Asparagus Cheddar, 144
Chicken Amandine, 165
Crab, 167
Deep Dish Tuna, 169
Herbed Mushroom, 174
Lorraine, 198
Mushroom Onion, 188
Salmon Mini (Appetizer), 13

Quiche cont'd
Spinach, 205
Tomato and Broccoli, 209
Quick Cashew Curry, 312
Quick Fish Bake with
Hot Tartar Sauce, 279

R
Ragout, Creamy Veal, 118
RAISIN
Cake, Orange, 447
Cheesecake, Chocolate
Rum and, 553
Cottage Pudding with Brown
Sugar Sauce, Apple, 493
Muffins, Peanut Butter, 468
Muffins, Pumpkin, 470
Rice Pudding, Crazy, 495
Scones, Scottish, 414
Ranch Style Potatoes, 385
Rarebit, Asparagus, 143
RASPBERRY
Cooler, 41
Cream, 520
Delight Cake, 454
Flan, 600
Frappé, 42
Frosty (Beverage), 36
Sauce, Danish Rum
Pudding with, 498
Squares Layered, 455
Trifle, Festive, 503
Red Pepper Frittata,
Mushroom and, 187
Regal Chocolate Sauce, 406
RHUBARB
Apricot Pudding Creamy, 494
Cake, 436
Cheesecake Pie, 581
Cobbler Strawberry, 548
Sauce, Whole Wheat
Crêpes with, 537

INDEX

RICE
and Apricot Dessert, 496
Beef and Rice Crisp, 108
Cabbage Rice Duo, 361
Chicken Rice Bake, 230
Chicken, Vegetable
and Rice Casserole, 235
Chicken Wild Rice Casserole, 233
Chocolate Rice Pudding, 487
Crazy Raisin Rice Pudding, 495
Creamed Peach Rice Mold, 540
Creamy Rice Pudding, 501
Golden Cheese
and Rice Bake, 172
Lemon Chicken Pilaf, 253
Likeable Liver
and Rice Casserole, 125
Parfait, Strawberry, 496
Peameal Bacon
and Rice Casserole, 328
Pears Condé Rice Pudding, 490
Seafood Mornay
with Parsley Rice, 290
Roulade with Mushrooms
and Bacon,Cheese, 156

RUM
and Raisin Cheesecake,
Chocolate, 553
Bavarian, 525
Cake, Orange, 453
Pudding with Raspberry
Sauce,Danish, 498
Sauce, Pina Colada Pancakes
with Creamy, 476
Rutabaga Soup, Cream of, 66

S
SALAD
Cabbage Fruit, 360
Dessert, Snow, 539
Dressing, Cabbage Fruit, 360
Dressing, Fruit, 407
Homemade Potato, 378
Mold, Crunchy Chicken, 246

Mold, Summer Tuna, 299
Party Special Fruit, 552
SALMON
à la Russe, Baked, 259
Cheddar Macaroni, 284
Easy Salmon Bisque, 75
Florentine, 283
Fresh Salmon with
Dilly Cucumber Sauce, 272
in Toast Cups,Saucy, 285
Loaf with Egg Sauce, 281
Mini Quiche (Appetizer), 13
Soufflé, 280
Supreme, 282
Sandwich, Turkey French Toast, 344
Sandwiches, Monte Cristo, 186
SAUCES
Brown Sugar, 493
Cheddar Sauce Supreme, 396
Cheese Sauce, 157
Cheese Sauce, Pork
Pastitsio with, 333
Cheese Sauce, Stuffed
Eggs in, 207
Cheese Sauce, Turkey French
Toast Sandwich, 344
Clam Sauce, Linguine with, 179
Clam Sauce, Spaghetti with, 203
Country Kitchen Pancake, 398
Cranberry Sauce,
Chicken Ring with, 238
Creamy Lemon Sauce, Fresh
Peach Cobbler with, 238
Creamy Onion, 399
Creamy Rum Sauce, Pina Colada
Pancakes with, 476
Creamy Vanilla Sauce, Streusel
Topped Baked Apples with, 541
Curry, 390, 400
Custard Sauce, 551
Custard Sauce, Easy Steamed
Christmas Puddings with, 492
Dill Sauce, Breaded Chicken
Breasts with Spinach and, 214

INDEX

Sauces cont'd

Dilly Cucumber Sauce, Fresh Salmon with, 272

Egg Sauce, Salmon Loaf with, 281

Fresh Fruit, 515

Fresh Fruit Sauce, Cheese Blintzes with, 158

Hearty Spaghetti, 401

Herbed Cheese Sauce, Spaghettini with, 202

Hot Tartar Sauce, Quick Fish Bake with, 279

Lemon Sauce, Seafood Kebabs with, 294

Lobster and Shrimp Sauce, Pasta with, 195

Mimosa Sauce, 375

Mornay, 402

Mushroom Sauce, 123

Mushroom Sauce, Easy, 129

Mushroom Sauce, Fish in a Pouch with, 271

Mushroom Sauce, Manicotti with, 185

Mushroom Sauce, Veal Burgers and, 140

Mustard, 403

Onion Sauce, Liver with Bacon 'n', 126

Orange Peach Sauce, 540

Pineapple Peanut, Oriental Barbecue Pork with, 327

Plum Sauce, 517

Raspberry Sauce, Danish Rum Pudding with, 498

Regal Chocolate, 406

Rhubarb Sauce, Whole Wheat Crêpes with, 537

Savoury Mushroom, 404

Smoke-House Bacon, 405

Sour Cream Sauce, Breaded Cauliflower with, 356

Spinach Yogurt Sauce, Poached Eggs in, 197

Sauces cont'd

Tomato Sauce, Spinach Balls in, 389

Vegetable Sauce, Poached Fish with, 276

White Sauce, Classic, 397

Wine Sauce, Chicken in, 223

Saucy Fruit and Chops, 334

Saucy Salmon in Toast Cups, 285

Saucy Short Ribs, 131

Sauerbraten Sauced Beef Loaf, 132

SAUSAGE

and Macaroni Supper, Skillet, 337

Fry Up, Pasta and, 194

Loaf, Savoury Stuffed, 335

Pizza, Baked, 313

Savoury Biscuits, 234

Savoury Cheddar Bread, 419

Savoury Cheese Muffins, 471

Savoury Coated Chicken and Sauce, 256

Savoury Franks and Cabbage, 336

Savoury Mushroom Sauce, 404

Savoury Stuffed Sausage Loaf, 335

Savoury Veal Casserole, 133

Scalloped Carrots, Cheese, 364

Scalloped Fish Casserole, 287

Scalloped Hamburger Casserole, 134

Scalloped Potatoes, Prestige, 384

Scalloped Potatoes with Cheese and Herbs, 386

Scalloped Scallops, 286

Scallops, Bubbly Baked, 260

Scallops on Toast, 288

Scones, Scottish Raisin, 414

Scottish Raisin Scones, 414

Scrambled Easter Egg Brunch, 199

Scrambled Eggs, Night-Before, 189

Sea Captain's Casserole, 289

Sea Shells and Cheese, 200

SEAFOOD

Bubbly Baked Scallops, 260

Buffet Chicken and Shrimp à la King, 216

Casserole, 296

634

INDEX

Seafood Cont'd
Cheesy Seafood Casserole, 261
Chowder A#1, 45
Clam Chowder with
Bacon and Croutons, 58
Clam Soup, 59
Coquilles St Jacques, 5
Cottage Curried Shrimp, 265
Crab Custard Baltimore, 267
Crab Imperial, 266
Crab on Melon Appetizer, 6
Crab Quiche, 167
Crabmeat Luncheon Dish, 268
Crêpes, 291
Delicate Curried Crab Soup, 73
Hearty Clam Soup, 83
Individual Clam Bakes, 273
Kebabs with Lemon Sauce, 294
Linguine with Clam Sauce, 179
Lobster Bisque, 90
Martha Washington's Cream
of Crab Soup, 92
Mornay with Parsley Rice, 290
Newburg, 293
Oyster Stew, 96
Pasta with Lobster
and Shrimp Sauce, 195
Pronto Shrimp Curry, 278
Scallop, 295
Scalloped Scallops, 286
Scallops on Toast, 288
Sea Captain's Casserole, 289
Seaside Potatoes, 387
Shrimp Fromage Casserole, 297
Spaghetti with Clam Sauce, 203
Springtime Crab
and Asparagus Tart, 206
Surprise Lasagne, 201
Thermidor en Casserole, 300
Shepherd's Pie, 135
Sherbet, Strawberry Banana, 609
Sherbet, Surprise, 611
Sherried Chicken and Mushrooms in
Toast Cups, 257
Shortcake, Peaches 'n' Cream, 452

SHRIMPS
à la King, Buffet Chicken and, 216
Cottage Curried, 265
Curry, Pronto, 278
Fromage Casserole, 297
Sauce, Pasta
with Lobster and, 195
Simply Seasoned
Casserole Bread, 420
Sippers, Dessert, 26
Cricket, 26
Grasshopper, 26
Skillet Sausage and
Macaroni Supper, 337
Smoke-House Bacon Sauce, 405
Snow Pie, 582
Snow Salad Dessert, 539
Soda Bread, Crisp-Crusted, 423
Sole Supreme, 298

SOUFFLES
Almond, 587
Asparagus Cheese
Soufflé Roll, 145
Cheddar Cheese, 152
Cheese Soufflé
in Pepper Cups, 161
Chilly Lime, 588
Company Broccoli, 366
Hot Chocolate, 589
Macaroni and Cheese, 182
Salmon, 280
Spinach, 391
Squash, 392
Swiss Cheese and Bacon, 208

SOUPS
A#1 Seafood Chowder, 45
Autumn Vegetable, 46
Bean and Bacon Potage, 48
Borscht, 47
Broccoli, 49
Canadiana Cheddar Cheese, 50
Cauliflower Bisque
with Cheesy Croutons, 51
Cheese Cauliflower, 52

INDEX

Soups cont'd
Cheese Zucchini, 53
Cheesy Broccoli, 54
Cheesy Vegetable, 55
Chilled Vichyssoise Supreme, 56
Chunky Chicken Chowder, 57
Clam Chowder with
  Bacon and Croutons, 58
Clam Soup, 59
Cold Beet Borscht
  with Cucumbers, 61
Cold Pink Beet, 60
Corn Chowder, 63
Corn 'n' Beef Soup, 62
Cream of Asparagus (Diet), 65
Cream of Rutabaga, 66
Cream of Tomato and Leek, 64
Creamed Pumpkin, 67
Creamy Broccoli Potage, 68
Creamy Mushroom, 72
Creamy Italian Minestrone, 70
Creamy Onion, 69
Creole Bisque, 71
Curried Avocado, 74
Delicate Curried Crab, 73
Easy Salmon Bisque, 75
French Onion Soup au Lait, 76
Fresh Cauliflower
  and Ham Chowder, 77
Garden Turkey, 78
Golden Carrot, 79
Green Split Pea, 80
Harvest Chowder, 81
Hearty Clam, 83
Hearty Ham Bone Chowder, 82
Herbed Fresh Tomato Bisque, 84
Hungarian Goulash Soup, 85
Iced Cucumber, 86
Iced Spinach, 87
Kettle of Fish Chowder, 88
Lettuce, 89
Lobster Bisque, 90
Lo-Cal Cream of Zucchini, 91
Martha Washington's
  Cream of Crab, 92

Soups cont'd
Melon and Potato, 93
Mushroom and Leek, 94
Onion Soup au Lait, 95
Oyster Stew, 96
Potage Jardiniere, 97
Potato-Bacon Harvest, 98
Protein Power Cheddar, 99
Swanky Franky Chowder, 100
Watercress Vichyssoise, 101
Zucchini Buttermilk Bisque, 102
Sour Cream Baked Potatoes, 388
Spaghetti Casserole, Wiener and, 142
Spaghetti Meat Bake, 136
Spaghetti Sauce, Hearty, 401
Spaghetti with Clam Sauce, 203
Spaghettini with Herbed Cheese
  Sauce, 202
Spanish Cream, 522
Spice Islands French Toast, 482
Spicey Apple Filling, 474
Spicy Beef Casserole, 137
SPINACH
  and Dill Sauce, Breaded
    Chicken Breasts with, 214
  Balls in Tomato Sauce, 389
  Flan, 204
  Florentine, Salmon, 283
  Quiche, 205
  Ring, 390
  Roll, Cheese, 162
  Soufflé, 391
  Soup, Iced, 87
  Yogurt Sauce, Poached
    Eggs in, 197
Springtime Crab
  and Asparagus Tart, 206
Sponge Pudding, Baked Lemon, 486
SQUASH
  Beef Stuffed, Acorn, 107
  Ring, Baked Winter, 355
  Soufflé, 392
  Stack-Ups, Bacon and Tomato, 147
  Strata, Baked Cheddar, 148

INDEX

Strata, Cheese and Vegetable, 155
Strawberries 'n' Cream Cake, 451
STRAWBERRY
Banana Shake, 24
Banana Sherbet, 609
Bavarian, 526
Cream, 521
Frost (Blender Breakfast), 16
Ginger Cream Pie, 584
Grilled Cinnamon Pears
with Fresh Strawberries, 551
Ice, Buttermilk, 602
'n' Cream Cake, 451
Parfaits, Chocolate, 508
Parfaits, Summertime, 509
Rhubarb Cobbler, 548
Rice Parfait, 496
Velvet Shake, 20
Yogurt Pops, 610
Streusel Topped Baked Apples
with Creamy Vanilla Sauce, 541
STEW
Home Style Chicken, 251
Lamb, 309
Oyster, 96
Stroganoff, Peppery Beef, 130
Stuffed Acorn Squash, Beef, 107
Stuffed Eggs in Cheese Sauce, 207
Stuffing, Cranberry Nut, 367
Summer Tuna Salad Mold, 299
Summertime Strawberry Parfaits, 509
Sunrise Starter (Beverage), 43
Sunshine Citrus Pie, 585
Super Creamy Pecan Parfaits, 506
Surprise Sherbet, 611
Swanky Franky Chowder, 100
Swedish Meat Balls, 138
Swedish Tea Ring, 430
Sweet and Sour
Pineapple Meat Balls, 139
Sweet Cheese Crêpes, 535
Sweet Noodle Pudding Deluxe, 500
Sweet Potato and Apple Pie, 586

Sweet Potato Combo, Turkey, 350
Swiss Cheese and Bacon Soufflé, 208
Swiss Cheese Bacon Bread
(Appetizer), 421

T
Tangy Tomato Dressing, 408
TARTS
Bacon, Cheese and Onion, 146
Colonial Maple, 590
Maple Fudge, 590
Mud Pie, 591
Springtime Crab
and Asparagus, 206
Tartar Sauce, Hot,
Quick Fish Bake with, 279
Tea Loaf, Carrot, 424
Tea Ring, Swedish, 430
Tender Pork au Lait, 338
Tetrazzini, Ham, 324
Tetrazzini, Turkey, 351
Thermidor en Casserole, 300
Tipsy Bread Pudding, 491
Toast Cups, 12
TOMATO
and Broccoli Quiche, 209
and Leek Soup, Cream of, 64
Bacon Wraps, 393
Herbed Fresh Tomato Bisque, 84
Pie, Cheesy, 164
Sauce, Spinach Balls in, 389
Stack-Ups, Bacon and, 147
Tangy Tomato Dressing, 408
Tuna, Egg, Mozzarella
Casserole, 301
Topsy-Turvey Custards, 516
TORTE
Black Forest, 592
Devil's Food, 595
Fresh Peach, 596
Frozen Neapolitan, 606
Mocha Meringue, 593
Orange Walnut, 594

INDEX

Tote Along Casserole, 210
Trifle, Festive Raspberry, 503
Trifle, Fresh Fruit, 504
Triple Treat Lemon Cake, 437
TUNA
à la King, 304
Broccoli Casserole, 303
Casserole Deluxe, 305
Casserole Macaroni, 274
Celery Crunch, 302
Curried, 270
in a Tub, 306
Quiche, Deep Dish, 169
Salad Mold, Summer, 299
Tomato,Tuna,Egg, Mozzarella
Casserole, 301
TURKEY
Buffet Casserole, 341
Casserole, 342
Divan, 343
French Toast Sandwich, 344
Leftovers with Almonds, 345
Noodle Casserole, 346
Pastry Rolls, 347
Pie, 348
Pot Pie with Herbed
Dumpling Crust, 349
Soup, Garden, 78
Sweet Potato Combo, 350
Tetrazzini, 351
Turnip Apple Bake, 394

V
Vanilla Cream French,
Profiterole Pyramid with, 544
Vanilla Fondue, Creamy, 542
Vanilla Ice Cream, French, 603
Vanilla Sauce, Creamy
Streusel Topped Baked
Apples with, 541
Veal Burgers
and Mushroom Sauce, 140
Veal Ragout, Creamy, 118
Casserole, Savoury, 133

Veal Supreme, 141
Vegetable Crêpes, 211
Versatile Chicken and
Ham Pâté (Appetizer), 14
Versatile Freezer Pork Dinners, 339
Basic Pork Mixture, 339
Curried Pork, 339
Deep Dish Pork Pie, 340
Vichyssoise Supreme, Chilled, 56
Vichyssoise, Watercress, 101
Vol-au-Vent, Chicken, 222

W
Walnut Torte, Orange, 594
Water Chestnuts,
Green Beans and, 377
Watercress Vichyssoise, 101
Wheat Muffins, Carrot, 462
Whirly-Twirly Pinwheel Cookies, 458
White Sauce, Classic, 397
Whole Wheat Crêpes
with Rhubarb Sauce, 537
Whole Wheat Sesame Biscuits, 411
Wiener and Spaghetti Casserole, 142
Wine Sauce, Chicken in, 223
Winter Squash Ring, Baked, 355

Y
Yogurt Flip (Beverage), 38
Yogurt Sauce, Spinach,
Poached Eggs in, 197
Yogurt Pops, Strawberry, 610

Z
ZUCCHINI
Buttermilk Bisque, 102
Casserole, 395
Creamy Broccoli and, 371
Soup,Cheese, 53
Soup, Lo-Cal Cream of, 91

NOTES

NOTES

NOTES

NOTES

NOTES